D0742721

# OPTICAL AND ELECTRONIC PROCESS OF NANO-MATTERS

# Advances in Optoelectronics (ADOP)

ADOP   Advances in Optoelectronics

# OPTICAL AND ELECTRONIC PROCESS OF NANO-MATTERS

Edited by

Motoichi OHTSU
*Tokyo Institute of Technology, Yokohama*

KTK Scientific Publishers / Tokyo

Kluwer Academic Publishers
Dordrecht / London / Boston

C.I.P. Catalogue record for this book is available from the Library of Congress.

ISBN 0-7923-6987-4 (Kluwer)

Published by KTK Scientific Publishers, 2002 Sansei Jiyugaoka Haimu, 27-19 Okusawa 5-chome, Setagaya-ku, Tokyo 158-0083, Japan / Kluwer Academic Publishers, P.O. Box 17, 3300 AA Dordrecht, The Netherlands.

Kluwer Academic Publishers incorporates
the publishing programmes of
D. Reidel, Martinus Nijhoff, Dr W. Junk and MTP Press.

Sold and Distributed in the U.S.A. and Canada
by Kluwer Academic Publishers,
101 Philip Drive, Assinippi Park, Norwell, MA 02061, U.S.A.
in Japan by KTK Scientific Publishers,
2002 Sansei Jiyugaoka Haimu, 27-19 Okusawa 5-chome, Setagaya-ku,
Tokyo 158-0083, Japan.

In all other countries, sold and distributed
by Kluwer Academic Publishers,
P.O. Box 322, 3300 AH Dordrecht, The Netherlands.

Printed in Japan

# Preface

Sizes of electronic and photonic devices are decreasing drastically in order to increase the degree of integration for large-capacity and ultrahigh-speed signal transmission and information processing. This miniaturization must be rapidly progressed from now onward. For this progress, the sizes of materials for composing these devices will be also decreased to several nanometers. If such a nanometer-sized material is combined with the photons and/or some other fields, it can exhibit specific characters, which are considerably different from those of bulky macroscopic systems. This combined system has been called as a mesoscopic system. The first purpose of this book is to study the physics of the mesoscopic system.

For this study, it is essential to diagnose the characteristics of miniaturized devices and materials with the spatial resolution as high as several nanometers or even higher. Therefore, novel methods, e.g., scanning probe microscopy, should be developed for such the high-resolution diagnostics. The second purpose of this book is to explore the possibility of developing new methods for these diagnostics by utilizing local interaction between materials and electron, photon, atomic force, and so on.

Conformation and structure of the materials of the mesoscopic system can be modified by enhancing the local interaction between the materials and electromagnetic field. This modification can suggest the possibility of novel nano-fabrication methods. The third purpose of this book is to explore the methods for such nano-fabrication.

Several articles on mesocopic systems have reviewed their physics, diagnostics, and fabrication separately from the viewpoint of electronic interaction. Some of them are the reviews on the application of scanning tunneling microscopy. In contrast with them, the unique feature of this book is to introduce the concept of nano-optics, i.e., near field optics into discussions on mesoscopic systems.

Intensive discussions on the three purposes presented above, including the methods of scanning tunneling microscopy, near field optics, and atomic force microscopy, have been made by the 17 members of a working group in the Institute of Electrical Engineers of Japan from April 1996 to March 1998, to which I have served as chairman. This book summarizes the main topics of the discussion, where most of the co-authors have been the members of this working group. Chapters 1–5 are devoted to physics. Chapters 6, 7, and 9–11 are for diagnostics, and Chapter 8 is for fabrication. Among them, reviews on electronic systems are dealt with in Chapters 2–6, and 11. Those topics on near-field optical systems and on atomic force systems are mentioned in Chapters 6–8 and in Chapters 9–11, respectively. All the chapters are carefully organized so that the readers can obtain a deep understanding on the contents. I hope that this book will enable undergraduate and graduate students, junior scientists, and engineers to systematically study the physics, diagnostics, and fabrication of nano-sized materials and devices.

M. Ohtsu
March, 2000

# Contents

# List of Authors

Chapter 1    H. Hori
             Department of Electronics, Yamanashi University,
             4-3-11 Takeda, Kofu-shi, Yamanashi 400-8511, Japan

Chapter 2    S. Tarucha[1,2], D. G. Austing[2], T. Fujisawa[2] and L. P. Kouwenhoven[3]
             [1]Department of Physics and ERATO Mesoscopic Correlation
               Project (JST), University of Tokyo,
               7-3-1 Hongo, Bunkyo-ku, Tokyo 113-0033, Japan
             [2]NTT Basic Research Laboratories,
               3-1 Morinosato Wakamiya, Atsugi-shi, Kanagawa 243-0198, Japan
             [3]Department of Applied Physics and ERATO Mesoscopic Correlation
               Project (JST), Delft University of Technology,
               P.O. Box 5046, 2600 GA DELFT, The Netherlands

Chapter 3    J. Bae and K. Mizuno
             Research Institute of Electrical Communication, Tohoku University,
             2-1-1 Katahira, Aoba-ku, Sendai-shi, Miyagi 980-8577, Japan

Chapter 4    H. Nejo and Z.-C. Dong
             National Research Institute For Metals,
             1-2-1 Sengen, Tsukuba-shi, Ibaraki 305-0047, Japan

Chapter 5    M. Tsukada, N. Sasaki and N. Kobayashi
             Department of Physics, Graduate School of Science,
             University of Tokyo,
             7-3-1 Hongo, Bunkyo-ku, Tokyo 113-0033, Japan

Chapter 6    T. Murashita
             NTT Photonics Laboratories,
             3-1 Morinosato Wakamiya, Atsugi-shi, Kanagawa 243-0124, Japan

Chapter 7    T. Saiki
             Kanagawa Academy of Science and Technology,
             KSP East 408, 3-2-1 Sakado, Takatsu-ku, Kawasaki-shi,
             Kanagawa 213-0012, Japan

Chapter 8    M. Ohtsu[1,2] and G. H. Lee[2]
             [1]Interdisciplinary Graduate School of Science and Engineering,
              Tokyo Institute of Technology,
              4259 Nagatsuta-cho, Midori-ku, Yokohama-shi,
              Kanagawa 226-8502, Japan
             [2]ERATO Localized Photon Project,
              Japan Science and Technology Corporation,
              687-1 Tsuruma, Machida-shi, Tokyo 194-0004, Japan

Chapter 9    S. Morita and Y. Sugawara
             Department of Electronic Engineering, Osaka University,
             2-1 Yamadaoka, Suita-shi, Osaka 565-0871, Japan

Chapter 10   T. Hattori, H. Nohira and K. Takahashi
             Department of Electrical and Electronic Engineering,
             Musashi Institute of Technology,
             1-28-1 Tamazutsumi, Setagaya-ku, Tokyo 158-0087, Japan

Chapter 11   M. Hara[1] and K. Kudo[2]
             [1]Frontier Research System, RIKEN,
              2-1 Hirosawa, Wako-shi, Saitama 351-0198, Japan
             [2]Department of Electronics and Mechanical Engineering, Chiba University,
              1-33 Yayoi-cho, Inage-ku, Chiba-shi, Chiba 263-8522, Japan

# Chapter 1

# Electronic and Electromagnetic Properties in Nanometer Scales

## 1.1 Introduction

Recent developments in nano-fabrication techniques based on self-organization and other techniques have put us into the position of being able to start investigations on novel functions in nanometer sized electronic devices. The nanometer region is the stage where both electronic states and electronic transport properties show mesoscopic natures. As the size of device elements becomes closer to the electron de Broglie wavelength, the electronic states and electronic transport properties manifest their quantum natures, so that the electronic behaviors depend strongly on the size and shape of the device element via boundary conditions for electron wave functions. The features of electronic devices are also determined by the character of the electromagnetic field associated with electronic motions, since their function is to control the transport of the electrical signal. In macroscopic conditions, such electromagnetic considerations correspond to of circuit design problems which are usually independent of the microscopic construction of each element. In nanometer scales, however, the construction and function of devices which determine the electronic and electromagnetic properties might be strongly related, so that the size and shape of each device element has a strong and immediate influence on the signal transport according to the electromagnetic response of matter and associated boundary conditions. An extensive study on interactions between electronic systems and electromagnetic fields is indispensable in order to understand the overall functions of devices. Such a study would let us take advantage of the peculiar properties of electronic and electromagnetic phenomena revealed at the nanometer scale for the realization of devices with novel functions.

During this decade, remarkable progress has been made in near-field optics, providing us with one of the key techniques for diagnosing and controlling the electromagnetic properties of nanometer-scale electronic systems. The techniques are being further extended not only to an optical method for nano-fabrication but also to the realization of opto-electronic devices with novel functions. The nanometer scale also corresponds particularly to the mesoscopic region for optical phenomena in the sense that it lies in the middle of the macroscopic scale of optical wavelengths and the microscopic scale of atomic constituents. As a result of these characteristics, the optical near-field would bring us extensive degrees of freedom in physical quantities which we can utilize for controlling states and motions of nanometer-sized objects.

As is noted, it is important in the realization of nano-electronic/opto-electronic devices to understand not only the independent properties of electronic systems and optical fields but also their interactions in confined spaces. However, these microscopic considerations alone are not enough to understand the overall properties of practical devices. Another indispensable key would be to implement a scheme by which microscopic phenomena are transformed into a macroscopic signal which is observable from the outside, and through which one could transport and control any kind of information. This would require us to be a composite analysis of the microscopic and macroscopic natures of devices in respect of all of the quantum mechanical, electromagnetic, and thermodynamical properties which produce the overall functions of electronic and opto-electronic devices.

The following chapters of this book present a number of recent developments in ideas, materials, fabrication techniques, diagnosis, analysis, and theories of electronic, optical, and composite processes in nanometer space. A reader who is engaged in research within related fields might proceed immediately to the concrete topics presented in the following chapters according to individual interests and then back to reconsideration of the basic problems set out in this chapter. Those who are not familiar with those subjects might start with getting an understanding of the general aspects of nano-electronic and optoelectronic devices before investigating the rapidly extending fields of nanometer technology.

In this chapter, we will first survey the basic features of electronic systems, optical fields, and interactions between these, where importance, peculiarity, and connections are investigated from the view points of the quantum mechanical, thermodynamical, and electromagnetic behavior of nanometer scales. We will discuss details of transport properties of electrons and electrical signals, near-field phenomena, quantum mechanical behaviors and observations accompanied by several instructive examples. Many of examples are those which we will encounter in the practical studies of nano-electronic and opto-electronic devices in the following chapters.

## 1.2 Basic Features of Electronic and Optoelectronic Systems

### 1.2.1 Quantum behavior of electrons in the mesoscopic region

In nano-electronic devices, one encounters the quantum nature of electronic systems such as the discrete character of electric charges and energetical excitations, the interference effects of electron waves, and quantum fluctuations. They have the potential to bring unusual functions into nano-electronic devices through their individual characteristics and mutual relationships. Quantum mechanical behavior is generally related to two different aspects of either the quantum state or quantum observation [1]. The former corresponds to an isolated system and the latter to a coupled system of a microscopic system with a classical observation scheme [2]. In this section, we will investigate these properties using some examples.

First, the discrete nature of a physical quantity in general often manifests itself as a result of quantum wave interference under certain boundary conditions. However, in some situations, it might arise as a result of observation process corresponding to a decoherence process or the destruction of a quantum state. One of the examples is the constructive interference of scattered electronic waves producing discrete energy states such as found in quantum wells and dots. This arises as a result of the properties of isolated quantum systems. Another is the decoherence process of electronic waves in a double-barrier tunneling which results in the blockade of the electron transport, called "Coulomb blockade", as a reflection of the discrete nature of the electric charge [3].

In some devices, a system with well-defined discrete energy states such as a quantum mechanical harmonic oscillator exhibits quantum fluctuations corresponding to zero-point energy in its low-excitation limit. We might be reminded of a device operating with fundamental excitations or a coupled mode of electronic system with electromagnetic field. This property is related to an isolated microscopic system. On the other hand, these quantum fluctuations are suppressed or squeezed by introducing certain quantum correlations and coherence between two or more coupled quantum systems.

These different properties in similar systems depend on the degrees of coherence time or relaxation time of the quantum system and on the degrees of isolation or correlation to reservoirs or other systems.

We can make use of this quantum mechanical behavior in nano-electronic devices with novel functions by actively controlling the degrees of coherence/decoherence and isolation/correlation of quantum systems. A more detailed discussion on coherence and decoherence process will be presented later in this chapter. We will firstly investigate the relationship of quantum systems to information transfer in nano-electronic/photoelectronic devices. Basically, this corresponds to a thermodynamic consideration of these devices.

### 1.2.2 Thermodynamic considerations in respect of electronic devices

Electronic devices in general serve as switches, amplifiers, and controllers or transformers of signals or information in certain well-defined directions, so that they inevitably have the properties of non-equilibrium open systems. Some examples are shown schematically in Fig. 1.1. A switching device turns on and off the flow of electrical signals from a source to a load which serves as a detector of electromagnetic signal. An amplifier controls the flow of electrical energy from a power source to a load according to an incident signal. A transformer or controller modulates signals coming from a source with an external control and this results in an output to a load. Each of these device elements produces a well-defined macroscopic flow of electrical energy from sources to electrical loads under the control of microscopic electronic behaviors.

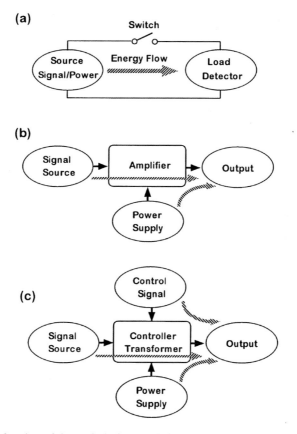

Fig. 1.1. Basic functions of electronic devices and their relationship to the flow of electric energy and signals.

With respect to these properties a careful consideration should be made both in relation to quantum mechanical behavior and of the thermodynamic properties of electronic systems which transform microscopic electron motion into macroscopic transport of signals. Here, we have two ways of evaluating the temporal evolution of mesoscopic electronic systems, one in terms of the Schrödinger equation and the other in thermodynamic terms, as schematically shown in Fig. 1.2.

The thermodynamic property is related to the irreversibility of the electronic process which regulates a unidirectional flow of both electrons and signals. There are two apparent categories; entropy increase and energy dissipation. The former is related to diffusive electron and signal transport from filled to empty spaces, and the latter arises when an electron goes across a chemical potential difference or a part of an electromagnetic signal is consumed and transformed into heat under the influence of an environmental reservoir. However, the difference between these two is subtle when the quantum

### (a) Quantum mechanical Evolution

### (b) Thermodynamical Evolution

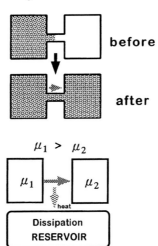

Fig. 1.2. Quantum mechanical and thermodynamical evolutions of systems with reversibility or irreversibility.

behavior of electrons is also taken into account. To see this, we will consider the quantum process of an electron tunneling through a potential barrier separating two closely spaced conducting electronic states extending to electron reservoirs with different values of chemical potential, as is schematically shown in Fig. 1.3. From a quantum mechanical point of view, this problem is related to the decoherence process of the quantum system due to observation taking place under the influence of an electromagnetic reservoir. From thermodynamic view point it is related to an entropy increase through which the electromagnetic field associated with the electron transport is lost into environmental reservoirs.

We note firstly that quantum coherence survives only for a short period when a macroscopic number of atoms are involved with the process under consideration. This is because even if a weak perturbation is applied to the system, the quantum coherence or information is lost into an enormous degree of freedom of the macroscopic system according to entropy increase. That is, there is a much greater chance of excitation transfer from the system to the reservoir than in the reverse process. Even though the quantum mechanical tunneling effect is a coherent process, its final-state wave function is immediately dephased due to fluctuations of macroscopic numbers of atoms which constitute

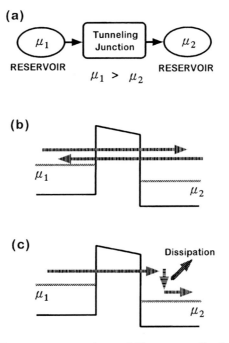

Fig. 1.3. Electron transport and reversibility at a tunneling junction.

the conducting material. The final state of the tunneling electron is related to extended states in conducting matter with lower chemical potential. Most frequently, the electron tunneling system under consideration is coupled with an environmental reservoir as shown in Fig. 1.4(a), so that for the establishment of thermal equilibrium in the final state, the electron energy corresponding to the chemical potential difference is dissipated via interactions with a reservoir, such as a reservoir of electromagnetic radiation or a reservoir of lattice vibrations. This process, in turn, is due to the macroscopic number of the reservoir states into which electromagnetic radiation or acoustic vibrations is lost according to entropy increase.

In contrast to these, there are several types of coherent nano-electronic systems composed of small numbers of electronic states, such as is found in resonant tunneling devices with narrow intermediate states in which the quantum coherence survives for a certain period, depending on environmental conditions such as low temperature. Provided that the system is perfectly isolated from external systems, the whole process is reversible until a load or detector consumes electron energy into heat, as shown in Fig. 1.4(b). Then the

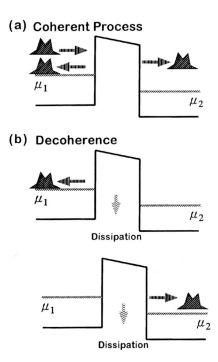

Fig. 1.4. Quantum mechanical behavior of an electron tunneling through a potential barrier and its decoherent final states.

temporal evolution of such an isolated system is completely due to the entropy increase between invalanced electron reservoirs with different chemical potential.

Sometimes the electromagnetic dissipation mechanism results in photon emission observable by means of both near-field and far-field photonic detection schemes. In this event, the decoherence time of the final state of the tunneling electron might be long enough to allow coherent interaction between the electron excitation and radiation or photons in a specific mode. This process is often referred to as a *photon-assisted tunneling*, and will be discussed later. Here we would remind the reader that investigations of electron interaction with electromagnetic field show us a number of important features of electron transport and their relationship to signal transport. This also implies that we can make use of such an electron-photon interaction to do a diagnosis with functional analysis.

It should be stressed that the nature of electromagnetic radiation also depends on the shapes and sizes of the surrounding matter as well as the method of detection. Therefore, a change in the environmental electromagnetic mode, towards which an excited electronic system would radiate, has a direct effect on the electronic energy dissipation lifetime. This aspect becomes especially important in the study of electromagnetic or optical near-field diagnoses of nano-electronic devices. Such an effect of the observation process on the result relates to one of the basic properties of quantum measurements. Further, in a mesoscopic phenomenon we will encounter the partly destructive but partly nondestructive nature of measurements.

We will now proceed to an investigation of the electromagnetic interaction of electrons in mesoscopic systems, one of the most important subjects dealt with throughout this book.

### 1.2.3 *Electromagnetic reactions to electronic behavior*

In the above, we saw that the functions of electronic devices are not directly concerned with electronic transport but with control of electromagnetic signals associated with electronic behavior. In other words, an electromagnetic field under the control of electronic behavior brings us information about the microscopic events taking place in the electronic system. Electromagnetic fields are also affected by the nature and shape of the surrounding matter, which in turn has an influence on electronic behavior. Therefore, a detailed consideration also has to be made of the self-consistent properties of the electromagnetic field which actually dominates information transport.

Regarding electromagnetic considerations, the nanometer scale also corresponds to the mesoscopic region between the macroscopic scale of the optical wavelength and the microscopic scale of an atomic size. We should take account of nanometer scale variations of the electromagnetic field, which oscillates at a frequency up to the optical region. This corresponds to near-field optical problems or quasi-static treatments of electromagnetic fields. At the nanometer scale, the macroscopic quantities set for the optical responses of matter, such as the refractive index, would lose their relevance since each of these represents an average of the microscopic responses of constituents over a space as large as the optical wavelength considered. However, every microscopic event of atomic scale is still hidden behind the behavior of mesoscopic electronic systems of the nanometer size.

Before proceeding to nano-electron devices, we will survey the basic instructive properties of an electromagnetic field in several typical material environments. So far, we have been familiar with electromagnetic fields in free space, in which we have considered optical or photonic processes of short wavelength limit at least with respect to photo-detection schemes. That is, even when a light wave is injected into a material medium such as a dielectric, microscopic interactions of the field with the medium are hidden by establishing a picture of light waves in a medium using the idea of a macroscopic refractive index which represents a macroscopic average of microscopic light-matter interactions over a spatial range corresponding to the optical wavelength considered. These pictures hold as long as we are interested in a long-range correlation for light-matter interaction describable in terms of propagating waves. At this stage, however, we should extend our investigations into subwavelength regions. Before starting our investigation into optical processes in nanometric space, we should remind ourselves of several basic properties of light-matter interaction in order to understand the many aspects of the properties of light and how light behaves.

### 1.2.3.1 Cavity electromagnetic mode and electromagnetic interaction

Interference of light waves sometimes results in discrete energy levels of photonic modes. For example, a discrete energy state of photonic modes is introduced using a cavity which consists of a material system exerting a coherent light scattering frequently that constructive interference results in the discrete cavity modes. It should be noted that the interactions of matter with such a discrete mode electromagnetic field drastically alters the radiation properties of the electronic system under consideration. One of the most remarkable effects is a suppression of spontaneous radiation in a high-quality optical cavity which causes an extinction of the photonic mode at which excited electronic systems would otherwise radiate. This is the so-called cavity quantum electrodynamics (cavity-QED) effect which has been under extensive study aimed at demonstrating QED-related phenomena for the concrete verification of fundamental physics as well as applications to control quantum states of electronic systems [4–9]. It should be noted that under a strong coupling of an electronic system with a cavity mode, the radiation process even alters the electronic states and results in a shift of the electronic energy levels. This is an example the electromagnetic boundary conditions exerting a strong effect on the interacting electronic states. It has been recently demonstrated experimentally that even a cavity electromagnetic mode in its vacuum state has a significant effect on the state of atoms in the cavity [10,11]. This might be quite natural if we remind ourselves of the QED treatments of electrons whose observed mass and charge are renormalized to include all possible interactions with the free-space electromagnetic modes of a vacuum, so that a modification of the electromagnetic vacuum requires us to correct the electronic process. We are also reminded of another profound problem to do with the mechanism of radiation, often described as no radiation being able to be emitted in the absence of an absorber [12].

### 1.2.3.2 Transition probability and final state density

We will now recall the golden rule of Fermi which governs the quantum mechanical transition probability per second $P_{\text{Transition}}$ of a system going from an initial state |initial> to a final state |final> with the state density $\rho_{\text{final}}$ under a perturbation operator $\hat{T}$ driving the transition;

$$P_{\text{Transition}} = \frac{2\pi}{\hbar} \left| \langle \text{final} | \hat{T} | \text{initial} \rangle \right|^2 \cdot \rho_{\text{final}}. \tag{1.1}$$

This formula suggests the importance of the final-state density investigated above for the cavity QED effects [13].

As an example, we will recall the spontaneous radiation process of an excited atom in free space via electric dipole interaction, in which the initial state corresponds to an atomic excited state plus electromagnetic vacuum and the final state to a lower atomic state plus a single photon state. The perturbation operator in this case is that of the electric dipole interaction described as a product of the atomic dipole operator and the electric field operator. The final-state density depends on both the spin state of the atom and the electromagnetic mode density. The final state of transition is established when the coherence in atom-photon interaction is destroyed by fluctuations in the relevant electromagnetic modes. Using a photo-detector which provides us with some visible results via electronic instruments, atomic radiation will be observed when we find a single phoon being consumed. Therefore, in practice, one should account for the final state density according to the observation scheme to be used and regardless of the compatibility of the mode description. We will go back to such quantum mechanical considerations of measurements in the following section.

Here, it should be noted that a photo-detector for a quantum mechanical system is not necessarily a single independent one but sometimes composed of two or more correlated or anti-correlated photo-detectors. An example is a pair of photo-detectors put after an optical beam splitter, where, in the weak intensity limit, a single photon is separated into correlated states of two output ports. These photo-detectors are equivalent to each other with respect to the quantum mechanical nature of the interaction with a correlated single photon, however either one of the detectors receives a photon as the result of observation in a counting mode. The important point is that the final state density should be evaluated with respect to the correlated or quantum measurement process, not to the result of classical observation processes of a stochastic nature.

### 1.2.3.3 States and operators in mesoscopic problems

At this stage, we should ask ourselves about the choice of initial and final states and the perturbation operator in our consideration of nano-electronic devices. In a mesoscopic problem, each of the initial and final states is a composite of interacting electronic systems and electromagnetic fields, and therefore the perturbation also corresponds to a specific portion of entire microscopic interactions. That is, the interaction of mesoscopic systems corresponds to a characteristic portion of the interaction taking place describing the overlapping or mixing of the initial and final states of composite systems of mesoscopic nature. Therefore, such separations into "initial", "final" and "perturbation" between these are dependent on the way in which our modeling and understanding of the phenomenon is introduced. According to Heisenberg's assertion, we should discuss physics on the basis of clearly established states and operators. However, mesoscopic problems retain extensive freedom in choice of states and operators according to individual interests. It is therefore important in any consideration of mesoscopic natures to pay attention to several different aspects of the problems. We will return to this point with some examples later in this chapter.

### 1.2.3.4 Multiple scattering in ordered material systems

Here, in order to obtain a clearer understanding of the mesoscopic nature of electromagnetic fields, we will survey another instructive process of coherent light matter interaction with an example of so-called photonic crystals resulting from a multiple scattering process in ordered light-scattering objects. We have recently noticed that multiple scattering light processes exhibit strong resonance behavior not only for optical cavities but also for spatially ordered scatters in the same way as an electronic system in crystals results in band structures in its dispersion relation. Extensive studies are being made of photonic crystals and photonic bands [14,15,16]. At the same time, we have found that a small disorder introduced into a photonic crystal results in a three-dimensional Anderson localization of optical modes [17].

Provided that an excited local electronic system is resonantly coupled to such a localized optical mode, its radiation characteristics would be strongly modified. This example is instructive for two reasons. Firstly, if the long-range order of the scattering medium is disturbed by any cause we will observe the light or photon emission which was inhibited in the original configuration.

Secondly, even if the original configuration is maintained to inhibit light emission into far field, there is still an optical near-field of a corresponding frequency in the subwavelength vicinity of the localized photonic state. Summing these up, we can observe an initially inhibited light emission by means of the destruction of symmetry or order in the material system or by a near-field observation with a local probe. Investigations of this system from different aspects of control and observation of electronic and photonic states provide us with a number of important ideas related to mesoscopic problems and the near-field and far-field diagnosis of electronic states. These problems will be discussed in detail in the following sections.

## 1.3 Mesoscopic Electromagnetic Processes and Coupled-Mode Descriptions

### 1.3.1 Models and understandings of mesoscopic systems

As the size of devices shrinks down to nanometer range for the purpose of dense integration or novel functions, the electromagnetic field associated with electronic systems also manifests unusual characteristics in confined space [18]. An consequence is that the field distributions will be strongly dependent on the local properties of material system, such as the shapes of material boundaries, and also on the non-local electronic responses in nanometer-scale systems [19,20]. In such a system, accounting for the signal transfer via interactions of the electromagnetic field with the electronic system should be done carefully. That is, for nano-electronic devices, no trivial or empirical relationships can be assumed between the transport of electronic carriers and the flow of electromagnetic energy as signal or information. Of course, for some of these systems, an electromagnetic scaling law might still be available on account of the spatial variables. However we must, at least, be able to identify those scalable problems from inscalable ones, which would exhibit strong size-dependent peculiarities. To this end it is useful to establish several model descriptions of electronic system in coupled mode with electromagnetic fields, and to signify the interactions between these coupled modes. In this section, we will briefly survey electromagnetic problems using this approach.

Here, we will investigate how we can describe and understand mesoscopic processes in nano-electronic devices. We should extract some important features of subsystems relevant to the functions of devices from the complex interaction processes taking place in the whole system. In general, our understanding of a composite system is gained by descriptions of each element and its interaction with other parts of the system under consideration. The ways of separating fundamental elements and coupling them to each other are neither explicit nor unique in nano-electronic devices of a mesoscopic nature. Therefore the way of describing and understanding the nature and functions of the system are dependent both on our model descriptions and questions. We should at least get answers to the following basic questions:

1.  What is the set of electronic, photonic, and coupled states in discussion?

(This is related to definitions of each element.)

2.  What is the perturbation to be considered?

(This relates to the interaction between elements.)

3.  What is the signal expected to be obtained?

(This relates to the global feature of the system.)

In order to understand the nature of each problem, we will survey some examples of model descriptions and understandings of electronic processes with some well-established ideas found in standard solid state physics or the physics of condensed matter as it relates to the macroscopic order of electronic systems.

### 1.3.2 Electronic modes in crystalline fields

In crystals such as metals and semiconductors, electronic behavior is determined by the interactions of enormous numbers of electrons with atomic lattices. Each electron is affected by the local potential field of each lattice atom as well as the long-range potential fields produced by the cooperative motions of electrons and their interactions with crystalline lattices with a specific order. In considering the macroscopic behaviors of electrons, it is convenient to make up a theoretical description which pays special attention to the long-range cooperative motions of electrons in several characteristic states. In doing this, every short-range interaction is included or renormalized into some adequately re-defined physical quantities such as effective mass and charges. The effective quantities will then describe the macroscopic behavior of electrons in certain collective modes with long-range order in response to an external field which is applied to control or observe electronic behavior in the mode. Bloch wave descriptions, electronic band structures, and so on are introduced into solid state physics in this way. These models and ideas provide us with a concrete framework for our understanding of electronic behavior in condensed matter.

One of the other important ideas that we should remind ourselves of is the hole description of electronic behavior in semiconductors. When we consider the electronic mode of a crystal which is almost filled with huge numbers of electrons which are able to approximate a state with certain distinctive features, it becomes convenient to start a theoretical description with the corresponding mode in a perfectly filled state. The absence of electrons in the mode could then be treated in terms of a positively charged electronic mode consisting of holes representing the overall responses of electrons, including their interactions in the mode under the influence of an external field. It is also well known that a similar description was once introduced by Dirac in his positron theory in relation to relativistic quantum mechanics. These ideas provide us the comprehensive theories in relation to many body problems.

Keeping these examples in mind, we will proceed to investigating descriptions of electronic systems in a coupled mode with electromagnetic fields. This will show us several important features arising out of the description and understanding of nanometer size devices. This is related also to the principles of near-field diagnosis.

### 1.3.3 Electrons in a coupled mode with electromagnetic fields

Let us firstly consider the concrete example of plasmon polariton to introduce the features of electronic systems in a coupled mode with the electromagnetic field and the interaction and observation processes relating to

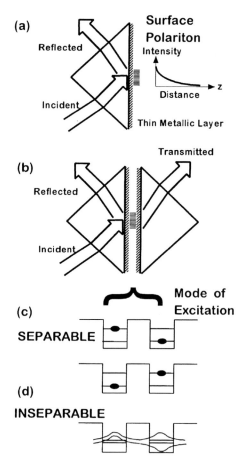

Fig. 1.5. Coupled mode description and its relation to interaction picture viewed from separability and inseparability of subsystems. Surface plasmon polariton is considered as an example.

coupled modes. This will provide also us with an introduction to quasi-particle descriptions of coupled modes [21].

We will take the example of a glass prism with a thin uniform layer of metallic material, outside of which is a vacuum or air, as shown in Fig. 1.5. Electrons in the metal conducting state behave like negatively charged gas in a positively charged homogeneous medium. A fluctuation in the electron density produces an electric field in its neighbor. When the density fluctuation is a local event within a distance called the Debye length, the reactions of neighbors to the electric field screen the long-range influence of the local fluctuation. In contrast, when the density fluctuation occurs over a macroscopic region beyond the screening distance, it produces a long-range electromagnetic interaction in the electron gas. This results in the collective motion of electrons showing a macroscopic oscillation at a characteristic frequency called the plasma frequency $\omega_p$ [22].

Usually the plasma oscillation in electronic gas is treated in the framework of second quantization as "plasmon" with the single quantum excitation energy of $\hbar\omega_p$. The plasmon mode accompanies an oscillating electric field in its vicinity, which sometimes produces a co-propagating electromagnetic wave with the plasmon mode depending on electromagnetic boundary conditions. Such a state in which electron-density waves are coupled with electromagnetic waves is called the plasmon-polariton mode in the second quantization context [21,23,24]. This signifies that a wavelike space-time correlation is produced in quantum process of plasmon excitation/deexcitation and associated photon annihilation/creation. In the case under consideration of thin metallic film fabricated on a planar dielectric surface, the metal-vacuum and metal-dielectric boundaries allow only waves propagated along the planar surface due to spatial translational symmetry. That is, the plasmon polariton mode satisfies both the electronic and electromagnetic boundary conditions at the same time.

### 1.3.4 Optical excitation of coupled modes

The co-propagating electromagnetic field in a plasmon mode cannot radiate energy into the vacuum-side half-space because the dispersion relation of plasmon-polariton cannot coincide with that of electromagnetic waves or photons in free space. This means that the wavelength of plasmon-polariton is shorter than the optical wavelength in a vacuum at the oscillation frequency of $\omega_p$. As the result, the electromagnetic field in this mode shows an exponential decay in the direction normal to the metallic surface in the vacuum side, referred to as an evanescent wave in wave terminology.

The plasmon-polariton mode can however be excited by an optical wave in a dielectric to the extent that the plasmon-polariton wavelength lies between the optical wavelength in a vacuum and that in a dielectric. This implies that a light wave in a dielectric also represents a electromagnetic field in a coupled mode with the electric polarization field in the medium and corresponds to another kind of polariton. These different types of polaritons will then interact resonantly provided that both the oscillation frequencies and the phases along the metal-dielectric boundary coincide. This is described as a coincidence between the dispersion relations. In contrast, if the plasmon-polariton wavelength is shorter than that of optical waves or polaritons in the dielectric medium, the electromagnetic field of the plasmon polariton in the dielectric side also shows an exponential decay, and therefore this mode is localized or confined in a two-dimensional space along the thin metallic film.

It should be noted that one of the restrictions imposed on interactions between these modes is due to the translational symmetry of the planar metal-dielectric boundary. Therefore even those plasmon polaritons with shorter wavelengths could be excited by perturbing the planar boundary by introducing small protrusions or corrugations such as a grating with an elemental size as small as the plasmon polariton wavelength [25]. We will go back to this point later in relation to the near-field diagnosis of the localized mode. We will now firstly investigate the interaction between coupled modes.

### 1.3.5 Interactions of coupled modes and subsystem descriptions

We will consider a coupling or excitation transfer between two of these plasmon-polariton modes defined for two electronically isolated similar thin metallic layers on glass prisms as described above. It will be assumed that one is excited by a light incidence with a resonant frequency and appropriate incident angle to satisfy the dispersion relation related to the metal-dielectric boundary condition. When two systems are put at a distance as close as the decay length of evanescent electromagnetic waves associated with plasmon polariton as shown in Fig. 1.5(b), the electromagnetic portions of plasmon-polariton modes overlap with each other and the interaction between plasmon-polariton modes results in an excitation transfer from an illuminated to a non-illuminated mode.

Here we might notice that the above descriptions of interactions depend on the way of our setting problems and asking questions. Indeed, we firstly assumed two independent but similar plasmon-polariton systems and then asked some questions about the interaction between two initially assumed well isolated modes, when the two systems are put closer to make an overlap of electromagnetic fields. The electromagnetic part of the plasmon polariton then will become an approximate concept in contrast to the electronic state which remains isolated.

In order to treat this problem in a rigorous way, we should rather consider the double-layered plasmon-polariton mode because the boundary conditions in the vacuum side of the metallic layers are no longer those of free half-space. In this double-layered model, the concept of evanescent waves in the vacuum side disappear from the problem as long as we consider far-field problems, in which we only have a double-layered thin metallic film with a narrow vacuum gap sandwiched by two glass prisms.

However our approach to this problem, based on interaction between two different plasmon-polariton modes, provides a good solution to questions about the excitation transfer, as long as the interaction is weak and multiple back and forth excitation transfer does not cause a drastic change in the plasmon-polariton modes relevant to the two separate electronic systems. For this weak coupling case we can interpret the double layered system in the following way. We will let one metallic layer be the left layer and the other the right layer. We can then describe the excitation transfer process as a transition from the initial state with an excited left layer and an unexcited right layer into the final state with an unexcited left layer and an excited right layer. This internal transition in the double layered plasmon polariton mode then deflects the incident light wave into a transmitted light wave. In this sense, the left and right layer could be separated with respect to their excitation so that our initial description of this problem as an interaction of two plasmon polariton modes holds as a good approximation of the entire process.

In contrast, for strong interaction conditions, we should directly solve the double-layered problem in a non-perturbative way, in which excitation features of the left and right layers is no longer separable. Fortunately, for this type of system the issue of excitation transfer between two well-defined layers is no longer applicable. The crucial point is how we identify these weak and strong interaction conditions. However, the mesoscopic region of these limits often gives rise to interesting problems.

### 1.3.6 Varieties of coupled modes

The discussion above shows us one important aspect of mesoscopic problems: that a model description is not only one involving alternatives but depends on the nature of the system one would like to consider. This also applies to near-field diagnosis of electronic and photonic states, which is discussed in the following section.

Indeed it is always possible in principle to construct a first-principle based description of the entire system of electronic states and interactions with electromagnetic fields, but this is effective only for limited cases which are concerned with a system composed of small numbers of electrons and well defined boundary conditions, environments, sources and detectors, and so on, from the view point of our computational ability as well as our understanding of underlying physics.

There are numbers of different types of fundamental excitations and their polaritons, such as the exciton-polariton as a coupled mode of electromagnetic field with excited pairs of electrons and holes behaving like quasi-particles in response to external fields. A coupled mode of electron spins and electromagnetic fields such as magnon polariton on a material surface might be one of the most interesting modes to be investigated further. We also find a special kind of coupled mode in superconductors as so-called Cooper pairs which represent the couplings of huge numbers of relieving or alternating pairs of electrons interacting with lattice distortion producing an unusual ground state of a coupled mode with superconducting properties. All of these coupled modes of electronic systems with fields have been potential candidates for the basic elements of electronic and photo-electronic devices realizing novel functions.

In order to make the most of these coupled modes in nanometer-sized devices, we should understand the properties of the localized mode and its coupling to an electromagnetic field. This is related to the problem which was briefly discussed above with respect to the excitation of shorter wavelength plasmon polariton modes. That is, even a subwavelength-sized structure under illumination produces a characteristic field in its vicinity as narrow as its spatial size. This involves polaritons having a shorter wavelength with a spectral bandwidth as wide as the inverse of the spatial size of the structure. This is one of the basic principles of near-field diagnosis with spatial resolution of subwavelength range. One of the comprehensive approaches to this important problem will be discussed later in this chapter on the basis of the angular spectrum representation. We will find numbers of novel theoretical treatments of electronic modes with localized modes of electronic systems in the following chapters of this book.

### 1.4 Quantum Measurements and Interpretations

In the study of nanometer-scale electronic and photoelectronic devices, we often encounter quantum observation properties as well as those of matter and field. These are related not only to diagnosis techniques but also to the functions of devices. We will consider these problems briefly from the viewpoint of nano-electronic devices and then proceed to investigation of the general features of measurements and diagnosis at the nanometer scale.

Quantum measurements are concerned with the process of deriving output signals from electronic devices operating with small numbers of electrons. Especially when the function of the device is concerned with electronic spin states, the problem requires a fully quantum mechanical treatment. Quantum features provide us with novel functions in signal processing with electronic and optoelectronic devices. For instance, an entangled state of several quantum systems and their coherent operation in an electronic device would allow us a parallel processing of signals as a coherent superposition of quantum states. Such quantum signal processing is under extensive study as quantum computing, quantum cryptography, and so on [26]. At the same time a continuous quantum observation process called quantum Zeno effect blocks a decoherent process [27]. We will briefly discuss the quantum nature of observation in electronic and photonic devices.

### 1.4.1  Features of quantum measurements

Basically, a quantum measurement is a repetition of destructive observations of identical quantum systems [1,2,28]. As shown in Fig. 1.6, an ensemble of similar quantum systems is prepared to be systems isolated from external world. They are then sent one by one into an observation scheme where each quantum system makes an interaction with a classical system to provide an observable result for us. Each one of the identical systems is made to come into contact with an observation scheme which provides a classical quantity as the result of measurement. The observed quantum system is then projected onto one of the eigen states of observed quantity. During this process, a part of the quantum coherence is destroyed. With a set of repeated observation procedures we can obtain enough information to estimate the initial state of the quantum nature. With respect to the flow of information, the measurement scheme should be considered as being a non-equilibrium open-system including a classical system with reservoirs which dissipate the quantum coherence. Aspects of the source of identical quantum systems are different depending on the Fermionic or Bosonic nature of the system under investigation. It is noted that a frequently repeated set of observations for a single quantum system still has a very interesting property referred to as the quantum Zeno effect and corresponding to an inhibition of transition of the quantum system from its initial state due to the continuous observation process projecting the system onto the initial state before a perturbation drives the system to establish a certain degree of final state amplitude.

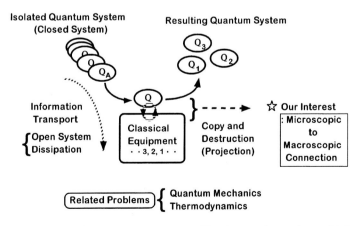

Fig. 1.6.  The concept of quantum measurement with respect to interaction and information transport between quantum and classical systems.

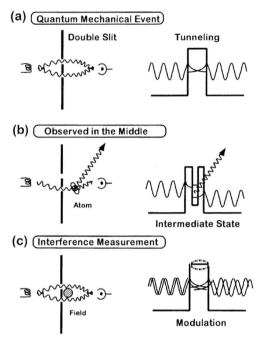

Fig. 1.7. Quantum mechanical processes (a) and possible perturbations on these with destructive (b) and coherent (c) natures.

As examples of the quantum observation process, we will investigate double slit and tunneling problems as shown schematically in Fig. 1.7. The quantum motion of a particle is described in terms of the transition amplitude between an initial state at one point of space-time to the final state at another point. It is assumed that the observation procedures are allowed only for the initial and final states, and therefore the quantum evolution should not be left alone. We can apply a potential field such as a double slit and barrier to control the quantum behavior of electron.

The quantum evolution of electrons is described in terms of quantum mechanical waves with the Schrödinger equation. It is sometimes described by a particle in the path-integral formulation where contributions from all possible paths between initial and final states are summed up with a phase factor imposed on each path. In any description, the quantum motions add up coherently to form a final state. The transition probability per second is described in the form of Fermi's golden rule, in which the absolute square is taken after summing up all the coherent contributions into transition matrix elements, which results in interference between different classical paths or superposed states. Here, a detector plays the role of a projection operator for putting the quantum system into one of the final states with classical identity. The observation process then should be repeated as many times as is necessary to establish the result of the measurement as an interference pattern or a number of particle counts in spatial or time domain, respectively.

### 1.4.2 Requirements for detectors

We will now pay some attention to the characteristics of detectors required to observe quantum effects in the double-slit and tunneling systems based on Heisenberg's uncertainty principle.

For the double-slit problem, a detector used for observing interference patterns must be well localized to allow electrons coming into the detector with a spread of lateral momentum $\Delta p_x$ along the slit displacement $\Delta x$ as wide as to satisfy the Heisenberg uncertainty principle $\Delta x \Delta p_x \geq \hbar$. This looks rather trivial for ordinary experiments, however it should be kept in mind as we consider nanometer-sized electronic and optoelectronic devices. We encounter an instructive example of this in near-field microscopy in which an ultrahigh resolution is achieved by using a highly localized probe tip allowing optical modes to pass with an extremely wide spread for the lateral momentum or wave number, being much larger than that of light in free space. This near-field character will be discussed in the following sections.

For the tunneling problem, a detection system for electron tunneling should operate in a duration $\Delta t$ long enough to identify the electron energy so as not to let it transmit over the potential barrier of height $\Delta E$. This satisfies the uncertainty relation $\Delta t \Delta E \geq \hbar$, corresponding to the shortest limit for this problem. In fact this discussion is concerned with the problem of tunneling time across the potential barrier, for which we still have numbers of questions, including the definition of tunneling time itself. The problem of kinematic understanding of tunneling process shows us the limitation and range of our understanding of quantum mechanical behavior as well as of the observation processes of quantum mechanical effects, which might be essential for a deep understanding of nano-electronic devices. However, we must resort to numbers of specialized articles and volumes for detailed discussions on this problem. These discussions which might be exhaustive would provide some useful suggestions for the realization of novel functions in nano-electronic and nano photo-electronic devices.

### 1.4.3 Destruction of quantum coherence

We will now consider what happens if we try to observe an electron undergoing its quantum motion. To do this for the double-slit problem, let us put an atom which scatters electrons in the close vicinity either of the slits and makes some effects observable to us. As we find in several detailed discussions in quantum mechanics textbooks, this observation destroys the interference pattern at the observation plane. Here we note that for microscopic devices of the double-slit type, the quantum properties still remain until observation of the scattered atom is accomplished. In the duration between the scattering and the observation of either electrons or atoms, a quantum correlation having the nature of superposition of the scattered and unscattered electron states still remains. This type of state could be used as an entanglement of quantum systems for the realization of novel functions of nano-electronic devices.

An observation scheme implemented in a tunneling barrier is instructive in showing us that if we want to observe electrons in the barrier with incident energy below the barrier height, we should either bore a hole in the barrier to produce real electronic states in the barrier region or excite the electrons to the state above the barrier potential, as shown in Fig. 1.7(b). Though the former still maintains the tunneling problem, what we will observe is not the electronic state in the tunneling barrier but that just after the tunneling of the first half of the double potential barriers. Therefore in this sense, we cannot ask that the electron be observed during tunneling. However we can disturb electronic behavior during its tunneling through the barrier. This is the third problem, discussed in the following.

### 1.4.4 Quantum perturbations and measurements

Let us consider a possible coherent perturbation on the quantum state of the electron under its quantum motion through a double-slit scheme or potential barrier. This corresponds to a phase modulation of the electronic state (see Fig. 1.7(c)).

In the double-slit problem, one can introduce a solenoidal vector potential field around the double slit which results in a phase difference between electronic states coming through one of the double slit to the other according to the nature of vector potential as the gauge field with respect to the phase of electron wave functions known as the Aharonov-Bohm effect [29]. This produces a change in the interference pattern on the observation plane.

In the case of the tunneling problem, a coherent perturbation corresponds to the introduction of another double-slit problem for the quantum behavior of electrons. Let us firstly consider a spatial modulation of the potential barrier keeping the average height the same as before. The electron wave function after tunneling is then spatially modulated in a similar way to in the double slit problem, though in this case no real path is assumed in the barrier region. The modulations are only due to continuity relations at the edge of the barrier between the wave functions of propagating modes and those in the barrier region with a different penetration depth depending on the barrier height modulation. This would alter the tunneling probability of the electron as the sum of the interference patterns on the observation plane.

A more interesting problem is posed when we introduce electronic spin states into the tunneling problem. When we apply a static magnetic field inside the potential barrier in a direction perpendicular to the incidental electronic spin, the potential barrier for electronic spins with its projection onto the direction of the magnetic field, up or down, results in two different barrier heights for incidental electrons in a superposition state of up and down spin states with respect to the applied magnetic field. In this way we have produced another type of double-slit problem for electron tunneling process. This problem can be treated from the kinematic viewpoint of quantum behavior as a Larmor precession of spin taking place in the potential barrier as a result of an applied magnetic field. One of the tunneling time considerations is based on this phenomenon.

### 1.4.5 Electron spin measurements

In the above we survey the properties of electronic quantum behavior, their destruction and observation of them, and the coherent way of perturbation. These examples might serve as a basis for considering the functions of electronic and photoelectronic devices of nanometer size. As a summary of this section, we will make some comments on spin devices and optical near-field interaction from a quantum mechanical point of view.

Electronic spin brings another degree of freedom to electronic devices. Most of the spin-dependent features so far utilized in our devices are related to magnetism corresponding to the macroscopic order in electron spin systems. We might of course be able to extend some of these concepts relating to spin magnetism down to the nanometer region by miniaturizing a domain of correlated spin. However, the real quantum nature of electron spin would manifest itself in a much more delicate way. We can find the quantum nature of electron spins in a degenerated multiple of electronic states, or in their interaction with orbital angular momentum. The former results in selection rules like those for chemical bindings and for transitions between electronic states in external magnetic fields, and the latter provides various ways of making indirect measurements of electron spins.

For instance, the electron spin resonance technique for atoms and molecules reveals the multiple nature of electronic states under the influence of external magnetic fields, and an optical spectroscopy provides information about the electronic spin via optical transitions between electronic states with different orbital angular moments, each of which interacts with electronic spin via spin-orbit (L-S) coupling. We then have no direct way of making spin observations of the electron in its free motion. According to the purely quantum nature of electronic spin, we should pay special attention to local fields and local angular orbital motions of electrons in considering the local spin state of electrons and its application in electronic and optoelectronic devices. This is because the local characteristics of these fields and electronic motions are strongly dependent on the shapes and sizes of the surrounding materials, especially at the nanometer scale.

Here, we should also take great care that we are not prejudiced by our understanding of macroscopic phenomena. Both experimental and theoretical studies on spin-related microscopic or mesoscopic phenomena are at the starting point in the search for applications for devices with ultimate novel functions.

A number of fundamental problems to do with the relationship between the quantum mechanical motion of electrons and electromagnetic fields still remain, some of which we will discuss in more detail in the following sections.

## 1.5 Measurements and Diagnosis at a Nanometer Scale

Developments in science and technology have been strongly promoted by innovations in techniques of measurements and diagnosis. The extensive progress in research into electronic and photoelectronic devices at a nanometer scale during this decade have actually been largely driven by progress made in spectroscopy, such as using low energy electron diffraction and novel sources of light, and ultra-high resolution probe microscopy such as scanning tunneling microscopy, atomic force microscopy, near-field optical microscopy, and variations and composites of them. A number of these innovations have been introduced into many reference books and further extensive developments of these techniques will be presented throughout this book.

Of course, a number of diagnosis procedures for microscopic material structure, such as X-ray and electron microscopy, were developed in the 20th century. However, these high-energy oriented methods aiming at localized probing are not adequate for doing functional analyses of nano-electronic devices working in the energy range of up to several 1 eV. We need to find some diagnostic procedure aimed at the nanometer-resolution locality keeping moderate or low probing energy. In this regard, one of our great achievements is the STM spectroscopy method [30,31,32,33]. Also for the purpose of functional evaluation of nano-electronic devices in terms of both electronic and electromagnetic behavior, optical near-field diagnosis provides other useful features. We will now briefly survey the basic ideas and characteristics of atomic and nanometer resolution probe microscopy. As has already been mentioned, we should pay attention to both electronic and electromagnetic behaviors, even for electronic devices.

### 1.5.1 Near-field measurements and diagnosis

The diagnosis of structures and functions of nano-electronic devices requires a composite measurement of material shapes, electronic states, their excitation and relaxation processes, and interactions with electromagnetic fields. Though the size of each element in devices and the probe tips of our microscopes are microscopic, they should also be equipped with a detector of a size remaining still in the macroscopic region, since it would provide information to us of a macroscopic size. Systems which connect microscopic events to macroscopic signals, in order to accomplish local probe techniques including spectroscopy should also therefore be implemented. Such near-field measurement systems may be regarded as a kind of electronic and optoelectric device, in respect of which careful attention should be paid to information theory or thermodynamics, as well as the nature of destructive measurements.

The basic system involved in near-field diagnosis is putting the tip of a microscopic probe close to a measurement object and enable them to make a local coupling which would finally result in a qualitative or quantitative change in the signals detectable from outside the system as a whole. This device consists of several elements of different characteristic sizes, depending on their functions, interacting with objects of a microscopic size and providing connections between microscopic subsystems and macroscopic sources and detectors. In such a system, we must know the properties of the probe tip and object in coupled mode and the mechanism whereby the local response of the coupled mode is transformed into a macroscopic signal to the detector. Let us investigate these properties in terms of three essential functions of near-field measurements; "local coupling", "source-detector isolation", and "near-to-far connection" [18].

### 1.5.1.1 Scanning tunneling microscopy, STM

As an example we will firstly study scanning tunneling microscopy (STM) according to the three essential functions mentioned above. The microscopic element is an electron tunneling effect through a potential barrier which is formed by an objective atom on a conducting surface and an atom at the apex of a metallic STM probe tip. The tunneling probability per second is determined by the local density of states around the objective and probe-tip atoms, which is described in terms of Bardeen's tunneling current. This corresponds to "local coupling" in the STM process. As a result of electron tunneling, a macroscopic current is driven in an external electric circuit, resulting in an electric (electromagnetic) flow of energy from a power supply to a load, giving us a certain amount of evidence that a microscopic electron tunneling event is taking place. This corresponds to a "near-to-far connection" in the STM process, which is implemented in a rather trivial way in STM, without detailed considerations. By counting the number per second of the events via macroscopic current we will know the "local density of state" of electrons coupled to the specific single atom on the probe tip. The coincidence between microscopic electron tunneling and macroscopic current is guaranteed by the fact, or assumption, that the STM system has been implemented so as not to drive any current in the circuit except for electron tunneling via the electronic state at the specific single atom of the probe tip. This feature corresponds to "source-detector isolation" in the STM process. In order to produce an STM picture as a two-dimensional map of the observed current, one needs auxiliary information about the position of the probe tip scanned on a plane above the objective surface. Instead of doing this, the control variable corresponding to the vertical position of the probe tip is often mapped, resulting in a constant current in the external electric circuit, which also provides information about the local density of the relevant electronic state.

Here we have made observations about the subtlety in descriptions of the STM process, which actually involves a number of delicate problems in respect of both microscopic and macroscopic behaviors. There will be detailed discussions on each subject in the following chapters in this book. Here, we will continue trying to extract some general features of near-field probe microscopy and make several comments about interpretations of these processes.

Before proceeding to these discussions, we will look at the properties of local coupling utilized in atomic force microscopy (AFM) and near-field optical microscopy (NOM) in order to introduce different aspects of near-field measurements.

### 1.5.1.2 Atomic force microscopy, AFM

The idea of AFM relies on the local measurement of attracting or repulsing forces between the surface of the object and the tiny probe tip. This corresponds to a potential gradient measurement. Force corresponds to the momentum exchange per second between the object and the probe and can be describable as the exchange process of particle or excitations in a field theoretical framework. Therefore, a force measurement has the quality of being partly compatible with transmissions of excitations or particles.

A microscopic to macroscopic transformation scheme in AFM is usually obtained by using cantilevers for its motional change due to the local force on the tip. For a relatively soft cantilever, even a weak force exerted on the tip could result in an observable bending of the whole cantilever, which is often observed as a directional change of light beam reflected by a mirror installed to cantilevers. However such a large bending of the cantilever causes a nonlinear modification of the probe-surface interaction so as to involve large numbers of atoms on the objective surface around the interaction region. This looks quite natural if we remind ourselves that the local force on the tip of cantilever results in a macroscopic bending involving a large number of composite atoms. The result of this is that the AFM image involves a wider range mechanical correlation on the sample surface, which sometimes results in the image having a periodic nature, showing a quality different from those due to the locality of interaction.

On the other hand, an AFM measurement with a relatively hard or high-stuffiness cantilever avoids the disadvantages of long range effects at the expense of simplicity in the probe bending detection system. Instead, a phase-sensitive measurement could be adopted for sensing change in the vibrational motion of the cantilever around its resonance frequency due to the minute effect of force exerted on the tip. This would realize a real local measurement of the mechanical interaction and actually provide an atomic resolution image, as is presented in the following chapter of this book. Such a long term resonant measurement with high-Q system or low-noise resonance circuit is an example of resonance enhancement and repetition measurements of microscopic events.

### 1.5.1.3 Near-field optical microscopy, NOM

The NOM corresponds to the general scanning probe microscopy techniques based on excitation transfer between local modes of object and probe tip. When these excitations are described in terms of quasi particles, the NOM process is quite similar to that of STM. The types of the excitations utilized in NOM measurements are very varied, and include photons (virtual), plasmons, excitons, magnons, and their polaritons. All of these represent the electromagnetic field in a coupled mode with matter. The problem of light matter interaction in a nanometer space is one of the main subjects throughout this book. We will present a detailed discussion on NOM and related systems later in this section.

### 1.5.1.4 Compatibility between force and transmission measurements

We will now make some observations about the compatibility between excitation transmission and force. In general, both transmission and exchange of particles or excitations exert mechanical force. For example, an electron exchange exerts force on the STM tip, which might enhance the tunneling current under certain conditions [34], and optical near-field interaction exerts force between the probe and the object. Similarly, one can see compatibilities in the electrostatic force and charge exchange between electronic states or excitation exchange between electric polarizations, magnetic force and spin exchange, and so on. The property important to near-field microscopy is the dependency of those transfer and exchange processes on the shapes, sizes, and qualities of electrons, photons, and excitations in coupled modes with environmental matter.

### 1.5.2 General requirements for near-field microscopes

As we have looked at examples of near-field microscopy, let us proceed to investigations into the fundamental requirements involved in general near-field measurements.

As we have seen from the discussion of plasmon polariton transfer, the process of near-field measurement is in a rigorous sense an inseparable system in which the object and the probe tip are in a coupled mode under external excitation. From this view point, a near-field measurement corresponds to a destructive measurement of the objective field by putting a probe tip providing some information about near-field effect into macroscopic detectors placed at the far field. However, as the process in the microscopy, we should be able to state, by any means, that an object or objective field of such a nature can be observed by a probe of such a nature. To do this, separability of objective and probing systems must be established, at least to a good approximation. An example of such separation has been already discussed for the transfer of plasmon polariton between two closely spaced identical metallic systems. We should establish well-defined descriptions of the object and probe fields corresponding to the classifications of local interactions we have studied above.

### 1.5.2.1 Range of interaction and identity of local probe measurements

An important point about the range of interaction which determines the local coupling of sample with probe tip should now be mentioned. That is, we should identify whether the field distribution at the probing point could be attributed, in a good approximation, mainly to the local nature of the object. Sometimes the local field strongly depends on the result of longer-range interaction with its neighbors. An example is seen in STM, where electronic local density of state at each probing position strongly reflects the band structures of the atomic layers near the surface layer determined by multiple scattering of electron waves in the crystalline lattice. Because of this fact in an interpretation of the STM image we should state that the STM probe tip has measured the local density of the electronic state of certain energy through the magnitude of the tunneling current. We could neither state that an STM image shows the local shape of the potential barrier nor that it provides an image of atoms on the objective surface. For several specific material surfaces, the atomic scale topography coincides with the observed STM image, verified by means of a elaborate numerical analysis of the electronic band structure as well as the interaction of certain definite atomic structure with STM tip, as will be discussed in detail in the following chapters.

Now we should at least be able to clearly identify the periodicity and locality of observed near-field images and their relation to the object structure and surface modes. That is, the idea of periodicity belongs to the macroscopic category, since it is related to symmetry under spatial parallel displacement along the observed surface. On the other hand, the idea of locality is reflected in the broadness of the spatial frequency spectrum in a Fourier analysis of observed images. This nature will be clearly formulated for near-field observation later in this section by using an angular spectrum representation of scattered field.

One of the empirical ways of identifying the nature of the locality from the long-range correlation is giving a numerical estimation of the ratio of the size of the local object under consideration to the characteristic wavelength of the source field utilized for measurement. In the case of STM, this ratio is of the order of unity because the object corresponds to atomic size, which determines both the size of a local object and a typical distance between the object and the probe tip, and the source field wavelength corresponds to the electron de Broglie wavelength for the usual energy range considered in STM. In the case of NOM, the ratio is much smaller than unity because of the nanometer size both of the probe tip, and the sample-object distance is much smaller than the optical wavelength of illuminated light of around one micron. Generally speaking, when the ratio is small enough, the wave nature of the source is negligible, so that the near-field approximation holds as a good approximation but when it is close to unity the long-range scattering and resulting interference effect are significant when considering the results of near-field measurements.

### 1.5.3 Local probe measurement and macroscopic spectroscopy of local mode

It is instructive to compare the qualities and advantages of local probe measurements with those of spectroscopic observation of local phenomena from the far field. Observations of local modes such as surface electronic states and impurity states is realized by spectroscopy techniques using far-field observations, provided that resonant behavior arises in the scattering of electron, light, and other particle beams. Because of having weak signals compared with bulk originated modes, such a spectroscopic method is useful only when the frequency of the local-mode resonance falls on a bandgap in the bulk-mode spectrum. Otherwise we should install an ultra-sharp band-path filter which only lets the local resonance signal come through its spectral window, which is actually infeasible in ordinary situations.

On the other hand, a local probe makes a direct connection with the local mode by means of spatially selective properties, so that the macroscopic bulk-mode signals can scarcely be transmitted through the tiny area of the local probe tip. Such a filtering character or isolation system for signals from macroscopic modes is one of the greatest advantages of local probe measurements. A combination of this filtering mechanism with mesoscopic phenomena having shape dependent properties for local probe tips with objects in coupled mode then provides a basic system for near-field observation and diagnosis at a nanometer scale. We will discuss this in detail with practical examples of near-field optical microscopes in the following sections.

To conclude this general near-field study, we note that the introduction of resonance properties in the local probe method also increases the selectivity of local events and provides local probe spectroscopy or enhancements of signals. In addition, injection of excitation into nanometric electron systems is one of the most important ways that the local responses of nanometric semiconductor devices can be investigated [35].

## 1.6 Electromagnetic Signal Transport as Circuit Design at a Nanometer Scales

### 1.6.1 Transformation of local effects into macroscopic signals

An electronic device is installed with several basic functions, including switches, amplifiers, or controllers of electromagnetic signals or power. These functions become active when they are connected to an external circuit, including sources of electric signals or power and load-consuming electric energy as a detector for outputs. The overall function of the device is therefore described in terms of energy transport between reservoirs in different equilibrium states. In a standard electronic device of macroscopic size, the transport of electrical signals is considered in terms of the circuit design, which provides knowledge about the correspondence between microscopic electronic behaviors and macroscopic signal transport via an electromagnetic field. On the other hand, each device element is often studied in solid state electronics in terms of its individual character, apart from any circuit configurations. Because of this, we are sometimes inclined to signal transport properties and electronic behaviors in the same class. However, each of these basically belongs to a different aspect of electromagnetic interaction, and therefore the functions of electronic devices, such as switches, amplifiers, and so on, are realized only in connection with circuit configurations. For example, in a transistor circuit such as an emitter follower, a similar system of carrier injection and transmission provides control of signal or power transmissions in various ways according to each circuit design. We should reminder this crucial point especially when we are considering signal transmissions in nano-electronic devices.

In contrast with ordinary electronic devices of macroscopic size nano-electronic devices require us a careful consideration to carefully consider how electron and electromagnetic properties relate to each other because of the mesoscopic nature. For instance, at the microscopic limit, it is not trivial to ask whether a single electron transport produces any electromagnetic signal, which would allow us to extract some information about the microscopic process. As we have discussed in the previous sections, the mesoscopic behavior of electrons and electromagnetic fields are both under the influence of the shapes and properties of surrounding matter. Detailed investigations into the conversion of microscopic processes of electrons into macroscopic electromagnetic signals provides us the principles of circuit design using elements at a nanometer scales. It might now be instructive to remind ourselves of fundamental processes of electron tunneling and resulting macroscopic current as seen in STM, in which the microscopic process of each electron tunneling produces an observable current in an external circuit controlling the flow of electromagnetic energy from a power supply to a load as evidence of the tunneling event. Here, we can either establish or assume that there is a simple one-to-one correspondence between the microscopic tunneling process and the macroscopic current due to the fact that no other transport channel can be allowed in the system configuration. Here, we again encounter the isolation system discussed for near-field observations. Indeed, there always is a local electromagnetic field in the close vicinity of any electronic system in microscopic structures, but it can be observed only when a specific system is implemented to convert the microscopic field into a macroscopic signal providing some information for us. In the following section, we will investigate several examples.

### 1.6.2 Directionality of signal and transport of electronic carriers

With respect to the transport properties of carriers or electronic charges, in general, an electronic device has three terminals corresponding to input, control, and output, respectively. The function of each terminal is therefore determined with respect to the external connections to power supplies and loads, i.e. the directionality of electromagnetic energy and signal. In contrast, a microscopic electronic process is governed by quantum mechanics and shows coherent behavior of a reversible nature. In order to obtain any information from such system, one has to install an observation system which produces an one-directional flow of signals at the cost of energy dissipation and decoherence. In general, such a quantum electronic observation system is produced by making a copy of electronic behavior in terms of an entangled electromagnetic field, through which the quantum coherence is destroyed by means of electromagnetic energy dissipation so as to provide an output signal and to make the entire process irreversible. In principle, a sufficient number of repeated observation cycles is required to obtain any information about a quantum ensemble of identical electronic systems.

In order to understand such a coherent/decoherent feature and the observation process, it might be instructive to compare electronic circuits with optical circuits of conventional type. In general, an optical circuit utilizes the scattering, diffraction, and resulting interference of optical fields both in the spatial and temporal domains. We often count this parallel processing feature as a major advantage. In such a system, an observable signal is obtained only by consuming the optical output by a photo detector, so that the entire process is reversible until the optical field is dissipated at one of the photo detectors, a basic feature of any coherent device. In this system, it can be said that the signal transport, i.e. the direction of temporal evolution, is determined in terms of entropy increase in the photonic system. That is, a signal is transmitted in one directional from a source being filled with photons to a detector which is empty. In contrast to such optical circuits, the usual electronic circuit provides an electromagnetic signal apparently in a different way, where electrons produce a bi-directional flow between electron sources and sinks, and a direct observation is not performed for the electrons but for the electromagnetic field via the interaction of the charge current with the electromagnetic environment. Since electronic circuits are in general composed of current loops, a flow of electrons will therefore only serve to control the unidirectional flow of electromagnetic energy. In this process, electromagnetic dissipation takes place due to electron relaxation passing across the chemical potential difference and results in the irreversible transport of electronic carriers and electromagnetic signals. Sometimes an electronic circuit is described in terms of electron transport between two or more electron reservoirs with different chemical potentials. In this case, a real signal transport is accomplished when an electron exerts relaxation between the chemical potential differences with the help of an interaction with electromagnetic field, which is usually hidden behind the action of the reservoirs. That is, to the extent that the electron transport results in a charging of the capacitor formed by each pair of electron reservoirs, the process of electron transport is, in principle, reversible. In electronic devices, each element is biased and produces a unidirectional flow of electromagnetic signals according to external controls so as to contribute to the function of the entire circuit.

The recent development of mesoscopic electronic devices make it feasible to produce a coherent electronic device in which the signal flow is reversible until an observation scheme causes destruction of the quantum coherence of the electronic system. A decoherence process is implemented in an usually rather trivial way in macroscopic systems consisting of a macroscopic number of quantum elements, since the decoherence time is so short that the system is sure to behave like a classical system. For a mesoscopic system composed of rather small numbers of atoms or molecules one should be aware of the importance of the decoherence mechanism. In order to provide either quantum or classical logic, according to the purpose to which it will be put, one can implement a mesoscopic system with either a coherent or incoherent function.

### 1.6.3 Scanning tunneling microscopes and electric signals

We will now investigate the dissipation mechanisms of electronic systems and the character of signal transport for STM. Here STM is viewed as a device for regulating the flow of the electromagnetic signal according to atomic or nanometer scale control parameters. The following example might help our understanding of the interaction process between electric carriers and electromagnetic fields.

An STM probe-tip is considered to have an atomic protrusion at its apex and it interacts with the electronic system of a sample surface via overlap of the electronic wave-functions of the tip atom and the surface system to be observed. The magnitude of interaction is therefore dependent on local variables, i.e. local density of state of the electronic system at the sample surface. Depending on bias voltages applied between the sample and the tip, the resulting electron tunneling events are observed frequently enough to provide information about the microscopic process in the form of a current in the macroscopic electrical circuit connected to the sample-probe system.

Although is rather trivial in this case, it is worth noting that an STM probe-tip realizes these functions not only with a protruded atom exerting local interaction but also with all others providing a micro-to-macro connection for the transfer of the tunneling electron into a conducting state in a macroscopic wire carrying electrical current. We can consider a protruded atom at the STM tip as being a microscopic element of the electronic device regardless of whether it was produced by chance or by a controlled process such as ion beam lithography. This element exhibits its function when it forms an electronic circuit by being put in an atomic-scale vicinity of object surface and being connected to an external ampere meter or a load serving as a detector of the electric signal. The observed quantity, i.e. electrical signal carrying information, is not each electron but electrical energy flow observed in its function of biasing voltage and the position of the tip atom, as shown in Fig. 1.8.

The quantum coherence of electrons in each tunneling phenomenon is lost via relaxation of electron or electron energy dissipation between the chemical potential difference between the tip and the sample surface driven by the electromagnetic environment, regardless as to whether it is observed by us or not. Before the detector or load dissipates electrical energy, the quantum coherence is already partly lost and the flow of information is made irreversible. Any dissipation processes contribute to make the flow of information irreversible. Performance and applications of this device is then evaluated in terms of the efficiency or yield of information that is obtained in the form of an electrical signal per each single quantum process of electron tunneling. Fortunately for STM, such a yield is obtained to a perfect degree because the

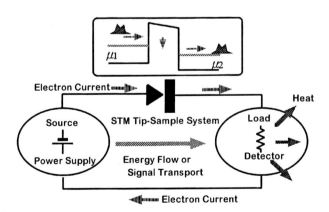

Fig. 1.8. The relationship between electron transport and electromagnetic energy flow in the case of scanning tunneling microscope, STM.

tip plus surface system is very well isolated and only via electron tunneling can the electric charge flow in the external circuit. This corresponds to one important feature of system isolation discussed as one of the general requirements for near-field probings.

The degree of isolation is usually expressed in terms of the tunnel conductance which should be highest in the overall circuit configuration. This fact makes the function of STM devices rather trivial. However we stress that the STM process was nor trivial when it showed the first atomic scale picture of $7 \times 7$ structure of Si surface [36].

## 1.7 Electrical Signals in Coherent/Incoherent Electronic Devices

### 1.7.1 Coherent and incoherent electronic devices

In nano-electronic devices, the coherent behavior of electronic Schrödinger waves is expected to play an important role in realizing novel functions, such as performed by quantum wells, quantum wires, quantum dots, resonant-tunneling devices, and so on. According to quantum mechanics, in order to take advantage of coherent properties, we should repeat a considerable number of cycles in which the system is first kept in free evolution for a certain period to enable quantum results to be obtained, and then it is observed via a decoherence procedure to obtain classical results. Such a controlled set of coherent and decoherent procedures is one of the basic requirements for measurements of quantum mechanical effects.

We will now investigate an example of both coherent and incoherent behavior in electronic systems and signal transport characteristics by comparing two typical types of electron tunneling processes, Coulomb-blockade and resonant tunneling phenomena. We will consider a double-barrier tunneling device which has an intermediate region bounded by the double potential barriers and the extended regions outside the barriers, as shown in Figs. 1.9(a) and (b), and assume that the intermediate electronic state consists of several discrete energy levels. When an electron bound in the intermediate state has certain lifetime, or in other words it is kept coherently for a duration long enough to reflect back and forth between the potential barriers, the tunneling event through the double barrier takes place in a coherent way and the interference effect due to multiple scattering between the barriers will result in a sharp resonant transmission dependent on the incidental electron energies.

On the other hand, when the decoherence time is considerably shorter, i.e. shorter than the characteristic time for establishing a resonance in the barrier and also shorter than the shortest range of the tunneling time through the double barrier, the decoherence process forces the electrons into either of the states "be there" or "not be there" in the intermediate electrode, which corresponds to an implicit observation of electrons at the intermediate electrode. This in turn results in a change in the electrostatic energy of the intermediate

**(a)  Resonant Tunneling**          **(c)  Equivalent Circuit**

**(b)  Coulomb Blockade**          **(d)  Equivalent Circuit**

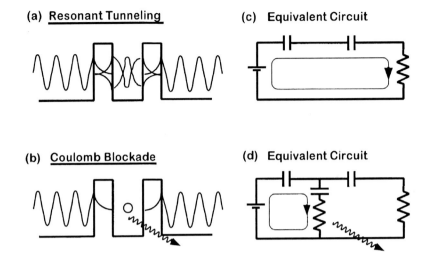

Fig. 1.9. Electron tunneling through a double barrier system and its equivalent-circuit description showing either coherent or incoherent behavior in the intermediate state.

electrode by an amount corresponding to a single electronic charge and this blocks the tunneling of another electron coming into the intermediate electrode provided that the bias voltage is not high enough to charge the intermediate electrode with two electronic charges.

These examples clearly show us that two different aspects of the electron transport problem for the double barrier configuration arise, depending on the decoherence time of the electronic quantum state in the intermediate electrode. The decoherence process is due to interaction of the electron with environmental systems which therefore receive information about electron tunneling through the first barrier. In contrast, in the case of resonant tunneling, the electron tunneling through the double barrier system should be perfectly isolated until it is observed to have come through both the potential barriers, so that no information can be obtained about the electronic existence in the intermediate electrode. In the latter case, what we can do is to estimate the probability of the electron sitting in the intermediate electrode if we dare do such observation of it in a destructive manner.

In Fig. 1.7, we have already provided examples of such an understanding of coherent/decoherent properties, comparing two quantum mechanical processes and related observation processes for the electron tunneling through potential barrier and describing the well known gedanken experiment of the double-slit experiment with electrons.

It would now be especially instructive to compare the quantum behaviors of electronic de Broglie waves with an optical analogue. Let us take the example of a Fabry-Perot resonator which consists of a set of two optical mirrors with almost perfect reflectivity, and which reflect back and force the incidenta optical wave between the mirrors to build up an interference between incident and reflected field, resulting a strong resonant character in the entire transmission/reflection properties as shown in Fig. 1.10.

Under certain conditions this corresponds to an optical analogue of the resonant electron tunneling through the double barrier system which we considered above. The most significant feature of this system is, of course, the strong enhancement of the transmitted light field for the resonance wavelengths that almost all of the optical field is transmitted through set of two mirrors of ~100% reflectivity. The performance of this system, the so called Q-factor or finesse, is determined by the decoherence time or the decay rate of the optical field inside the resonator. When the decoherence time is long enough for the optical field inside the cavity to build up a strong interference enhancement, this results in a constructive interference in the transmitted field and a destructive interference in the reflected field of the resonator system. The sharp resonant transmission behavior is observed only for this condition. In contrast, when a minute loss is introduced inside the resonator, the optical field in the cavity cannot establish a field intensity strong enough to cause resonant behavior. With perfect reflectivity of the two cavity mirrors, the light transmission through the Fabry-Perot resonator can be compared with the resonant electron tunneling through the double barrier system. Here, any

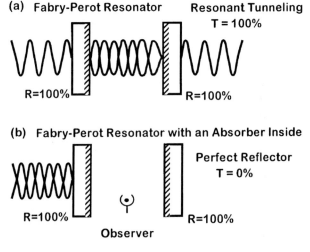

Fig. 1.10. Light transport through a Fabry-Perot resonator with or without observer or loss inside the cavity.

minute loss introduced inside the cavity completely blocks the light transmission through the resonator. In other words, a continuous observation with infinitesimally weak interaction could force the state of the field inside the cavity to being empty, so that for every short period, the system is again put into its initial state with no photon inside the cavity. This kind of effect due to continuous observation or decoherence of a coherent system is referred to as the quantum Zeno effect for quantum mechanical systems.

### 1.7.2  Features and equivalent circuit understandings of tunneling devices

It would be instructive to introduce equivalent circuit descriptions of these coherent/decoherent processes for double barrier tunneling, The equivalent circuit configulations reflect the way by which we understand the characteristics of resonant tunneling and the Coulomb blockade. For the resonant tunneling process, it is not allowed to make any observation of the intermediate electronic state. We must consider the system as a two-terminal tunneling device having a single current loop, as shown in Fig. 1.9(c). Only in terms of the current through the double tunneling junction can information about the electronic transport be obtained. On the other hand, the Coulomb blockade devices, in which the intermediate electronic state should be observed in order to destroy the quantum coherence, should be consider as a three-terminal tunneling device making up an electric circuit with two independent current loops, as shown in Fig. 1.9(d). In the latter, a resistance is connected to the intermediate electrode via a stray capacitance, which consists of another current loop. This resistance dissipates electromagnetic energy into heat to make the tunneling through the first potential barrier deterministic and irreversible. This process is usually irrelevant to our observation systems and described in terms of an environmental reservoir which is coupled to the electron via electromagnetic interaction.

With Coulomb blockade devices the third terminal is explicitly connected not, but via so-called stray capacitance (in circuit terminology), which actually represents a radiation process due to the change in the electric charge on the intermediate electrode and represent electromagnetic environment or reservoir around the tunneling barrier. This implies that the decoherence of the tunneling electron is due in part to the dissipation of the associated electromagnetic field via interaction with its environment. This in turn suggests that to produce a resonant tunneling device, we have to maintain both the coherence in the electronic state and in the electromagnetic field associated with the tunneling process until the electron completes the tunneling through the second potential barrier. This indicates that the coupling via the stray capacitance around the junctions is small enough at the corresponding frequency region, which actually depends on the shapes and electromagnetic responses of the electrodes and material surroundings of this device. We should remind ourselves of these features in designing electronic devices at a nanometer scale.

As is seen above, the problem of the isolation and connection of nano-electronic devices to electromagnetic reservoirs is a crucial point and it determines the function of the device. In this regard, it is instructive to note photon assisted electronic processes which take place in the intermediate region of double barrier tunneling systems, where a high frequency field or optical field exerts a resonant excitation or de-excitation of electronic states [37,38,39,40,41,42]. Such photonic injection or emission serves as a third terminal for double-barrier devices which depend on the properties of the electromagnetic environment. That is, as long as coherent interaction between the optical field and electronic states is maintained, an optical input or output acts to exert coherent control of the tunneling process, otherwise it provides a third terminal through which electronic coherence is destroyed. In these cases, the coherence and decoherence of the optical field have to be taken into careful consideration in investigation of the function of the nano-electronic device, since they are directly relevant to the observation of quantum process, electron tunneling.

Some more details of these photon assisted processes will be discussed later in this chapter in terms of electromagnetic interaction in nanometer-scale electronic systems. We will proceed to study on the relationship between controls and output or cause and results in mesoscopic devices, which is also related to the directionality of signal.

### 1.7.3 Separability and inseparability and physical pictures of interactions

Let us investigate another important aspect of signal transfer in terms of the separability and inseparability of an interacting system into several meaningful subsystems and interactions between these. This problem is concerned with unidirectionality or bidirectionality of the interaction as well as the quasi-particle description of interactions. We will consider this using the example of optical systems.

In general, an optical system is a measurement system or optical responses of an object through the observation of light transmission from a light source to a photo detector. Let us take the example of two electronically distinguishable dielectric matter interacting via an optical field, as shown in Fig. 1.11.

An incidental optical field drives oscillating electric polarizations on the dielectric objects, which in turn produce an electromagnetic field whose properties depend on the shape and optical properties of the dielectrics as well as the character of the waveguide or space configuration through which the optical radiation is propagated to the far field region. We will let the second dielectric object be put into the scattered field of the first one so that it produces a scattered field. Three typical effects will be shown: (a) "far-field interaction", where the spatial separation between the objects is much larger than the optical wavelength so that the interaction between these is due to the propagating components of the field scattered from the first one, (b) "near-field" interaction,

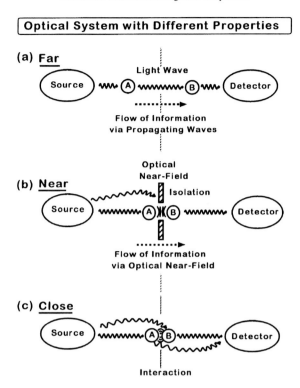

Fig. 1.11. Optical systems with different characteristics with respect to separability into subsystems, intermediate mode descriptions, and signal transport and isolation.

where the spatial separation is smaller than the optical wavelength so that the interaction is due to the near-field part of the field scattered from the first one, (c) "close" interaction in which two objects are electromagnetically coupled even without any illuminating field so that two objects should be treated as a single system.

These systems, however, exhibit no difference from the viewpoint of rigorous theories of the electromagnetic interaction of matter. A difference arises only in the way the observer understands the systems and how comprehensive theoretical descriptions and approximations are introduced. Such differences also become crucial from a technological point of view. Let us study several aspects of these problems.

The first point is the separability of the entire optical system into subsystems corresponding to each scattering object. The descriptions (a) and (b) can be distinguished from (c) depending on whether it is seen as being adequate to separate the system into two interacting modes; the mode relating to the first object being excited by the optical source and the other to the second object being excited by the scattered field of the first one which results in a radiation field being extended to the photo detector. This corresponds to the question of whether a unidirectional flow of signal and a certain causal relation can be set or not for the optical system under consideration. Due to the strong interaction between the objects, system (c) is difficult to separate into two subsystems, even though the multiple scattering process is taken into account and should be treated as a unique scattering system. In this case it is also actually difficult to obstruct the direct coupling of the second object to the light source and that of the first one to the photodetector. Of course, even in (a) and (b) such cross couplings spoil the identities and meaning of the mode separation. Such an isolation system is of fundamental importance in (b) which operates through an optical near-field interaction. Provided that an isolation system is established in (b) one can signify a light coming into the detector as being a direct result of excitation transfer from the first to the second object via near-field interaction.

The second important point which distinguishes case (a) from (b) is the nature of the modes dominating the interactions between the first and second objects. In the case of the far-field interaction (a), the optical mode mediating the interaction corresponds to the propagating mode of light which satisfies the dispersion relation in free space and has a real wavevector. In contrast to (a), the near-field interaction (b) is due to a local mode which can be represented as a superposition of evanescent waves showing exponential decay in the direction normal to the object surface, i.e. waves with imaginary components in wave number. Such interaction arises since the two objects are spaced within a subwavelength distance and the propagating components of the light field are blocked in the middle of these objects. Here we clearly identify the separability and inseparability as well as the meaning of isolation systems. These determine model descriptions and understandings of interacting systems.

Here we should note that even in a macroscopic problem there arise inseparable systems of which we already have examples in the optical interferometers and photonic crystals discussed in the previous section.

## 1.8 Near-Field Optical Microscopes and the Micro-Macro Connection

In this section, we will investigate the general features of electronic devices from an electromagnetic view point with the instructive example of near-field optical microscopes, NOM for short, which function via interaction of the optical field with electronic systems in condensed matter at the nanometer scale [18,35,43,44,45]. The investigation of the NOM process will provide a clear understanding of the electronic interaction with the optical field in nano-electronic devices which consist of several characteristic subsystems with qualitatively different size-dependent natures including microscopic interaction, microscopic to macroscopic signal transfer, and intermediate-scale connection between these subsystems.

### 1.8.1 Construction and function of near-field optical microscopes

A typical NOM configuration can be described as follows; [18]. First an electronic system with a size much smaller than the optical wavelength is illuminated to excite oscillating electric polarizations on it, where a light source serves as an optical-frequency power supply to it. As a result of internal interaction within the object, distribution of oscillating electric polarization depending on the shape, size, and optical response of atomic or molecular constituent is established. This polarization field produces a characteristic electromagnetic field relevant to the internal electromagnetic interaction and radiation. Especially, there is a strong optical near-field in the narrow vicinity of the object, showing strong localization within the spatial range as wide as the size of the electronic system scattering the optical field. In NOM systems, a probe tip is installed with an apex manufactured as small as the size of the object, and which is put close to the object to make an optical near-field interaction. It is stressed that the optical near-field is in general the characteristic field intermediating the internal electromagnetic interactions of the electronic system, so that it is strongly related to the local structure and optical nature of the objectives. When a system causes distinctive resonant behavior at the excitation frequency, the characteristic field is described in terms of an electromagnetic field in a coupled mode with electronic system. Such a coupled mode is often interpreted as a quasi particle representing fundamental excitations in the second-quantization framework, such as plasmons, excitons, and so on. Further more, for the excitations showing strong spatial correlations described as the surface waves of coupled electromagnetic field with excitation, these modes are treated as polaritons of fundamental excitations, such as the plasmon polariton, the exciton polariton, and so on in the second quantization framework [48,49].

Since the optical near-field itself, resulting from near-field interaction, cannot be observed directly, the near-field probe must be fabricated to have the function of transforming information about the near-field into some electrical signal extending to a photo detector placed in far-field region. One of the most effective ways to this end is to fabricate a tapered waveguide extending gradually from the probe-tip to the optical fiber or waveguide, and having the appropriate shapes, metal claddings, and guiding mode suitable for the near-field interaction conditions required for specific purposes [18]. In some cases, it is more efficient to implement a wavelength transformer such as fluorescent dye molecules or photo-current devices which convert the near-field signal into other types of electrical signal.

We are reminding of the microwave version of the near-field microscope, in which a microwave diode converts the results of near-field interactions into electric signals transmitted through a pair of current-carrying wires to a load used as an observation device. Such a microwave near-field measurement can be viewed as a microwave device. In the same sense, the functions of near-field optical microscopes basically correspond to opto-electronic devices in the usual sense, that is an optical fiber probe causes a near-field interaction and converts information about the optical near-field into propagating waves in the optical fiber with a help of a conical waveguide and the metal cladding of the sharpened probe, and transfers an electrical signal into an external observer, i.e. a photodetector. The sharpened optical fiber-probe in this case serves as a near-field sensor with a character of a spatial frequency filter and a converter of electromagnetic near-field modes into the propagating field coupled to an outside optical circuit. The basic function of the near-field devices are to make a microscopic interaction, to make a efficient transmission bass connecting the microscopic event to the macroscopic signal, and to isolate any direct coupling between the source and the detector of light. Because of its small size, a nano-electronic device should not necessarily be constructed of well-defined elements such as switches, amplifiers, filters, and so on, but consist of inseparable and multifunctiond basic elements such as we found in the sharpened optical-fiber probe.

### 1.8.2 Angular-spectrum representation and the properties of near fields

We will study one of the most important bases in describing near-field problems, the angular spectrum representation of scattered fields, in which an arbitrary field is expanded in terms of a series of plane waves specified by the directional angle of the propagation vector analytically extended to the complex region. Of course, plane waves are the basis for representing the invariance or symmetry of the system under spatial translational transform or parallel displacement. However, even an arbitrary field configuration exhibiting no such invariance can be expanded by introducing so-called "inhomogeneous" or "evanescent" waves with complex directional angles of propagation [46,47], in addition to propagating plane waves, so-called "homogeneous" waves. This corresponds to the mathematical procedure of analytic continuation with respect to the angular parameter extended into the complex region. By this means, we can describe scattering processes and interactions in terms of "penetration depth" of the field normal to an assumed boundary plane and the "spatial frequency spectrum" of the field in a direction parallel to the boundary. Investigating the behavior in the inhomogeneous part of the angular spectrum such as the central spatial frequency at which the spectrum takes the maximum, spectral intensity, spectral width, and so on, we can evaluate the character of the scattered field and the nature of interaction within it in terms of range or effective distance of the field as well as its locality. These features provide us with understanding and suggestive theoretical descriptions of near-field optical microscopes, NOM, and of general optical near-field problems.

### 1.8.2.1 Angular spectrum representation of a scattered field

We will firstly consider the scattering problem of a scalar field and study the mathematical framework and physical meaning of angular spectrum representation. The scattering process of a scalar field by matter, or equivalent scattering potential, is treated in the framework of the time-independent scattering problem based on the Helmholtz equation written in the following way:

$$\left[\nabla^2 + K^2\varepsilon(r)\right]\varphi(r) = 0, \tag{1.2}$$

where $K = \omega/c$ and $\varepsilon(r)$ are the wave number of the incidental field in a vacuum and the dielectric function of the scattering object, respectively, and $V(r) = K^2[1 - \varepsilon(r)]$ corresponds to the scattering potential. The Helmholtz equation describes the process in which the incident and scattered fields interact with an induced source field in a material system.

Let us assume that the scatterer is localized in a finite spatial domain $D$ and assume tentative planar boundaries $\Sigma^+$ and $\Sigma^-$ attached to both ends of the scatterer, as shown in Fig. 1.12. Then the whole space is separated into right, $R^+$, and left, $R^-$, half spaces outside the scatterer. The problem is to describe the scattered scalar field $\varphi$ in terms of plane waves in both of the half spaces. Even for a scattering system with no spatial translational symmetry, one can describe the scattered field in terms of plane waves if one resorts to an analytic continuation of the directional angle of wave vectors into the complex region [50,51,52]. This corresponds to an angular spectrum representation of the scattered field,

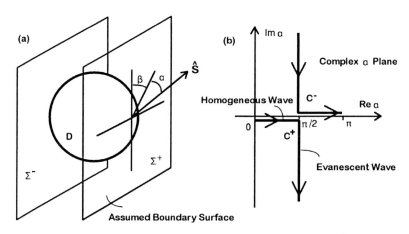

Fig. 1.12. Coordinate system for an angular spectrum representation of scattered waves and related complex contour of the propagation angle.

$$\varphi^{(SC)}(\boldsymbol{r}) = -\frac{1}{4\pi}\int_D G_0(\boldsymbol{r},\boldsymbol{r}',\omega)V(\boldsymbol{r}')\varphi(\boldsymbol{r}')d^3r', \tag{1.3}$$

based on the Weyl transform of the Green's function [46] given by

$$G_0(\boldsymbol{r},\boldsymbol{r}',\omega) = \frac{e^{iK|\boldsymbol{r}-\boldsymbol{r}'|}}{4\pi|\boldsymbol{r}-\boldsymbol{r}'|} = \frac{iK}{2\pi}\int_{-\pi}^{\pi}d\beta\int_{c^{\pm}}d\alpha\sin\alpha e^{iK\hat{v}\cdot(\boldsymbol{r}-\boldsymbol{r}')}. \tag{1.4}$$

As shown in Fig. 1.12, here, the $C^+$ contour of integration takes 0 to $\pi/2$ on the real axis and $\pi/2$ to $\pi/2 - i\infty$ on an imaginary path in the complex $\alpha$ plane for the transmitting component of the scattered wave into the right half-space, whereas, the $C^-$ takes $\pi/2 + i\infty$ to $\pi/2$ on the complex region and $\pi/2$ to $\pi$ on the real axis. The analytic nature of the angular spectrum representation has been studied to show that the angular spectrum is the entire function of $\alpha$ [53]. This results in the instructive theorem that any scattered field includes both propagating (homogeneous) and evanescent (inhomogeneous) waves at the same time. For an evanescent wave, the real part of the complex wavenumber forms a vector which represents the symmetry under spatial translation in the direction parallel to the assumed boundary plane. On the other hand, the imaginary part represents the exponential decay of the field with respect to the direction normal to the assumed boundary plane. On account of the NOM process, the large wavevector of evanescent waves being propagated parallel to the assumed boundary is the source of the ultra-high spatial resolution beyond the diffraction limit imposed on homogeneous wave measurements.

### 1.8.2.2 Origin of locality in sphere-sphere interaction
Next, we will study the optical near-field interaction of subwavelength-sized dielectric spheres put at a subwavelength distance empirically known to be localized within a spatial range corresponding to the sphere distance in the near-field conditions under consideration. The origin of the locality can be understood in terms of an angular spectrum representation of an electric dipole propagator. Here, we consider the electric dipole interaction between two small dielectric spheres of subwavelength size. The angular spectrum of the dipole propagator shows us the locality and range of interaction between the spheres. In Fig. 1.13, we show the angular spectrum representing the electric dipole propagator corresponding to the wave components in the imaginary part of the integration contour $C^+$, which dominates the inteaction so far as the sphere-sphere distance is less than the optical wavelength, as a function of the lateral wavenumber of the spectral component described by evanescent waves,

$$K_{EV} = K\cosh(\mathrm{Im}\{\alpha\})$$

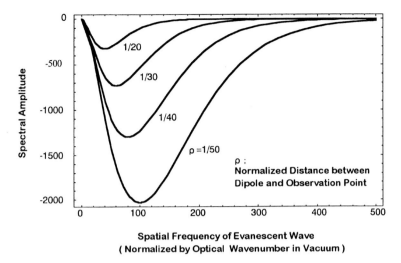

**Spatial Frequency of Evanescent Wave**
**( Normalized by Optical Wavenumber in Vacuum )**

Fig. 1.13. Angular spectra of a scattered field from an oscillating electric dipole represented on the assumed observation planes lying at distances $\rho$ from the point dipole (normalized by the wavenumber in vacuum). Only those for inhomogeneous or evanescent waves are shown as a function of the real wavenumber parallel to the assumed surface (normalized by the wavenumber in vacuum).

for several values of the sphere-sphere distance $d = |R \cdot z|$ in the near-field regime $Kd \ll 1$. In the figure, $\rho = Kd$ and $\xi = \cosh(\text{Im}\{\alpha\})$ are used. Note that $K = \omega/c$ is the wavenumber in vacuum, and the penetration depth is given by the inverse of the normal wavenumber [54,55],

$$|K_z| = |K \sinh(\text{Im}\{\alpha\})|$$
$$= \left| \sqrt{K^2 - K_{EV}^2} \right|.$$

As one can see in Fig. 1.13, the angular spectrum takes its maximum at the lateral wavenumber, $K_{EV} = K\cosh(\text{Im}\{\alpha\})$, approximately equal to the inverse of the sphere-sphere distance, i.e. $2/d$, or $\xi = 2/\rho$. That is, the dipole interaction between spheres is dominated by the angular spectrum components with the lateral wavenumber approximated by the inverse of the distance of the interacting electric dipoles. As the spheres approach, the lateral wavenumber becomes higher dominating the interaction and also the strength of the interaction. It can be understood why the amplitude diverges in the short distance limit by remembering the fact that the field intensity has singularity around a point charge in the static limit of a Coulomb interaction.

One can also see in Fig. 1.13 that the spectral width becomes broader as the sphere-sphere distance decreases. The full-width at the half-maximum (HWHM) of the spectrum is also approximated by the inverse of the sphere-sphere distance, i.e. $2/d$. It can be numerically shown that even if one cuts off the spectrum at the FWHM frequencies and reconstructs the electric dipole interaction by integrating over the dominant parts, one can reproduce the principal part of the electric dipole interaction which shows the radial dependence scaled by the third power of the inverse distance, $1/d$, corresponding mainly to the Hankel function $h_2^{(1)}$ involved in the general point dipole field [56]. The part that is omitted n dropping corresponds to the far-field relevant component described by the Hankel function $h_0^{(1)}$ with asymptotic behavior of the spherical wave in the far-field limit.

In conclusion, when we consider the electromagnetic interaction between spheres in the near-field regime, it is dominated by the interaction channel via short-ranged terms with a very large lateral wave number, $K_{EV}$, which is approximated by the inverse of the sphere-sphere distance, i.e. $2/d$. The broadening of the angular spectrum implies in turn that the lateral range of the corresponding interaction is also localized due to the interference of the evanescent waves spreading over a wide band.

## 1.9  Electron Interaction with Electromagnetic Fields in Nanometer Scale

### 1.9.1  Fundamental process of electron-photon interaction

As a fundamental process, photon-electron interaction in an otherwise free environment is not allowed to the extent that linear interaction is involved because of the difference between the electron and photon dispersion relations as free particles. That is, the conservation rule cannot be satisfied in any simultaneous transfer of energy and momentum between electron and photon (Fig. 1.14(a)). Photon-electron interactions in general are therefore allowed only in association with an environmental material field from which an excess momentum is borrowed to compensate for the difference in dispersion relations. This corresponds to the fundamental process known as "bremsstrahlung". The electron-photon interaction takes place provided that an electron scattered by a local potential field of environmental matter such as atomic nuclei and ionic crystals is in a virtual state having energy-momentum relation displaced from that of free electrons, i.e. in the so-called the state with "off-shell" energy-momentum relation.

Optical near-field interaction can be considered as being the counterpart of the bremsstrahlung in the sense that the optical fields or photons are scattered by environmental matter such as a dielectric or metallic system, so that their energy-momentum states are displaced from the free-photon dispersion relation and come into interaction with otherwise free electronic systems (Fig. 1.14(c)). One of the most interesting and instructive phenomena is the effect

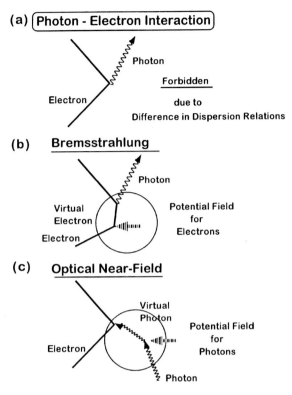

Fig. 1.14. Electron-photon interaction and effects of potential fields from surrounding material systems on both electrons and photons.

once reported by Schwarz and Hora, the so-called Schwarz-Hora effect [57,58], in which a direct modulation of an electron beam by a strong laser field was said to be observed with the help of a thin dielectric film providing a waveguide for the optical field, although this effect is still under experimental reconfirmation [59,60]. Another concrete example is Cerenkov radiation which is equivalently describable both as the result of bremsstrahlung in a dielectric medium and the result of an optical near-field interaction of electrons propagating in an infinitesimally thin hollow in a dielectric medium [61,62]. Further more, it is quite possible to utilize local excitations and the associated electromagnetic near-field for functional probing and analysis of general nano-electronic devices. In doing this, we must establish the theoretical background in order to make comprehensive descriptions of electromagnetic interactions in nanometer-sized electronic systems, in which both electronic and electromagnetic behavior are taken equally into account. Such a theoretical background would reveal a number of instructive ideas for realizing nanometer-scale electron-photon devices based on novel concepts.

### 1.9.2 Conserved quantities and local symmetry in interacting systems

It should be stressed that in the electron-photon interaction assisted by an environmental matter field, the shapes and properties of the surrounding matter strongly affect the quantities conserved in the process of interaction, such as energy, momentum, and angular momentum. These pseudo-conservation laws or conservation laws for coupled systems allow us to utilize novel types of interactions such as forbidden transitions in otherwise free space by using specially fabricated nano-structures [34,63]. This point likely to be especially important for electronic devices which make full utilization of electronic spins manipulated via coupling with pseudo-angular momentum.

Another important application of optical fields is also under extensive study aimed at the utilization of forces exerted by the optical field on a material system via resonant interaction. The laser cooling technique has made particularly great progress in the last decade in establishment of the novel field of atom optics and condensed state of Bose particles, and making it possible to utilize material waves in our device technology. Important progress has also been made in optical method of atom manipulation, especially by using the optical near-field, such as atom waveguiding and photo-chemical vapor phase deposition of semiconductor materials with nanometer resolution, and so on [35]. As mentioned in the above, in optical near-field regime, we can utilize the phenomena which are macroscopically forbidden but allowed for the case involving interactions of fields with mesoscopic electronic systems.

The mechanical effects exerted by the optical field on a material system are classified into those originating from the electrical energies of electric polarizations induced in the material system and those from magnetic interaction. The electric part is due to momentum transfer from the optical field to matter and described in terms of energy gradient of the induced electric polarizations of matter in the optical field [65]. One of the two characteristic effects is momentum transfer with dissipative properties associated with resonant light scattering or absorption. The other is the force of conserving nature usually referred to as dipole force and due to the potential energy of induced electric polarizations of matter which depends both on the intensity gradient of the optical field and frequency detuning of the optical field from the resonance frequency of the material system. The electrical part of the light-induced force has been widely used for laser cooling and trapping of atoms as well as for manipulations of atomic and microscopic particles as optical tweezers. The magnetic contributions are via the transfer of orbital angular momentum from the optical field to matter and thereafter, spin-orbital coupling taking place in the electronic system.

Because of the nature of the optical near-field as being a coupled mode of the electromagnetic field with matter, the definitions and conservation rules for physical variables such as momentum and angular momentum might become approximate ones, depending strongly on the shapes and sizes of coupled material systems. For this reason, these physical quantities are often referred to as pseudo-quantities [64]. Using the example of evanescent waves at planar dielectric surfaces discussed previously in this section, the wavevector along the surface corresponding to the pseudo-momentum of the surface wave has a value $k_{\parallel}$ larger than the wavenumber $\omega/c$ of the light field being propagated in the vacuum. This property results in the characteristic near-field mechanical effect that since $k_{\parallel}$ is proportional to the pseudo-momentum of the surface field, an atom exerting resonant absorption of an evanescent optical field will acquire a recoil momentum corresponding to the pseudo momentum of the evanescent wave, though the energy transfer involves same amount as the resonant absorption of free photons in a vacuum. This effect has been demonstrated experimentally by using high resolution laser spectroscopy by Matsudo *et al.* [66,67]. On the other hand, due to the exponentially decaying aspect of evanescent waves along the normal to a dielectric surface, a strong dipole force is exerted on atoms via off-resonant interaction, and this is efficiently utilized to waveguide atoms in a frequency selective way in a thin hollow optical fiber [68]. One remarkable aspect of atom manipulation by evanescent waves is that the wavevector along the surface and the intensity gradient along the normal to the surface are directing orthogonally to each other. This is in marked contrast to what happens when an optical wave is propagated in a vacuum, when the intensity gradient is surely directed to the wavevector of propagation. Furthermore, the angular spectral representation suggests that a highly localized optical near-field involves evanescent waves with an extremely large pseudo-momentum and would exert a significant mechanical effect. In fact, the force measurement of the optical near-field is under extensive study as well as several applications in this direction, such as single-atom manipulation. A mesoscopic electronic system showing strong resonant behavior in its interaction with optical field, such as plasmon resonance, would strongly enhance these mechanical effects. In addition, in a microscopic limit where objects are placed much closely, there arises a mechanical effect due to near-field electromagnetic interaction via vacuum fluctuation of the electromagnetic field and material polarizations, such as a van der Waals force. The mechanical effects of an optical near-field therefore could be considered as a kind of inter-atomic or molecular force under the control of external excitation and having the potential for application in a wide variety of research fields, such as the control and manipulation of microscopic matter, including biological samples.

Another important physical quantity is the pseudo-angular momentum of the optical near-field, determined by the local rotational symmetry of the interacting electromagnetic field plus the matter system [69]. For example, with resonant atomic interaction with an optical near-field, a theoretical consideration shows the transfer or conservation of angular momentum along the direction normal to the surface corresponding to the symmetry axis of the local interacting atom plus surface system. The pseudo-angular momentum is related to the magnetic effect of the optical field and would provide a number of interesting applications, such as for magneto-optical effects and the control of the electronic spin state via orbital to spin coupling in electronic systems. This can be briefly explained as follows. A transverse optical field being propagated in free space serves as a ladder operator for orbital angular momentum in the multipolar interaction. This can be clearly described on the basis of circular polarizations representing the rotational symmetry of the optical field along the axis of propagation. That is, the circular polarization of light in free space belongs to the rotational group in two dimensional Euclid space normal to the wavevector. This is the reason why the angular momentum of photons belongs to either of the spin states where $\ell_p = \pm 1 \hbar$ and also why they are easily polarized by only using wave plates. In contrast to this, the optical near-field corresponds to the coupled mode of the optical field with matter and exhibits an aspect found in the rotational group of coupled systems and provide qualitatively different properties in angular momentum. Several types of spin devices have recently been under extensive study especially for low dimensional electronic systems such in a top-most atomic layer with magnetic properties. Of these, the spin polarized STM with an optically pumped GaAs tip is under extensive study. As well as the usual optical pumping techniques for electronic spin states, the process consists of an optical excitation with associating increase or decrease in orbital angular momentum obtained from the optical field, after which there is orbit to spin coupling to polarize electronic spins. In order to prepare a spin-polarized probe-tip of subwavelength size, we should carefully investigate the pseudo-angular momentum in the related optical near-field around the tip.

### 1.9.3 Quantization of fields in confined space

In the study of devices at the nanometer scale we might encounter the weak limits of electromagnetic fields which would require a quantum treatment of the field in a confined space. Here we note two ways of approaching this problem.

Quantization of the field is often introduced in the second quantization framework, in which the field is described in terms of creation and annihilation operations of quanta in a certain mode in the momentum space. The mode description is in turn based on a Fourier analysis with respect to time, so that a dispersion relation is set for the field quantum under consideration. According to the second quantization procedures, it is not possible to consider any light quanta in a rigid confined space narrower than a half wavelength of the corresponding mode frequency [70]. Under this condition, we should consider the local interaction of an electronic system with an optical field as a process of creation and annihilation of a quantum in a photonic mode extended at least as wide as a half optical wavelength size via coupling of electronic excitation to the tail of the optical mode function. An example under this condition has been introduced as the quantization of evanescent electromagnetic waves based on the triplet mode description formulated by Carniglia and Mandel [71]. This approach could serve as a convenient framework in which to study problems involving only a single specific portion of a light scattering system with otherwise almost homogeneous space configurations.

In order to introduce a comprehensive quantum theory for a system composed of numbers of characteristic subsystems of a small size, we ought to resort to doing an approximate treatment accounting for only several specific features of each subsystem confined in a narrow space. This approach leads to the concept of local excitations and quasi-modes with a characteristic wavelength as short as the size of the confined space considered. This also corresponds to a virtual photon approach to interaction problems. These modes actually represent the coupled modes of the electromagnetic field with matter such as the polariton of plasmons, excitons, and so on, as we discussed in the previous section. Consequently, the dispersion relations of these excitations differ from those of free photons. The whole process is then approximated by successive interactions between subsystems and described in terms of local excitation transfer and interference to this. By adding up all the contributions in the entire system, we can evaluate the overall transition probabilities from the source to a detector.

Practical examples of this field quantization in a confined space can be found in optical near-field related quantum problems, many of which are, at this stage, under investigations.

### 1.9.4 Notes on optical near-field diagnosis of nano-electronic devices

Electronic devices in general serve as the field in which an interaction takes place between the nonequilibrium system for electron transport and electromagnetic flow as signals. It is therefore important to evaluate the structures and functions of nano-electronic devices at the same time, because observations of the coupled mode are the most important basis in near-field diagnosis. Here, we should be again aware that observation processes of nano-electronic devices corresponds to a destructive measurement for both electric and electromagnetic systems. For example, optical near-field diagnosis, i.e. a destructive measurement for the optical field, results in decoherence in electronic systems. The smaller the size of object, the larger the magnitude and degree of perturbation during diagnosis.

The properties of both electronic systems and electromagnetic fields are strongly under the influence of the structures and features of the surrounding material environment. In contrast to the wave nature of electrons, the wave nature of the optical field is no longer important in nanometer space and quasi-static behaviors take priority. Such a combination of electronic and optical fields showing significantly different properties has the potential to produce novel devices which take full advantage of combined resonant and nonresonant responses. Further research in this direction would open the way to creating functional devices operating at optical frequencies even without specific resonant conditions due to free-space dispersion relations. For such nano-electronic devices taking full advantage of quantum properties, it becomes more important to consider dissipation mechanisms which produce an irreversible one-directional flow of electrical signals from the reversible quantum behavior of electrons as well as doing functional analysis evaluating properties of nonquilibrium open-systems.

For issues related to the ultimately weak electromagnetic fields of microscopic devices up to the fast response range in optical frequency regions, a quantum theory of coupled modes of field and matter would be indispensable. We need to do further research in order to gain a much more profound nature of quasi-particle picture and develop the theory of fundamental excitation as well as that behind the fundamental processes of electromagnetic interaction with matter.

[References]

1)    J. A. Wheeler and W. H. Zurek, *Quantum Theory and Measurement*, Princeton University Press, Princeton (1983).

2)    M. Burne, E. Hagley, J. Dreyer, X. Maitre, A. Maali, C. Wunderlich, J. M. Raimond, and S. Haroche, Observing the Progressive Decoherence of the "Meter" in a Quantum Measurement, *Phys. Rev. Lett.*, **77** (1996) 4887–4890.

3)    H. Grabert and M. H. Devoret (eds.), *Single Charge Tunneling*, Plenum, New York (1991).

4)    G. S. Agarwal, Quantum Electrodynamics in the Presence of Dielectrics and Conductors. I. Electromagnetic-Field Response Functions and Blackbody Fluctuations in Finite Geometries, *Phys. Rev. A*, **11** (1975) 230–242.

5)    G. S. Agarwal, Quantum Electrodynamics in the Presence of Dielectrics and Conductors. II. Theory of Dispersion Forces, *Phys. Rev. A*, **11** (1975) 243–252.

6)    G. S. Agarwal, Quantum Electrodynamics in the Presence of Dielectrics and Conductors. III. Relations among One-Photon Transition Probabilities in Stationary and Non Stationary Fields, Density of States, the Field-Correlation Functions, and Surface-Dependent Response Functions, *Phys. Rev. A*, **11** (1975) 253–264.

7)    D. Mechede, W. Jhe, and E. A. Hinds, Radiative Properties of Atoms Near a Conducting Plane: An Old Problem in a New Light, *Phys. Rev. A*, **41** (1990) 1587–1596.

8)    W. Jhe and J. W. Kim, Atomic Energy-Level Shifts Neara Dielectric Microsphere, *Phys. Rev. A*, **51** (1995) 1150–1153.

9)    M. Janowicz and W. Żakowicz, Quantum Radiation of a Harmonic Oscillator Near the Planar Dielectric-Vacuum Interface, *Phys. Rev. A*, **50** (1994) 4350–4364.

10)   M. Burne, F. Schmidt-Kaler, A. Maali, E. Hagley, J. Dreyer, J. M. Raimond, and S. Haroche, Quantum Rabi Oscillation: A Direct Test of Field Quantization in a Cavity, *Phys. Rev. Lett.*, **76** (1996) 1800–1803.

11) C. J. Hood, M. S. Chapman, T. W. Lynn, and H. J. Kimble, Real-Time Cavity QED with Single Atoms, *Phys. Rev. Lett.*, **80** (1998) 4157–4160.

12) J. A. Wheeler and R. P. Feynman, Interaction with the Absorber as the Mechanism of Radiation, *Rev. Mod. Phys.*, **17** (1945) 157–181.

13) R. P. Feynman, *Quantum Electrodynamics*, Benjamin/Cummings, Reading, Mass. (1961).

14) E. Yablonovitch and T. J. Gmitter, Photonic Band Structure: The Face-Centered-Cubic case, *Phys. Rev. Lett.*, **63** (1989) 1950–1953.

15) K. M. Leung and Y. F. Liu, Full Vector Wave Calculation of Photonic Band Structures in a Face-Centered-Cubic Dielectric Media, *Phys. Rev. Lett.*, **65** (1990) 2646–2649.

16) Z. Zhang and S. Satpathy, Electromagnetic Wave Propagation in Periodic Structures: Bloch Wave Solution of Maxwell's Equations, *Phys. Rev. Lett.*, **65** (1990) 2650–2653.

17) S. John and J. Wang, Quantum Electrodynamics Near a Photonic Band Gap: Photon Bound States and Dressed Atoms, *Phys. Rev. Lett.*, **64** (1990) 2418–2421.

18) M. Ohtsu and H. Hori, *Near-Field Nano-Optics*, Kluwer Academic/Plenum Publishing Corp., New York (1999).

19) K. Cho, Nonlocal Theory of Radiation-Matter Interaction: Boundary-Condition-Less Treatment of Maxwell Equations, *Progr. Theor. Phys. Suppl.*, **106** (1991) 225–233.

20) K. Cho, Y. Ohfuti, and K. Arima, Study of Scanning Near-Field Optical Microscopy (SNOM) by Nonlocal Response Theory, *Jpn. J. Appl. Phys.*, **34** (1994) 267–270.

21) J. J. Hopfield, Theory of the Contribution of Excitons to the Complex Dielectric Constant of Crystals, *Phys. Rev.*, **112** (1958) 1555–1567.

22) C. Kittel, *Introduction to Solid State Physics*, 6th ed., John Wiley & Sons, New York (1986).

23) A. D. Boardman (ed.), *Electromagnetic Surface Modes*, John Wiley & Sons, Chichester (1982).

24) S. M. Barnett, B. Huttner, and R. Roudon, Spontaneous Emission in Absorbing Dielectric Media, *Phys. Rev. Lett.*, **68** (1992) 3698–3701.

25) M. Specht, J. D. Pedaring, W. M. Heckl, and T. W. Hänsch, Scanning Plasmon Near-Field Microscope, *Phys. Rev. Lett.*, **68** (1992) 476–479.

26) C. Monroe, D. M. Meekhof, B. E. King, W. M. Itano, and D. J. Winland, Demonstration of a Fundamental Quantum Logic Gate, *Phys. Rev. Lett.*, **75** (1995) 4714–4717.

27) W. M. Itano, D. J. Heinzen, J. J. Bollinger, and D. J. Winland, Quantum Zeno Effect, *Phys. Rev. A*, **41** (1990) 2295–2300.

28) V. B. Braginsky and F. Y. Khalili, *Quantum Measurement*, Cambridge University Press, Cambridge (1992).

29) A. Tonomura, N. Osakabe, T. Matsuda, T. Kawasaki, J. Endo, S. Yano, and H. Yamada, Evidence for Aharonov-Bohm Effect with Magnetic Field Completely Shielded from Electron Wave, *Phys. Rev. Lett.*, **56** (1986) 792–795.

30) C. J. Chen, *Introduction to Scanning Tunneling Microscopy*, Oxford University Press, Oxford (1993).

31) H. J. Güntherodt and R. Wiesendanger (eds.), *Scanning Tunneling Microscopy I*, 2nd ed., Springer-Verlag, Berlin (1994).

32) R. Wiesendanger and H. J. Güntherodt (eds.), *Scanning Tunneling Microscopy II*, 2nd ed., Springer-Verlag, Berlin (1995).

33) R. Wiesendanger and H. J. Güntherodt (eds.), *Scanning Tunneling Microscopy III*, 2nd ed., Springer-Verlag, Berlin (1996).

34) C. J. Chen, Attractive Interatomic Force as a Tunneling Phenomenon, *J. Phys.: Condens. Matter*, **3** (1991) 1227–1245.

35) M. Ohtsu (ed.), *Near-Field Nano/Atom Optics and Technology*, Springer-Verlag, Tokyo (1998).

36) G. Binnig, H. Rohrer, Ch. Gerber, and E. Weibel, $7 \times 7$ Reconstruction on Si(111) Resolved in Real Space, *Phys. Rev. Lett.*, **50** (1983) 120–123.

37) P. K. Tien and J. P. Gordon, Multiphoton Process Observed in the Interaction of Microwave Fields with the Tunneling between Superconductor Films, *Phys. Rev.*, **129** (1962) 647–651.

38) L. P. Kouwenhoven, S. Jauhar, K. McCormic, D. Dixon, P. L. McEuen, Yu. V. Nazarov, N. C. van der Vaart, and C. T. Foxon, Photon-Assisted Tunneling through a Quantum Dot, *Phys. Rev. B*, **50** (1994) 2019–2022.

39) L. P. Kowenhoven, S. Jauhar, J. Orenstein, P. L. McEuen, Y. Nagamune, J. Motohisa, and H. Sakaki, Observation of Photon-Assisted Tunneling through a Quantum Dot, *Phys. Rev. Lett.*, **73** (1994) 3443–3446.

40) P. Johansson, R. Monreal, and P. Appel, Theory of Light Emission from a Scanning Tunneling Microscope, *Phys. Rev. B*, **42** (1990) 9210–9213.

41) R. Berndt, J. K. Gimzewski, and P. Johansson, Inelastic Tunneling Excitation of Tip-Induced Plasmon Modes on Noble-Metal Surface, *Phys. Rev. Lett.*, **37** (1991) 3796–3799.

42) R. Berndt and J. K. Gimzewski, Injection Luminescence from CdS(11$\bar{2}$0) Studied with Scanning Tunneling Microscopy, *Phys. Rev. B*, **45** (1992) 14095–14099.

43) D. W. Pohl and D. Courjon (eds.), *Near Field Optics*, Kluwer Academic Publishers, Dordrecht (1993).

44) M. A. Paesler and P. J. Moyer, *Near-Field Optics: Theory, Instlumrntation, and Applications*, John Wiley & Sons, New York (1996).

45) J. P. Fillard, *Near Field Optics and Nanoscopy*, World Scientific, Singapore (1996).

46) M. Born and E. Wolf, *Principles of Optics*, 3rd ed., Pergamon Press, Oxford (1965).

47) C. Cohen-Tannoudji, B. Diu, and F. Laloë, *Quantum Mechanics, Vol. 1*, Chap. 1, p. 71, John Wiley & Sons, New York (1977).

48) K. Cho (ed.), *Excitons*, Springer, Berlin (1979).

49) V. M. Aharanovich and A. A. Maradudin (eds.), *Excitons*, North-Holland, Amsterdam (1982).

50) A. Sommerfeld, *Partial Differential Equations in Physics*, Academic Press, New York (1949).

51) P. M. Morse and H. Feshbach, *Methods of Theoretical Physics, Part 1*, McGRAW-HILL Book Comp., New York (1953).

52) P. M. Morse and H. Feshbach, *Methods of Theoretical Physics, Part 2*, McGRAW-HILL Book Comp., New York (1953).

53) E. Wolf and M. Niet-Vesperinas, Analyticity of the Angular Spectrum Amplitude of Scattered Fields and Some of Its Consequence, *J. Opt. Soc. Am. A*, **2** (1985) 886–890.

54) T. Inoue and H. Hori, Representations and Transforms of Vector Fields as the Basis of Near-Field Optics, *Opt. Rev.*, **3** (1996) 458–462.

55) T. Inoue and H. Hori, Theoretical Treatment of Electric and Magnetic Multipole Radiation Near a Planar Dielectric Surface Based on Angular Spectrum Representation of Vector Field, *Opt. Rev.*, **5** (1998) 295–302.

56) J. D. Jackson, *Classical Electrodynamics*, 2nd ed., John Wiley & Sons, New York (1975).

57) H. Schwarz and H. Hora, Moculation of an Electron Wave by a Light Wave, *Appl. Phys. Lett.*, **15** (1969) 349–351.

58) H. Hora, Coherence of Matter Waves in the Effect of Electron Waves Modulation by Laser Beams in Solids, *Phys. Stat. Sol.*, **42** (1970) 131–136.

59) J. Bae, H. Shirai, T. Nishida, T. Nozokido, K. Furuya, and K. Mizuno, Experiantal Verification of the Theory on the Inverse Smith-Purcell Effect at a Submillileter Wavelength, *Appl. Phys. Lett.*, **61** (1992) 870–872.

60) J. Bae, S. Okuyama, T. Akizuki, and K. Mizuno, Electron Energy Modulation with Laser Light Using a Small Gap Circuit: A Theoretical Consideration, *Nuclear Instruments and Methods in Physics Research*, **A331** (1993) 509–512.

61) G. Torardo di Francia, On the Theory of some Cerenkovian Effects, *Nuovo Cimment*, **16** (1960) 1085–1101.

62) D. A. Tidman, A Quantum Theory of Radiative Index, Cerenkov Radiation and the Energy Loss of a Fast Charged Particle, *Nuclear Phys.*, **2** (1956/1957) 289–346.

63) D. F. Nelson, Momentum, Pseudomomentum, and Wave Momentum: Toward Resolving the Minkowski-Abraham Controversy, *Phys. Rev. A*, **44** (1991) 3985–3996.

64) R. Peierls, *More Surprise in Theoretical Physics*, Sec. 2.4–2.6, pp. 30–42, Princeton University Press, Princeton (1991).

65) J. P. Gordon, Radiation Force and Momenta in Dielectric Media, *Phys. Rev. A*, **8** (1973) 14–21.

66) T. Matsudo, H. Hori, T. Inoue, H. Iwata, Y. Inoue, and T. Sakurai, Direct Detection of Evanescent Electromagnetic Waves at a Planar Dielectric Surface by Laser Atomic Spectroscopy, *Phys. Rev. A*, **55**, (1997) 2406–2412.

67) T. Matsudo, T. Takahara, H. Hori, and T. Sakurai, Pseudomomentum Transfer from Evanescent Waves to Atoms Measured by Saturated Absorption Spectroscopy, *Opt. Commun.*, **145** (1998) 64–68.

68) H. Ito, T. Nakata, K. Sakaki, M. Ohtsu, K. I. Lee, and W. Jhe, Laser Spectroscopy of Atoms Guided by Evanescent Waves in Micron-Sized Hollow Optical Fibers, *Phys. Rev. Lett.*, **76** (1996) 4500–4503.

69) M. Kristensen and J. P. Woerdman, Is Photon Angular Momentum Conserved in Dielectric Medium?, *Phys. Rev. Lett.*, **72** (1994) 2171–2174.

70) L. Mandel, Configuration-Space Photon Number Operators in Quantum Optics, *Phys. Rev.*, **144** (1966) 1071–1077.

71) C. K. Carniglia and L. Mandel, Quantization of Evanescent Electromagnetic Waves, *Phys. Rev. D*, **3** (1971) 280–296.

# Chapter 2

# Electron Transport in Semiconductor Quantum Dots

## 2.1 Introduction

Recent advances in epitaxial growth and processing technologies have enabled us to fabricate semiconductor nanostructures whose dimensions are comparable to the de Broglie wavelength of electrons. Transport measurements on these nanostructures have revealed a rich variety of phenomena associated with the effects of quantum mechanical confinement [1]. Conductance quantization in one-dimensional quantum point contacts, and resonant tunneling through quantum wires and quantum boxes are such examples. These properties directly reflect the quantization of energy. In addition, charge quantization is observed for tunneling through a small dot, which acts as an island for electrons. When tunneling occurs, the charge on the island suddenly changes by a quantized amount namely "$e$". This leads to the change in the electrostatic potential of the dot by the charging energy, $E_c = e^2/C$, where $C$ is the typical capacitance of the island. The one-by-one change in the number of electrons on the island gives rise to oscillations in the tunneling conductance (Coulomb oscillations) when the gate voltage is swept. These oscillations are usually periodic when the number of electrons is "large". However, in a small dot holding just a few electrons, the charging energy can no longer be parameterized in terms of a constant capacitance, and the Coulomb oscillations are significantly modified by electron-electron interactions and quantum confinement effects. Both the quantized energy level spacing, and the interaction energy become large when the dot size is decreased, and can be similar when the dot size is comparable to the electron wavelength. Thus, the addition energy needed to put an extra electron on the dot becomes strongly dependent on the number of electrons in the dot. Such a system can be regarded as an artificial atom. [2] We show in section 2.2 that the addition energy spectrum of an artificial atom reflects atom-like features such as a shell filling and the obeyance of Hund's first rule when the dot has a high degree of cylindrical symmetry. Such a dot has only recently been developed by using a sub-micorn diameter double barrier heterostructure in which a circular disk-shaped dot is located. [3,4] The shell structure arises from the single-particle level degeneracy imposed by the two-dimensional (2D) harmonic potential in the circular dot. In the presence of a magnetic field, the interaction effects become more important than the quantum mechanical effects in determining the configuration of the ground state (GS), and this leads to transitions in the GS configuration. We use an excitation spectroscopy technique in section 2.3 to study the contributions of the many-particle interactions in these transitions in the GS. The atom-like electronic properties described are all dependent on the cylindrical symmetry of the dot. By breaking the cylindrical symmetry, the atom-like properties should be significantly modified. In section 2.4 we present a new device which enables an *in situ* control of the symmetry in the confining potential. The remaining sections are devoted to transport studies on double quantum dot systems which behave like artificial molecules. When two dots are quantum-mechanically strongly coupled via a tunnel junction, an electron can oscillate coherently back and forth between the two dots. This gives rise to two delocalized electronic states, i.e. a symmetric and an anti-symmetric state, which are separated by the tunnel coupling energy. The number of electrons is uniquely defined for the whole two dot system. On the other hand, when two dots are quantum-mechanically weakly coupled, the electronic states are usually localized in each dot. The number of electrons is then defined for each dot. Nonetheless, the respective electron numbers are still regulated by the electrostatic coupling between dots. We discuss the energy spectrum of two dot systems with strong or weak quantum mechanical coupling when the two dots are vertically connected in section 2.5, and when they are laterally connected in section 2.6. We use a triple barrier heterostructure to fabricate a vertically coupled molecule. In this configuration, two identical circular dots are connected to each other via an abrupt thin heterostructure potential barrier (see Fig. 2.14). This vertical tunnel coupling produces sets of states, either delocalized throughout the two dots, or localized in each dot separately. Each of the vertical dots has the same lateral states confined by a 2D harmonic potential. Electron filling is well distinguished between for the strongly coupled case, and the weakly (or electrostatically) coupled case. On the other hand, we use a modulation doped 2D heterostructure to fabricate a laterally coupled double dot molecule. Two dots are separately confined by Schottky gates placed on the surface and connected in plane via a Schottky gate induced electrostatic potential barrier (see Fig. 2.21). The coupling between the two lateral dots can produce a number of coupled states, while there is a single electronic (occupied) state in the vertical direction of each dot. The coupling strength, and charge configurations in the two dots, can be tuned *in situ* using the Schottky gates. We use a microwave technique to probe the energy spectrum of the coupled dot states. We discuss the coherent or elastic tunneling between two delocalized states, while for two localized states, we discuss the inelastic tunneling which is energetically coupled to the environment.

## 2.2  Quantum Dot Atoms

### 2.2.1  Single electron tunneling spectroscopy

The energy spectrum of a quantum dot can be probed using the *single electron tunneling spectroscopy* technique. Let us consider the quantum dot depicted in Fig. 2.1(a). The dot is only weakly coupled to the reservoirs (source and drain). The gate electrode is electrostatically or capacitively connected to the dot, and can be used to tune the electrostatic energy of the dot. We measure the tunneling current flowing through the dot for an arbitrarily small voltage applied between the source and drain: $V_{sd} = (\mu_s - \mu_d)/e$, where $\mu_s$ and $\mu_d$ are the electro-chemical potentials in the source and drain, respectively. Tunneling current, $I$, can only flow when a dot state is between the Fermi energies of the source and drain: $\mu_s > \mu(N) > \mu_d$, where $\mu$ is the electro-chemical potential of the dot, and is defined as $\mu(N) = U(N) - U(N-1)$, where $U(N)$ is the total energy of the $N$-electron GS. Zero and non-zero tunneling current can be observed as a function of gate voltage $V_g$, and the resulting current versus gate voltage $(I - V_g)$ shows a series of peaks, corresponding to the one-by-one change in the number of electrons on the dot. Each current peak effectively measures $\mu(N)$, and the spacing between the peaks measures the increment of $\mu(N)$, i.e. $\Delta\mu(N) = \mu(N+1) - \mu(N)$.

If we assume that the quantum levels can be calculated independently of the number of electrons on the dot, and also that the interaction energy can be parameterized with a constant capacitance, then $\Delta\mu(N)$ is given by

$$\Delta\mu(N) = \Delta E + e^2/C, \qquad (2.1)$$

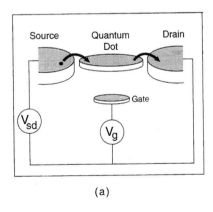 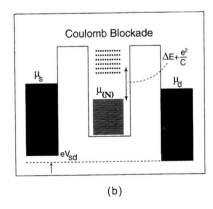

(a)                               (b)

Fig. 2.1. (a) Schematic representation of a quantum dot connected to source and drain contacts by tunnel junctions and capacitively coupled to a gate. (b) Energy diagram of a quantum dot when electron transport is blocked by the level spacing ($\Delta E$) and classical charging energy ($e^2/C$).

where $\Delta E = E_{N+1} - E_N$ is the single-particle level spacing between states with energies $E_{N+1}$ and $E_N$ (Fig. 2.1(b)). Note that $\Delta E = 0$ when $E_{N+1}$ and $E_N$ are for spin-degenerate states. This can lead to an even-odd asymmetry in the energy spectrum with respect to the number of electrons in the dot. In a many-electron system the level spacing is much smaller than the interaction energy, and a constant capacitance can be used to describe the interaction energy, so that $\Delta\mu(N)$ $(=e^2/C)$ is constant with $N$. On the other hand, in a few-electron system, the level spacing becomes comparable to the interaction energy, and the constant capacitance model no longer holds due to the contributions of the many-particle interactions. This can significantly modify the pattern of spacings between the Coulomb oscillations.

### 2.2.2  Planar and vertical single electron transistors

Single electron tunneling spectra have been extensively studied for small semiconductor dots using so-called single electron transistors (SETs) with many different geometries and configurations. This is particularly true for SETs with a planar geometry defined in a two-dimensional gas by surface Schottky gates (see Figs. 2.20(a) and 2.21). The lateral constriction forming the dot, the tunnel junctions between the dot and the reservoirs, and the plunger

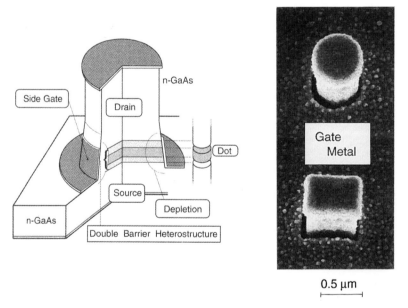

Fig. 2.2. Schematic representation of a quantum dot in a vertical device and scanning electron microscope images of the top contact and the gate metal for a circular and rectangular mesa. [3,4] The bottom (substrate) contact is not visible.

gate to tune the electrostatic potential of the dot are all made by applying voltages to Schottky gates. This enables a wide variation of device geometry and a large freedom of gate operation for tuning the transmission probability of the tunnel junctions as well as the electron number in the dot. In addition, the fabrication technology is well developed. However, there are a number of limitations. For example, the geometry of the dot is not so well defined, since the actual confining potential imposed by the Schottky gates can be significantly different from that of the gate geometry. In addition, it is difficult to fabricate a dot containing just a few electrons. In contrast, due to the presence of heterostructure barriers and vertical side walls, the dot in a vertical SET can have both good shape and high symmetry (see Fig. 2.2). Such a vertical SET enables a precise tuning of the electron number in the dot, *starting from zero*. The fabrication technology has only recently been developed. [3] In the following sections, we use a vertical SET to study atom-like electronic properties of artificial atoms. For quantum dot molecules, we use such a vertical SET containing a stack of two disk-shaped quantum dots to study the effect of quantum mechanical coupling. We also use a planar SET containing two coupled quantum dots defined by surface Schottky gates to perform special measurements, e.g. microwave excitation spectroscopy. In the vertical SET, some important parameters such as the tunneling probability between the dot and the nearest reservoir and the tunnel coupling between the dots are not tunable. In contrast, in the planar SETs both are tunable. In addition, it is possible to adjust the electrostatic potential of each dot independently by changing two gate voltages. We use this capability to study electron transport between two zero-dimensional (0D) states on-resonance and off-resonance.

### 2.2.3 Atom-like properties of a disk-shaped quantum dot

Clean quantum dots with a regular disk shape have only recently been fabricated in a semiconductor heterostructure. [3,4] This subsection presents the electronic properties of such disk-shaped quantum dots containing just a few electrons. We show that the GS energy spectrum reflects atom-like properties.

Electrons bound to a nuclear potential experience sufficiently strong quantum mechanical confinement and Coulomb interactions that they are well arranged in ordered states. This leads to the well-known ordering of atoms in the periodic table. The ionization energy has large maxima for atomic numbers 2, 10, 18, ... when shells are completely filled with electrons. In addition, for the filling of electrons in similar orbitals, parallel spins are favored until the set of orbitals is half filled, as would be expected from Hund's first rule. This also gives rise to secondary maxima in the ionization energy. [5] A good analogue to the three-dimensional shells in atoms can be realized for artificial atoms with the shape of a circular disk. The disk-shaped quantum dots we can fabricate are formed in a laterally gated sub-micron-sized double barrier structure, and contain a tunable number of electrons starting from zero. If the lateral confinement has the form of a harmonic potential, the eigen-energy $E_{n,\ell}$ for the lateral states is expressed using two orbital quantum numbers: the radial quantum number $n$ (=0, 1, 2, ...) and the angular momentum quantum number $\ell$ (=0, ±1, ±2, ...) to

$$E_{n,\ell} = (2n + |\ell| + 1)\hbar\omega_0, \tag{2.2}$$

where $\hbar\omega_0$ is the lateral confinement energy. Here we neglect the Zeeman effect, so each state is spin-degenerate. $E_{n,\ell}$ are systematically degenerate from the lowest level: including spin degeneracy, the first, second, and third shells respectively are two-fold degenerate with $E_{0,0}$ (1s orbital), four-fold degenerate with $E_{0,1} = E_{0,-1}$ (2p orbital), six-fold degenerate with $E_{0,2} = E_{0,-2}$ (3d orbital) = $E_{1,0}$ (3s orbital). For non-interacting electrons, these states are consecutively filled from the lowest, and complete filling of each set of degenerate states is attained for special electron numbers of $N = 2$, 6, 12, 20, etc. These are the "magic numbers" that characterize the shell structure. For interacting electrons, the degeneracy is lifted due to the Coulomb interactions. However, when the quantum mechanical confinement energy is comparable to or greater than the interaction energy, the above shells are still consecutively filled from the lowest, so that we can still see the same series of magic numbers as for the non-interacting case. In addition, for the filling of electrons in the same shell, parallel spins are favored in accordance with Hund's first rule. This leads to another series of magic numbers of $N = 4$, 9, 16, ... corresponding to the half filling of the second, third, fourth (and so on) shells, respectively.

We use a resonant tunneling structure to fabricate a disk-shaped quantum dot. The device configuration is schematically shown in Fig. 2.2. The scanning electron microscope image shows the actual shape of the device mesa. A single

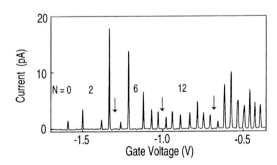

Fig. 2.3. Coulomb oscillations in the current versus gate voltage at $B = 0$ T measured at $T = 50$ mK. [4]

quantum dot is located inside each sub-micron cylindrical mesa made from a double-barrier structure (DBS), which consists of an undoped 12 nm- $In_{0.05}Ga_{0.95}As$ well and two undoped $Al_{0.22}Ga_{0.78}As$ barriers of thickness 9.0 and 7.5 nm. On either side of the double barrier structure there is an $n$-doped GaAs contact. The source and drain electrodes are placed on the top of the mesa and on the bottom of the $n$-GaAs substrate. The third electrode is a Schottky gate, which is wrapped around the mesa. The current $I$ flowing vertically through the dot is measured as a function of gate voltage $V_g$ in response to a dc voltage $V_{sd}$ applied between the contacts.

Figure 2.3 shows the current oscillations (Coulomb oscillations) observed for a device with a geometrical diameter of $D = 0.5 \, \mu m$. [4] A small bias of 0.15 mV is set for $V_{sd}$, so that only the GSs contribute to the current. The absolute values of $N$ can be identified in each zero-current region (Coulomb blockade region) between the peaks, starting from $N = 0$, because for $V_g < -1.6$ V no further current peaks are observed, i.e. the dot is empty. When $N$ becomes smaller than 20, the oscillation period depends strongly on $N$. In contrast, Coulomb oscillations observed for a large dot containing more than 100 electrons look very periodic. The current peak to the left of a Coulomb blockade region with $N$ trapped electrons measures the $N$-electron GS electrochemical potential $\mu(N)$, and the peak spacing labeled by "$N$" therefore corresponds to the increment of electro-chemical potential, $\Delta\mu(N + 1) = \mu(N + 1) - \mu(N)$ (see 2.2.1). $\Delta\mu(N + 1)$ can also be determined from measurement of the widths of the so-called "Coulomb diamonds" (see 2.5.2) .$\Delta\mu(N + 1)$ is

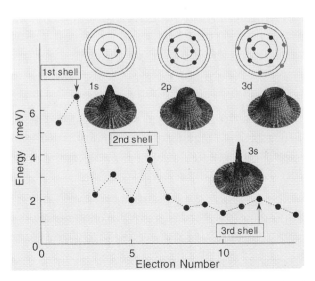

Fig. 2.4. Change of electro-chemical potential, $\mu(N + 1) - \mu(N + 1)$, as a function of electron number $N$. [4] Electron distributions for 1s, 2p, 3d, 3s orbitals are also shown.

plotted as a function of $N$ in Fig. 2.4. In correspondence with the spacings between the Coulomb oscillations, $\Delta\mu(N+1)$ is unusually large for $N = 2, 6$ and 12, and is also relatively large for $N = 4, 9$ and 16 (see arrows in Fig. 2.3). The values of 2, 6 and 12 arise from the complete filling of the first, second and third shells, respectively, while those of 4, 9 and 16 are due, respectively, to the half filling of the second, third and fourth shells with parallel spins in accordance with Hund's first rule. The 2-dimensional (2D) shell structure obtained in the disk-shaped quantum dot is also pictorially illustrated in Fig. 2.4. The addition energy spectrum is well reproduced by calculations of exact diagonalization method. The distribution functions for the 1s, 2p, 3s, and 3d orbitals in the 2D shell structure are included in Fig. 2.4.

## 2.3 Effects of a Magnetic Field

### 2.3.1 Fock-Darwin states

In real atoms, electrons are so strongly trapped that their quantum mechanical properties cannot be strongly modified under normal experimental conditions, for example, by applying a magnetic field. In contrast, the electrons in our quantum dots are bound in a relatively large region of the order of 100 nm. This allows us to use readily accessible magnetic fields, not only to identify the quantum mechanical states, but also to induce transitions in the ground states whose counterparts in real atoms can never be tested on earth.

The eigen-states for a 2D harmonic quantum dot are the Fock-Darwin (FD) states. [6] The eigen-energies at $B = 0$ T, i.e. $E_{n,\ell}$ in Eq. (2.1), are modified in the presence of a magnetic field ($B$-field) parallel to the tunneling current to

$$E_{n,\ell} = -\frac{\ell}{2}\hbar\omega_c + \left(n + \frac{1}{2} + \frac{1}{2}|\ell|\right)\hbar\sqrt{4\omega_0^2 + \omega_c^2}, \qquad (2.3)$$

where $\hbar\omega_c = eB/m^*$ is the cyclotron energy. Figure 2.5(a) shows $E_{n,\ell}$ versus $B$ calculated for $\hbar\omega_0 = 3$ meV, which is deduced from a comparison with the experimental data. Spin-splitting is ignored so each state is two-fold degenerate. [4] The orbital degeneracy at $B = 0$ T is lifted on increasing $B$, reflecting the first term of Eq. (2.3). As $B$ is increased further, new crossings can occur. The last crossing occurs along the bold line in Fig. 2.5(a). Beyond this crossing, the FD-states merge to form the lowest Landau level.

Figure 2.5(b) shows the $B$-field dependence of the position of the current oscillations shown in Fig. 2.3. [4] We take into account the interaction energy as well as the FD diagram when examining the experimental data. The current peaks generally shift in pairs with $B$. This pairing is due to the lifting of spin degeneracy. So from the shift of the paired peaks on increasing $B$, we assign quantum numbers to the respective pairs. For example, the lowest, second

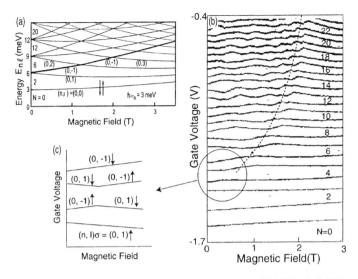

Fig. 2.5. (a) Fock-Darwin (FD) diagram calculated for $\hbar\omega_0 = 3$ meV. [4] (b) $B$-field dependence of current peak positions for the same device whose Coulomb oscillations are shown in Fig. 2.3. [4] (c) Modification of peak pairing in the second shell close to $B = 0$ T is schematically shown. The related F-D states are indicated by $(n, \ell)$ and $\uparrow$ or $\downarrow$, where $\uparrow$ or $\downarrow$ is a state with an up-spin or a down spin, respectively. [4]

lowest, and third lowest pairs correspond to the filling of electrons in the FD states $(n, \ell) = (0, 0)$, $(0, -1)$, and $(0, 1)$ with anti-parallel spins, respectively. Thus, the wiggles or anti-crossings between pairs of peaks correspond to the crossings of FD states. For example, the anti-crossing at the * point corresponds to the crossing of the FD states $(0, -1)$ and $(0, 2)$. However, from close inspection of the second and third lowest pairs of peaks in the vicinity of $B = 0$ T, we find that the pairing is modified in line with Hund's first rule (Fig. 2.5(c)). This is discussed in detail in section 2.3.3. The last wiggle of each pair of peaks appears along the dashed line, which corresponds to the bold line in Fig. 2.5(a). Beyond this line, all electrons are in the spin-degenerate lowest Landau level. This line also identifies filling factor $v = 2$. For $v < 2$, we see various other transitions associated with $B$-field enhanced Coulomb interactions. We now focus on such transitions.

### 2.3.2 Transitions in the ground state

Figure 2.6(a) shows the $B$-field dependence of the first five current peaks for $V_{sd} = 0.1$ mV. [7] Besides the overall smooth shift in the peaks, we see several kinks, which we have identified with different labels. These kinks are assigned to transitions in the GSs from a comparison to the calculations shown in Fig. 2.6(b). [8] This is an exact calculation of the GS electro-chemical potentials for $N = 1$ to 5 as a function of the $B$-field. The electro-chemical potentials for $N > 1$ show upward kinks; more kinks for the larger $N$. At each upward kink, the electro-chemical potentials for two different configurations with the same electron number cross one another. The configuration with the lower electro-chemical potential always forms the GS. That is, the GS undergoes a transition at each up ward kink. These transitions occur in such a way that the total spin, $S$, as well as the total angular momentum, $M$, are maximized in the consecutive transitions. For example, for the $N = 2$ GS, two electrons occupy the same orbital state $(n, \ell) = (0, 0)$ with anti-parallel spins implying $S = 0$ and $M = 0$ to the left of the kink, whereas they occupy the different orbital states $(0, 0)$, and $(0, 1)$ with parallel spins implying $S = 1$ and $M = 1$ to the right of the kink. This is the so-called spin singlet-triplet transition. For the $N = 3$ GS transition, the transition accompanies a change of $(M, S)$ from $(1, 1/2)$ to $(2, 1/2)$, then to $(3, 3/2)$. These transitions are indeed observed in the experiment. In this manner, for the regions between the kinks, we can identify the quantum numbers, including the spin configurations.

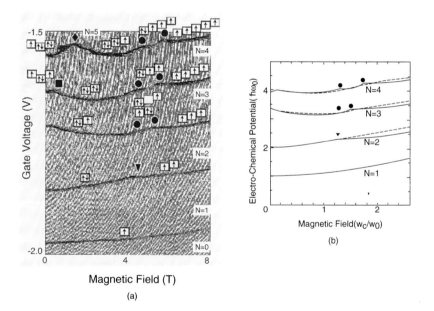

Fig. 2.6. (a) Evolution of current peaks for $N = 0$ to 5 with $B$-field. Ground state transitions are indicated by different symbols. The arrows in the boxes indicate the spin configurations. The lowest box corresponds to the FD state $(n, \ell) = (0, 0)$. For boxes to the right, $\ell$ increases by 1 with $n = 0$. For $N = 4$ and 5, near $B = 0$ T, also the $\ell = -1$ box is shown on the left of the $\ell = 0$ box. [7] (b) Exact calculation of electro-chemical potential for the $N = 1$ to 4 ground states. [8]

### 2.3.3  Spectroscopy of ground and excited states

To investigate the configurations of the GSs and ESs responsible for the kinks observed in Fig. 2.6(a), we use the excitation tunneling spectroscopy technique. [7] A large voltage of $V_{sd} = 5$ mV is now applied between the source and drain. This voltage opens a sufficiently wide transport window between the Fermi levels in the source and drain that both the GSs and ESs can be detected. As $V_g$ is made more positive, first the GS alone, and then the first few ESs can contribute to the current for any given $N$. $I$ versus $V_g$ and $B$ for the $N = 1$ and 2 GS and ESs is shown in Fig. 2.7. [7] For this particular value of $V_{sd}$, the two stripes just touch at $B = 0$ T. The lower edge of each stripe (indicated by the white dashed curve) is the GS. Inside the first stripe, a pronounced current change is observed near the black dashed curve labeled $(n, \ell) = (0, 1)$. This curve enters the upper edge of the first stripe at $B = 0.2$ T. This change identifies the position of the first excited state for the $N = 1$ dot. Note that at higher $B$ values, two higher excited states (black dashed curves) also enter from the upper edge of the stripe at 5.7 and 9.5 T, respectively. The energy separation between the GS (white dashed curve) and the first ES (black dashed curve labeled $(0, 1)$) can be read directly from the relative positions inside the first stripe, so the excitation energy is slightly larger than 5 meV at $B = 0$ T, and then decreases on increasing B. Note that even over this wide magnetic field range of 16 T, the first ES never crosses with the GS. In the second stripe in Fig. 2.7, however, we see the first ES crossing with the GS at $B = 4.15$ T, i.e. the first ES for $B < 4.15$ T becomes the GS for $B > 4.15$ T.

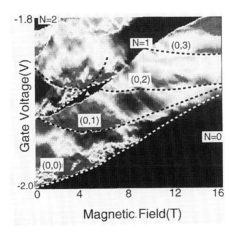

Fig. 2.7.  Surface plot of current amplitude for $N = 0$ to 2 measured with $V_{sd} = 5$ mV. Changes in the color near the dashed line indicate changes in the current amplitude. Black is the Coulomb blockade region where $I = 0$ A. The states in the $N = 1$ stripe (lower stripe) are identified by the quantum numbers $(n, \ell)$. [7]

Located exactly at this $B$-field is the kink attributed to the singlet-triplet transition in Fig. 2.6(a), so it is assigned to a crossing between the GS and the first ES. In a similar fashion, we are able to identify crossings between GSs and ESs at each kink in Fig. 2.6(a).

The last crossing we discuss is indicated by ■ in Fig. 2.6(a). This crossing is a manifestation of Hund's first rule as described in section 2.3.1. As the adjacent configuration diagrams show, there is a transition such that the third and fourth electrons go from having parallel spins to anti-parallel spins. When the states $(n, \ell) = (0, 1)$ and $(0, -1)$ are sufficiently close, the energy gain due to the exchange interaction between electrons with parallel spin favors a high-spin state. [4] When $B$ is increased, $(0, 1)$ and $(0, -1)$ move away from each other (see Fig. 2.6(a)), and at a particular value, a transition is made to an anti-parallel spin state where the third and fourth electrons both occupy $(0, 1)$. Figure 2.8 shows a surface plot of the $N = 4$ stripe measured at $V_{sd} = 1.6$ mV. [7] This surface plot clearly shows the $B$-field dependence of the single-particle states $(0, 1)$ and $(0, -1)$ including the crossing between the GS (white dashed curve) and the ES (black dashed curve) at 0.4 T.

### 2.3.4  B-N phase diagram for the few-electron ground states

Figure 2.9 summarizes pictorially the configurations of the few-electron and many-electron GSs in the plane of $B$ and $N$. At $B = 0$ T, Hund's first rule accounts for the high-spin states within a given shell. As the $B$-field is initially applied, these parallel-spin states disappear, and consecutive anti-parallel filling of electrons into spin degenerate states becomes widespread. As the $B$-field is increased still further, crossings of the FD states give rise to frequent changes in the GS configuration. Very recently, we have found that at each

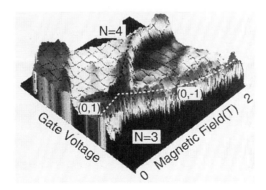

Fig. 2.8. Surface plot for the $N = 4$ current stripe measured with $V_{sd} = 1.6$ meV. The white and black dashed lines show the $N = 4$ GS, and first ES, respectively, from which transitions in the GS can be identified. [7]

crossing of FD states, a GS with parallel spins is favored in line with Hund's first rule. The last crossing of the FD states occurs at $v = 2$. For $2 > v > 1$, a sequence of $N/2$-spin flips is expected to lead to a sequential increase in the total spin of the $N$-electron GS from 0 to $N/2$, i.e. resulting in full spin polarization of the $N$-electron dot at $v = 1$. A self-consistent charge distribution in the dot results in a compressible center (second lowest Landau level partially filled) being separated from a compressible edge (first lowest Landau level partially filled) by an incompressible "ring" in which the lowest Landau level is completely filled. [9] Sequential depopulation occurs as electrons transfer across the ring from spin-down sites at the center to spin-up sites at the edge (see Fig. 2.5(c)). At $v = 1$, a maximum density droplet (MDD) is formed. [1,10] This is spatially the most compact state of spin-polarized electron droplet. All $N$-electrons are in the lowest Landau level, and occupy sequentially the up-spin states $(n, \ell) = (0, 0), (0, 1), ..., (0, N-1)$ without any vacancies. The detailed experiments on the spin-flip phase, and the MDD phase in our disk-shaped dot are discussed in Refs. 9) and 11).

## 2.4 Manipulation of the Lateral Potential Geometry of a Vertical Dot

In section 2.2.3 we described the atomic-like properties of *artificial atoms* observed by measuring Coulomb oscillations in disk-shaped dots. [4] Associated with the rotational symmetry of the circular dot, a "shell" structure, a pairing

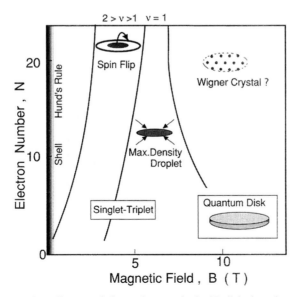

Fig. 2.9. *B-N* phase diagram of electronic states in the 2D disk-shaped quantum dot.

of the conductance peaks, and modifications predicted by Hund's first rule, are observed. This "conventional" single dot vertical structure is made from a DBS by surrounding an "isolated" etched sub-$\mu$m mesa with a single gate. [3] For a cylindrical mesa, this single gate "squeezes" the dot uniformly so that the rotational symmetry of the lateral confining potential is maintained.

In this section, we describe the characteristics of Coulomb oscillations when the mesa forming the vertical dot is non-circular. [12,13] Such non-circular mesas can be examined with a single gate, [12] but multiple independent gates for vertical dots are, in principle, more desirable as they offer greater control. We outline a new technology for fabricaring novel vertical SETs with separate gates which allows the effect of systematically manipulating both the extent and geometry of the lateral confining potential on the electronic states in the dot, and related phenomena, to be probed. [13] It is worth stressing that single-particle states in a quantum dot are very sensitive (far more sensitive than "classical" charging energy) to the geometry of the lateral confining potential.

Figure 2.10 shows a schematic view of a typical device. Full details of the concept, fabrication, and demonstrations of separate gate operation are described elsewhere. [3,13] Four separate Schottky gates (labeled A, B, C and D) surround the central square mesa at the center of the "cross". Current $I$ can only flow vertically from the substrate to the top contact metal pad up through the square mesa because the top contact metal pad is sitting on silicon oxide, and the semiconductor regions in the 0.2 $\mu$m wide top contact metal line mesas ($\alpha$), and gate "splitter" line mesas ($\beta$) are always "pinched-off". The specially designed DBS is the same as that shown in Fig. 2.2. [4] The configuration of one particular square mesa is also shown in Fig. 2.10. Coulomb oscillations are

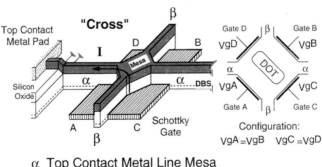

$\alpha$  Top Contact Metal Line Mesa
$\beta$  Gate 'Splitter' Line Mesa

Fig. 2.10. Schematic view and configuration of a square mesa with four separate gates. Drain current $I$ flows vertically through the quantum dot in the mesa at the center of the "cross". [13]

measured at small drain bias as a function of voltages $V_{gA}$, $V_{gB}$, $V_{gC}$, and $V_{gD}$ on the four gates. Gate A (C) and gate B (D) are diametrically opposite. We choose to alter the gate voltages such that $V_{gA} = V_{gB}$ and $V_{gC} = V_{gD}$. It should then be possible to controllably deform the geometry of the lateral confining potential from "square"-like (far from "pinch-off" when there are "many" electrons), or "circular"-like (close to "pinch-off" when there are only a "few" electrons), when $V_{gA}$ $(=V_{gB}) = V_{gC}$ $(=V_{gD})$, to "rectangular"-like or "elliptical"-like, when $V_{gA}$ $(=V_{gB})$ is more negative than $V_{gC}$ $(=V_{gD})$. If all gates are equivalent then the gate configuration we adopt should fix the dot center at the center of the "cross". In practice, the gates are not all equivalent, [13] but unless the gates are grossly different, this principle of operation is a good first approximation.

Figure 2.11 shows a "map" of the conductance peak positions of a 0.8 $\mu$m square device. Each horizontal slice of the map is obtained by sweeping $V_{gA}$ and $V_{gB}$ *together* from 0.4 to –2 V at constant $V_{gC}$ $(=V_{gD})$. [13] Each peak is represented by a black mark. As $V_{gC}$ and $V_{gD}$ are varied *together* from 0.4 to – 2 V, the dot is more strongly "squeezed", so the position of each peak shifts systematically to more positive $V_{gA}$ $(=V_{gB})$. To read the "map" and observe the corresponding peaks in each slice, the "map" should be viewed along the direction indicated by the two dotted parallel lines in the top right-hand corner. These two dotted lines actually mark the movement of two neighboring peaks. Because the voltage increment along the $V_{gC}$ $(=V_{gD})$ axis is not too large, it is straightforward to identify corresponding features in neighboring horizontal

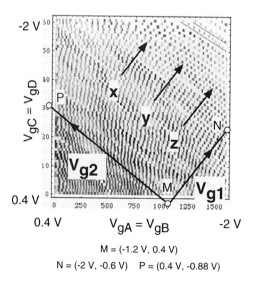

M = (-1.2 V, 0.4 V)

N = (-2 V, -0.6 V)    P = (0.4 V, -0.88 V)

Fig. 2.11. "Map" of conductance peak positions in $V_{gA}$ $(=V_{gB})$ – $V_{gC}$ $(=V_{gD})$ space. $\mathbf{V_{g1}}$ and $\mathbf{V_{g2}}$ are orthogonal voltage vectors. Voltage-vector scan along $\mathbf{x}$, $\mathbf{y}$, and $\mathbf{z}$ are shown in Fig. 2.13. [13]

slices from the patterns in both peak height and relative position, and the presence of unusually large gaps between some of the peaks. [13] It is thus possible to arbitrarily define sets of orthogonal voltage-vectors in $V_{gA}$ $(=V_{gB}) - V_{gC}$ $(=V_{gD})$ space: $\mathbf{V}_{g1}$ vectors, along which the dot is "squeezed" equally in the same sense by all four gates, and electrons are most efficiently removed from, or added to, the dot ($\mathbf{V}_{g1}$ vectors are perpendicular to the two parallel dotted lines); and $\mathbf{V}_{g2}$ vectors along which gate action for gates A and B, and gate action for gates C and D are in an opposite sense, without changing, at the first approximation, the area of the dot ($\mathbf{V}_{g2}$ vectors are parallel to the two parallel dotted lines). In other words, by applying voltages to the four gates along $\mathbf{V}_{g1}$ and $\mathbf{V}_{g2}$ voltage-vectors respectively, it is possible to probe separately how the dot size and dot shape affect the appearance of the current oscillations. Translation along a $\mathbf{V}_{g1}$ voltage-vector changes the area of the dot, but preserves the geometry, and is equivalent to the gate action of "conventional" vertical dots surrounded by a single gate. Generally, moving along a voltage-vector $\mathbf{V}_{g1}$ from the bottom left hand corner of Fig. 2.11 (number of electrons in dot $\approx 90$) to the top right hand corner (number of electrons in dot $\approx 10$), the *average* spacing between neighboring conductance peaks, which is related to the "classical" charging energy, increases steadily. This *average* spacing is not particularly sensitive to a translation along a $\mathbf{V}_{g2}$ voltage-vector, because the area of the dot remains almost constant. On the other hand, the *actual* spacing between neighboring conductance peaks, which is very much dependent on the nature of the single-particle states involved, can be very sensitive to such a translation, leading to deformation of the dot. A translation along a $\mathbf{V}_{g2}$ voltage-vector represents a new mode of operation that is not possible with a single-gated vertical dot.

Figure 2.12 shows a vector scan through the "map" in $V_{g1} - V_{g2}$ space over the region defined by points M, N, and P in Fig. 2.11. [13] With the systematic shift to more positive $V_{gA}$ $(=V_{gB})$ in Fig. 2.11 eliminated, the conductance peaks

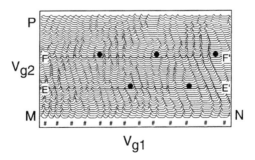

Fig. 2.12. "Map" of conductance peaks in $V_{g1} - V_{g2}$ space. Pairs of conductance peaks are identified by "#". Large gaps between peaks in these traces are marked by "●". [13]

generally run vertically as $V_{g2}$ is varied. Residual deviations or "wiggles" in the conductance peak positions can now be clearly observed and these reflect the sensitivity of the single-particle states to the dot geometry. The opening and closing of unusually large gaps between some conductance peaks signifies that the underlying "shell" structure is changing as the dot is deformed. We can distinguish, for example, along trace EE' a "shell" consisting of eight peaks bounded by two large gaps (large gaps are marked by "●"). Moving to trace FF', the positions of the gaps have changed. Since the number of electrons in the dot is more than 20, Hund's first rule alone cannot account for the large gaps. Because the dot is still quite "large", it is only possible to probe the "shell" structure when there are "many" electrons in the dot. Nevertheless, the effective dot size is comparable to the Fermi wavelength, so quantum effects are still important. Due to the non-circular nature of the lateral confining potential, and the presence of electron screening, there is as yet no theory that accounts for the exact details.

Another important observation related to Fig. 2.12 is that pairs of conductance peaks can be identified (each pair marked by "#"), i.e. the positions of the conductance peaks forming a pair are directly related as $V_{g2}$ is varied, and large gaps ("●") occur between pairs of peaks and not between peaks belonging to the same pair. This pairing of the conductance peaks is due to spin-degeneracy, and is insensitive to geometric distortions of the dot.

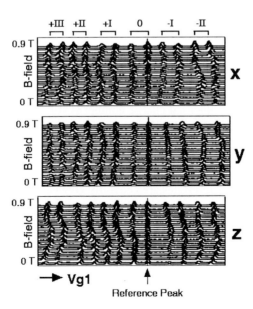

Fig. 2.13. *B*-field dependence up to 0.9 T of the conductance peaks along the three $\mathbf{V}_{g1}$ voltage-vectors **x**, **y**, and **z**. The "reference peak" is actually the same peak for each voltage-vector. [13]

Conservation of peak-pairing and the alteration of the "shell" structure with geometric distortion can be further demonstrated by following the $B$-field dependence of the positions of conductance peaks along different $\mathbf{V_{g1}}$ voltage-vectors. Figure 2.13 shows the $B$-field dependence up to 0.9 T along three different $\mathbf{V_{g1}}$ voltage-vectors $\mathbf{x}$, $\mathbf{y}$, and $\mathbf{z}$. These three vectors are indicated in Fig. 2.11. [13] All traces have been shifted horizontally so that an arbitrarily chosen "reference peak" is aligned, and its position is independent of the $B$-field and the choice of $\mathbf{V_{g1}}$ voltage-vector. Pairs of peaks are labeled + III, + II, + I, 0, –I, and –II. Analysis of the three sets of traces for $\mathbf{x}$, $\mathbf{y}$, and $\mathbf{z}$ confirms that peak-pairing is a basic intrinsic property, generally robust to geometric distortions of the dot. On the other hand, the pattern of "wiggles" of individual peaks, or equivalently pairs of peaks, are dissimilar, which indicates that the underlying dot geometry and thus the nature of the single-particle states is very different.

### 2.5 Quantum Dot Molecules

In section 2.2.3 we described the atomic-like properties of single *artificial semiconductor atoms* in high-quality disk-shaped vertical quantum dots containing a tunable number of electrons. [4] Knowledge of the attributes of a single quantum dot is invaluable for understanding single electron phenomena in more complex quantum dot systems. We now outline how vertically coupled disk-shaped dots can be employed to study the electronic states in quantum dot molecules.

Let us consider two quantum dots, in each of which charge quantization and energy quantization are well defined. If two 0D states, one state from one well and one state from the other well, are quantum-mechanically strongly coupled, an electron can oscillate back and forth coherently between the two dots. In this case, the charge number state of each dot is no longer a well-defined integer, while the total charge number is still a well-defined integer. The coupled states are delocalized over the two dots, and the eigen-functions are superpositions of the states in each dot, $\psi_1$ and $\psi_2$, in a symmetric and an anti-symmetric way, i.e. $\psi_S = (\psi_1 + \psi_2)/\sqrt{2}$ and $\psi_{AS} = (\psi_1 - \psi_2)/\sqrt{2}$ when the states are energetically aligned. If there is an energy mismatch, $\varepsilon$, between the two states, the delocalized states are generally given by $\psi_1 \sin\theta + \psi_2 \cos\theta$ (tan $\theta = (\Delta_{SAS} - \varepsilon)/2hT_c$), where $T_c$ is the coherent tunneling rate, and $\Delta_{SAS}$ $(=\sqrt{(\varepsilon^2 + 4T_c^2)})$ is the energy splitting between the symmetric and anti-symmetric states. In a vertical device, two disk-shaped dots are stacked on top of each other and are separated by a thin heterostructure potential barrier. The quantum mechanical coupling between the dots generates a symmetric state and an anti-symmetric state in the vertical direction. Both of the symmetric and anti-symmetric states are confined laterally by a parabolic potential. In section 2.5.2 we discuss the electronic properties of a quantum mechanically "strongly" coupled double dot molecule in the vertical configuration, and likewise in section 2.5.3 for a quantum mechanically "weakly" coupled double dot molecule. The electronic states in the latter are usually localized. Nevertheless, the "weakly" coupled dots are still coupled *electrostatically*, and this can lead to a pairing of conductance peaks.

### 2.5.1 Double dot molecules—vertical configuration

A vertically coupled quantum dot molecule can be realized in a geometry similar to that of a single disk-shaped quantum dot (see Fig. 2.2), but the DBS is replaced by a triple barrier structure (TBS). [14] The InGaAs wells are 12 nm wide, and the outer AlGaAs barriers are typically about 7 to 8 nm wide. The inset of Fig. 2.14 shows a schematic section through such a gated mesa of geometric diameter $D$. The drain current is measured as a function of drain voltage, $V_{sd}$, and gate voltage $V_g$ much like a single dot.

By changing the thickness of the central $Al_{0.2}Ga_{0.8}As$ barrier $b$ from 7.5 nm to 2.5 nm we are able to increase the energy-splitting between symmetric and anti-symmetric states, $\Delta_{SAS}$, from about 0.09 to 3.4 meV. *Quantum mechanically*, we consider the dots separated by the 7.5 nm barrier to be "weakly" coupled, and the dots separated by the 2.5 nm barrier to be "strongly" coupled. Figure 2.14 shows how $\Delta_{SAS}$ ($\varepsilon = 0$) varies with $b$ based on a simple flat-band model calculation with a material-dependent effective mass. As a rough guide, for the case of two electrons trapped in the system ($N = 2$), a typical "bare" lateral confinement energy $\hbar\omega_0$ of 4 meV is indicated for all values of $b$, a typical

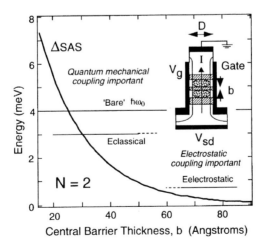

Fig. 2.14.  A simple calculation of $\Delta_{SAS}$ as a function of central barrier thickness, $b$. Inset shows a schematic section through the vertical double dot structure. [14]

average "classical" charging energy, $E_{classical}$ $(=e^2/C)$, of 3 meV is shown for $b < 5$ nm, and an electrostatic coupling energy, $E_{electrostatic}$, of 0.7 meV is marked for $b > 6.5$ nm. We stress that quantum mechanical coupling is not the only coupling mechanism in *artificial molecules*. In the regime where $\Delta_{SAS} \approx \hbar\omega_0$ ($>> E_{electrostatic}$), quantum mechanical coupling between the two dots is the dominant mechanism, but in the regime where $(\hbar\omega_0>) E_{electrostatic} >> \Delta_{SAS}$, it is electrostatic coupling between the dots which becomes important. Competition between the two mechanisms as $b$ is varied is expected to have a profound effect on the transport properties of the two-dot system.

The vertical quantum dot molecules described here are very different from other quantum dot molecules reported recently. The coupling strength can be tuned *in situ* in a planar double quantum dot as demonstrated by Waugh *et al.* (see also Fig. 2.19), but this type of transistor can only access the "many"-electron condition where $E_{classical} > \hbar\omega_0$. [15] Single-particle states are observed in an ungated vertical TBS by Schmidt *et al.*, but this type of device cannot accumulate electrons one-by-one at zero bias. [16]

### 2.5.2  Quantum mechanically strongly coupled double dot molecules—vertical configuration

Figure 2.15 shows a grey scale plot of $dI/dV_{sd}$ in the $V_{sd} - V_g$ plane for a $D = 0.56$ µm *quantum mechanically* "strongly" coupled ($b = 2.5$ nm) double dot transistor (outer barriers of TBS nominally symmetric). [14] Black (positive values of $dI/dV_{sd}$) and white (negative values of $dI/dV_{sd}$) lines criss-crossing the plot and running parallel to the sides of the diamonds identify bound and excited states. Well-formed Coulomb diamonds (grey regions where $I = 0$ pA) close to zero bias are evident. The symmetry of the diamonds with respect to bias direction confirms that the states responsible are delocalized over both dots. Notice that the $N = 2$ and $N = 6$ diamonds are unusually large compared to the adjoining diamonds. As with the single dots, the half-width of the $N$-th diamond is a direct measure of the energy needed to add one more electron to the system. This "addition energy" (or $\Delta\mu(N)$) contains information about the relevant lateral confinement, direct Coulomb, and exchange energies. The change of electro-chemical potential as a function of electron number for the "strongly" coupled double dot, and a $D = 0.5$ µm single dot is shown in the inset in Fig. 2.15. The familiar "magic" numbers 2, 6, 12, ... marking the complete filling of shells, and 4, 9, 16, ... marking the half filling of shells (Hund's first rule), for the single dot *artificial atom* are shown for comparison. [4] For the "strongly" coupled double dots, "magic" numbers 2 and 6 are very clear (12 is less so), although 4 is only faintly present. The 4 meV value for the "bare" $\hbar\omega_0$ in Fig. 2.14 is consistent with the addition energy for $N = 2$, and the $B$-field dependence of the first and second conductance peaks (data not shown). The 3 meV value of $E_{classical}$ in Fig. 2.14 is also in line with the average of the addition energies for $N = 1, 2$, and 3 in Fig. 2.15. Note that for $N > 15$, the addition energy

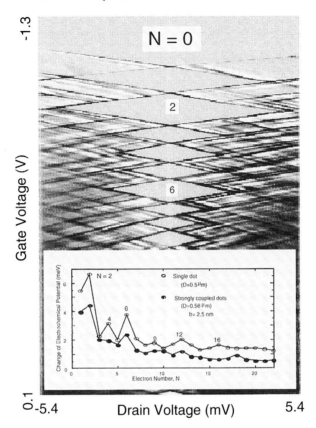

Fig. 2.15. Grey scale plot of $dI/dV_{sd}$ in the $V_{sd} - V_g$ plane for a *quantum-mechanically* "strongly" coupled double dot transistor. Inset: Change in electro-chemical potential as a function of electron number for the same "strongly" coupled double dot, and a $D = 0.5$ $\mu$m single dot. [14]

for this double dot is approximately half that for the single dot. This is reasonable, because the double dot occupies roughly twice as much volume as the single dot. For this *artificial molecule*, there is no evidence from Fig. 2.15 for $N < 20$ of the presence of anti-symmetric states, i.e. the lowest states are all symmetric states.

### 2.5.3 Quantum mechanically weakly coupled double dot molecules—vertical configuration

Figure 2.16 shows a grey scale plot of $dI/dV_{sd}$ in the $V_{sd} - V_g$ plane for a $D = 0.5$ $\mu$m *quantum mechanically* "weakly" coupled ($b = 7.5$ nm) double dot transistor showing what appear to be poorly formed and disrupted Coulomb diamonds close to zero bias from $N = 0$ to 7. [14] The black and white lines

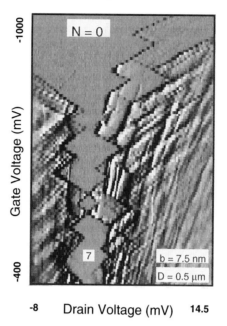

Fig. 2.16. Grey scale plot of $dI/dV_{sd}$ in the $V_{sd} - V_g$ plane for a *quantum-mechanically* "weakly" coupled double dot transistor. [14]

which partially cut across the Coulomb diamonds, i.e. lines not parallel to the sides of the diamonds, particularly those in the forward bias (the collector barrier of TBS is slightly thinner than the emitter barrier), are due to resonant tunneling between zero-dimensional states in the two dots (resonance width ≈ 0.3 meV ≈ $\Delta_{SAS}$ ($\varepsilon = 0$ meV)). The contrasting behavior between the "strongly" coupled dots in Fig. 2.15 and the "weakly" coupled dots in Fig. 2.16 is striking. Higher resolution measurements reveal that the disrupted diamonds are actually complete Coulomb kites (data not shown), and from the shape of these kites, we can deduce that the two dots are generally filled alternately by non-resonant processes, i.e. states are usually localized. The excitation voltage dependence from $V_{sd} = 50\,\mu V$ to $300\,\mu V$ of the $I - V_g$ characteristic for the "weakly" coupled double dot transistor is shown in Fig. 2.17. [14] When there are more electrons in the system, alternate filling of the dots can also be observed. Five consecutive pairs (each identified by a "●") of peaks are evident from $N = 7$ to 17. The pairing arises from electrostatic coupling between the dots. [17] From the related Coulomb kites (not shown), we can estimate the energy splitting between the peaks belonging to each pair. If $\varepsilon << E_{electrostatic}$, this energy splitting of about 0.7 meV is a measure of $E_{electrostatic}$ (hence the 0.7 meV value in Fig. 2.14).

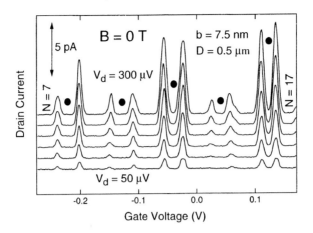

Fig. 2.17. Excitation voltage dependence from $V_{sd} = 50\ \mu$V to $300\ \mu$V of the $I - V_g$ characteristic for the *quantum-mechanically* "weakly" coupled double dot transistor. [14]

## 2.6 Double Dot Molecules—Planar Configuration

Compared with vertically coupled quantum dot molecules, as described in section 2.2.2, laterally coupled planar quantum dot molecules have greater tunability in coupling strength, and the device configuration is more varied. It follows that two charge states, namely an extra electron occupying one dot or an extra electron occupying the other dot, can be easily controlled in laterally coupled devices. This is true not only of the energy of the electrons, but also their charge states can be superposed. This is called a fully-controllable two-level system, and is used in this section to study the properties of molecular-like states, i.e. electrostatic coupling and coherent tunneling coupling. In section 2.6.3, a photon-assisted tunneling technique is employed to probe the energy spectrum. In addition, the effect of the environment around the quantum dots is discussed in section 2.6.4. Real atoms couple to the electromagnetic environment, while quantum dots couple to the lattice environment.

### 2.6.1 The Classical Coulomb blockade regime

We start with a molecule formed from two weakly coupled classical dots as the simplest example. The two charge states are only electrostatically coupled between the two dots. The device configuration and the equivalent circuit diagram are shown in Figs. 2.18(a) and (b), respectively. The gate voltage applied to each dot raises the electrostatic potential of the respective dots, and changes the number of trapped electrons one-by-one. We assume that there is no mutual influence between the two gate operations. Nonetheless, there is still an interaction between the dots through the coupling capacitance $C_i$, i.e. putting one more electron on the first dot raises the electrostatic potential of the second dot by fraction $C_i/C$ of the charging energy $e^2/C$. Here, $C \approx C_t + C_g + C_i$ is the total capacitance of the first dot. Thus, the electron occupancy in the two-dot system is a function of the gate voltages applied to the respective dots. Figure 2.18(c) shows the charging configuration for the electron numbers $n$ and $m$ on the two dots. Each $(n, m)$ configuration is represented by a hexagon, and hence we see a honeycomb structure in the charging diagram. Within the hexagons, the electron numbers are fixed at well-defined integers $n$ and $m$ and no electron transport occurs. On the boundary lines between two different configurations, an extra electron flows back and forth between one of the dots and the nearest reservoir, but it does not go through the other dot, so no actual current flows. A finite current only flows at the vertex where three different configurations meet. The number of electrons in each dot can then fluctuate by one, provided that the total electron number, $n + m$, can fluctuate only by one due to the influence of the electrostatic coupling between the dots. This leads to electron tunneling by two types of sequence. One is called an electron-like process in which an electron sequentially tunnels through the three barriers from the source to the drain. The charge configuration then changes in the sequence from $(n, m)$ through $(n + 1, m)$ to $(n, m + 1)$. The other is a hole-like process occurring in the reverse order, starting with $(n + 1, m + 1)$, then through $(n + 1, m)$ to $(n, m + 1)$, as if a positively charged particle, a hole, has moved toward the positively biased contact. These electron-like and hole-like processes appear at the vertex labeled P and Q, respectively in the figure.

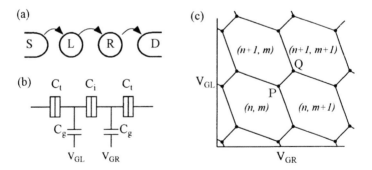

Fig. 2.18. (a) Schematic diagram of two quantum dots, L and R, coupled in series between the source S, and the drain D, contacts. (b) Equivalent circuit of the double dot system. (c) Charging diagram of the double dot system: $(n, m)$ denotes the stable charge configuration with $n$ electrons in the left dot and m electrons in the right dot. Electrons can pass through the double dot only at the triple points where three phases meet. Electrons tunnel sequentially in the normal order at the vertex P, and in reversed order at Q.

The effect of electrostatic coupling was previously studied for superconductor and semiconductor islands. Especially for the electron turnstile [15,18] and the pump operation, [19] it was observed that only one electron tunnels through the device during one period of the alternating voltage on the gates. For turnstile operation, sinusoidally alternating voltages are applied to two electrodes with a certain phase difference such that the charge configuration changes from $(n, m)$ through $(n + 1, m)$ to $(n, m + 1)$ to $(n, m)$. A single electron flows from the left to the right contact regardless of the bias voltage polarity. On the other hand, for pump operation, a single electron flows in the opposite bias direction on the phase of alternating voltage being reversed. This unique property of carrying one electron per cycle of the ac gate voltage might have an application as a current standard.

Figure 2.19 shows the current amplitude in the plane of the two gate voltages measured for a double dot device corresponding to Fig. 2.18(a). [20] In the limit of weak coupling, A, where $C_i$ is negligible to $C$, the current is seen as an array of spots and the electron-like and hole-like processes are hardly resolved. If the coupling is increased slightly to the case of B, the two processes are well resolved and the charge diagram becomes honeycomb-like. As the coupling is increased from B to D, co-tunneling process becomes distinct on the boundary of the hexagons. Finally, in the strong coupling limit, from E to F, the two dots merge into a large single dot and a periodic Coulomb oscillation characteristic of the large dot appears along the $V_{g1} + V_{g2}$ direction. When the tunneling conductance through the central barrier exceeds the value of $e^2/h$, charge quantization breaks down, so the electron numbers, $n$ and $m$, in each dot are no longer well defined as integers, but the total number, $n + m$, remains a well-defined integer. [20]

Note that we need two independent gates for understanding the full charging diagram of the double dot system. If one uses only one gate, the current can be measured only along one gate voltage axis in the charging diagram. The vertices corresponding to the configuration $(n, m)$ will rarely be cut, and the current modulation will be non-periodic and complicated. As seen in Fig. 2.19, the peak current amplitude depends significantly on the location in the charging diagram, and on the source-drain voltage. These dependencies have been explained in terms of a stochastic Coulomb blockade. However, measurement using a raster sweep over the whole charging diagram takes a lot of time and complicates understanding of the processes. One can sweep two gate voltages along the vector from vertex P to vetex Q (see Fig. 2.18(c)) to address this problem. [15]

### 2.6.2 Quantum regime

As described in section 2.2.1, tunneling current flows when the resonant 0D states in the double dot are between the Fermi energies of the source and drain. The overall tunneling rate is determined using the incoming rate from

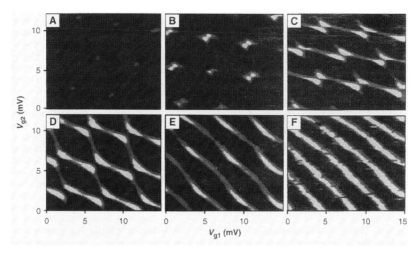

Fig. 2.19. Logarithmic plot of conductance in a double dot structure as a function of gate voltages, $V_{g1}$ and $V_{g2}$, whose zero is arbitrarily set. White indicates high conductance; dark regions represent low conductance. Interdot conductances are (A) $G_{int} = 0.22\ G_0$, (B) $G_{int} = 0.40\ G_0$, (C) $G_{int} = 0.65\ G_0$, (D) $G_{int} = 0.78\ G_0$, (E) $G_{int} = 0.96\ G_0$, and (F) $G_{int} = 0.98\ G_0$ (where $G_0 = 2e^2/h$); (F) is at the threshold of a larger value of conductance to accommodate a higher background conductance. [20]

the source $\Gamma_i$, the outgoing rate into the drain $\Gamma_o$, and the tunnel coupling between the dots $T_c$. This current is limited by the tunneling rate between the two 0D states, so that the resonance peak is not accompanied by thermal broadening. The coupling between the dots and the coupling to the reservoir broadens the quantum state in each dot. The tunneling current has a line shape given by the density matrix theory, $I(\varepsilon) = eT_c^2\Gamma_o/\{T_c^2(2 + \Gamma_o/\Gamma_i) + \Gamma_i^2/4 + \varepsilon\}$, where $\varepsilon$ is the energy difference between the two 0D states. [21,22] Note that $\Gamma_i$ and $\Gamma_o$ appear in an asymmetric manner, because the incoming tunneling builds up the single electron states and the outgoing tunneling breaks the states. The Lorentzian line shape has been clearly observed in Fig. 2.20(c). From the fit to the data for both positive and negative biases, the parameters, $T_c$, $\Gamma_i$ and $\Gamma_o$ can be determined. Note that the width of the current peak is narrower than the thermal energy. There are small deviations from the Lorenzian shape on one side in Fig. 2.20(c), which will be discussed in the next subsection.

### 2.6.3 Microwave excitation spectroscopy

Microwaves have a suitable energy range (4–200 $\mu$eV for 1–50 GHz) for exciting an electron from one state to another in quantum dots. This contributes to the tunneling current, so that the transition energy between the states can be

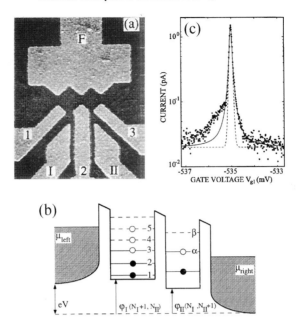

Fig. 2.20. (a) Scanning electron microscope images of the double dot structure with lithographic dimensions of $320 \times 320$ nm$^2$ (left dot) and $280 \times 280$ nm$^2$ (right dot). (b) Schematic potential landscape of the double quantum dot, where $\mu_{\text{left}}$ and $\mu_{\text{right}}$ denote the electro-chemical potential of the left and right reservoirs, and $V_{\text{sd}}$ is the bias voltage across the double dot structure. The 0D states in dot $I$ are denoted by levels 1 to 5, and in dot II by levels $\alpha$ and $\beta$. (c) The resonance of the two discrete levels measured with a bias voltage of 400 $\mu$V. The data points (black dots) can be fitted with a Lorentzian line shape (solid line). For comparison, we have plotted a thermally broadened resonance with a fit temperature $T = 35$ mK (dashed line). [21]

measured as a function of microwave energy. This is in the context of photon-assisted tunneling (PAT), which was experimentally studied for Josephson junctions, and analyzed in terms of time-dependent tunneling theory. [23] When the state $\psi_0 e^{iEt/\hbar}$ is placed in an oscillating potential $eVe^{i\omega t}$ the state takes on the form of a side-band structure: $\sum J_n(eV/\hbar\omega)\psi_0\exp(i(E + n\hbar\omega)t/\hbar)$. Here, $J_n(x)$ is an $n$-th order Bessel function of the first kind. These states are equally spaced in energy by $\hbar\omega$ and have probabilities of $J_n^2(eV/\hbar\omega)$. The total probability, $\sum J_n^2(x)$, is unity for any $x$. The photon-assisted tunneling is also described in chapter 6. We apply the PAT technique to investigate the energy spectrum in a double dot system.

The microwave field is irradiated via the central gate electrode as schematically shown in Fig. 2.21. [24] The double dot is made using the technique of ion implantation and deposition of finely patterned Schottky gates. This maximizes the discrete 0D energy spacing $\Delta E$. Typical values of $\Delta E$

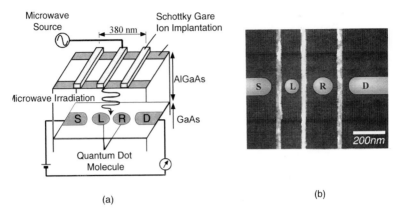

(a)                                                                    (b)

Fig. 2.21. (a) Double quantum dot device defined in the two dimensional (2D) electron gas of a GaAs/AlGaAs heterostructure by focused ion beam implantation. The narrow channel connects the large 2D source and drain leads. Negative voltages ($V_{gL}$, $V_{gC}$, and $V_{gR}$) applied to the metal gates (For $G_L$, $G_C$, and $G_R$; widths are 40 nm) induce three tunable tunnel barriers in the wire. The two quantum dots, L and R, respectively, contain about 15 and 25 electrons; charging energies are about 4 and 1 meV; and the measured average spacings between single-particle states are about 0.5 and 0.25 meV. (b) Scanning electron microscope images of the double dot structure.

range from 250 to 500 $\mu$eV in this sample. Figure 2.22 shows the effect of microwave irradiation on the 2D-raster sweep of the two gate voltages. [24] The sweep is taken only in the vicinity of the vertex M (electron-like process). The current peak at M is due to a 0D-0D resonance through the double dot when the energy levels, $E_L$ and $E_R$, are aligned at the Fermi energy. For microwaves of 50 GHz, additional positive and negative peaks appear on opposite sides of the primary peak, growing with microwave intensity. The two peaks are assigned to the electron excitation from the left-dot to the right-dot, and from the right-dot to the left-dot, respectively. We define two energy scales for the two dot states: the energy difference $\varepsilon = E_L - E_R$, and the average energy $\overline{E} = (E_L + E_R)/2$. Since the condition for the PAT is that the one level is located below, and the other level is located above the Fermi energy, the PAT peak has a width equivalent to the photon energy along the $\overline{E}$ direction. This technique can be used to precisely determine the energy scale on the plane of the gate voltages.

If $V_{sd}$ is zero or very small compared to the photon energy, as can be seen in Fig. 2.22, the electron occupies the lower energy state and is excited to the higher energy state by the microwave irradiation for both the positive and negative PAT peaks. This is the case for photon absorption. When the bias voltage is increased above the photon energy (i.e. $eV_{sd} > h\nu$), an electron can be injected into the higher energy state at the time when the lower energy state is empty (see Fig. 2.25). This leads to some inelastic tunneling from the higher

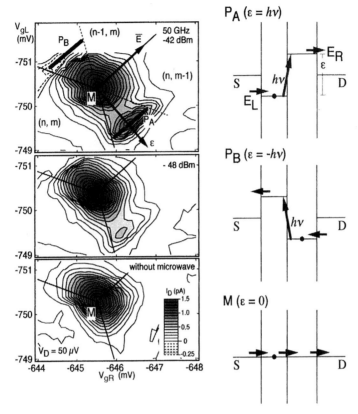

Fig. 2.22. Gray scale plot with contour lines of the current flowing through the double quantum dot structure measured with a small bias voltage of 50 μV. The three thin lines are boundaries of charge states $(n, m)$, $(n-1, m)$, and $(n, m-1)$. The main resonance peak is observed at the triple point (M) of the three charge states. The arrows indicate two energy directions along which the average energy, $\bar{E} = (E_R + E_L)/2$, and the energy difference, $\varepsilon = E_R - E_L$, can be changed. $P_A$ and $P_B$ is the conditions for PAT processes; $P_A$ for the excitation from the left to the right dot, and $P_B$ for the excitation from the right to the left dot. [24]

state to the lower state with the spontaneous emission of energy quanta. This will be discussed in section 2.6.4. By neglecting the spontaneous emission, the higher state will be populated while lower state is left empty ("population inversion" in a two-level system). [24] The application of a microwave field enhances the transition probability from the occupied higher state to the empty lower state, which is called stimulated emission. If one can integrate an array of double dots in to a cavity medium, laser operation is expected for the microwave or THz regime. [25]

    As time-dependent tunneling theory predicts, a non-linear optical regime or multiple photon-assisted tunneling should occur at an intense microwave amplitude. [26,27] Figure 2.23 shows how the higher order PAT processes evolve with microwave amplitude. [28] The peak amplitude plotted against the incident microwave amplitude (left inset) is found to follow the predicted Bessel function squared dependence well. The peak spacing for the one- and two-photon peaks from the main resonance are plotted against the frequency in the right inset. The clear linear dependence on frequency suggests the energy difference can be measured both from the gate voltage and from the frequency of the PAT spectrum. We note that spectroscopy using microwave excitation allows us to directly measure the energy spacing, while the gate voltages are only used to shift the electrostatic potential via the gate capacitance.

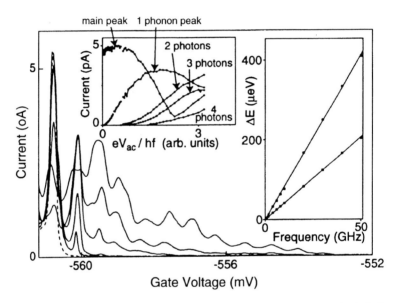

Fig. 2.23. Current versus gate voltage of a weakly coupled double dot structure. The dashed curve is for the case when no microwaves are applied and contains only the main resonance. The solid curves are taken at 8 GHz for increasing microwave powers resulting in an increasing number of satellite peaks. On the right side of the main peak, these correspond to photon absorption. The source-drain voltage is $V_{sd} = 700$ $\mu$eV and the photon energy is $hf = 32$ $\mu$eV at 8 GHz. At the highest power, we observe 11 satellite peaks, demonstrating multiple photon absorption. Left inset, height of the first four satellite peaks as a function of the microwave amplitude. The observed height dependence agrees with the expected Bessel-function behavior. Right inset, distance between main resonance and first two satellites as a function of the applied frequency from 1 to 50 GHz. The distance is converted to energy through $\delta E = \kappa \Delta V_g$, where $\kappa$ is the appropriate capacitance ratio for our device that converts gate voltage $V_g$ to energy. The agreement between the data points and the two solid lines, which have slopes of $h$ and $2h$, demonstrates that we observe the expected linear frequency dependence of the one- and two-photon processes. [28]

For the weakly coupled double quantum dot system just discussed, the effect of coherent tunnel coupling between the two dots is ignored. If the two spatially separated charge states are quantum-mechanically strongly coupled, two coherent states, i.e., a symmetric state and an anti-symmetric state, are formed throughout the two-dot system as described in section 2.5. The energy splitting between these symmetric and anti-symmetric states, $\Delta_{SAS}$ $\left( = \sqrt{\left( \varepsilon^2 + 4T_c^2 \right)} \right)$, can be measured using the PAT technique. Figure 2.24 shows the PAT spectrum against gate voltage (converted into energy difference $\varepsilon$) for different frequencies of 7.5–17 GHz. [28] The peak traces the expected hyperbolic function of the energy difference in $\Delta_{SAS}$ very well. The separation between symmetric and anti-symmetric states, $\Delta_{SAS} (\varepsilon = 0) = 2T_c$, is estimated to be 36 $\mu$eV in this case, and can be well controlled from 11 to 60 $\mu$eV by changing the central gate voltage or magnetic field perpendicular to the sample surface. [28,29]

The dynamics of the two levels are of fundamental interest in quantum physics. Consider the two unperturbed levels, $E_R$ and $E_L$, aligned in energy

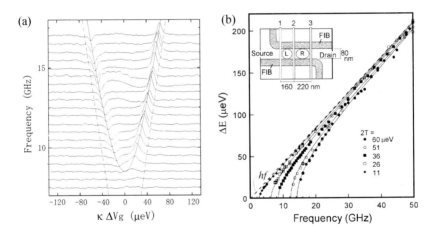

Fig. 2.24. (a) Measured photon assisted tunneling current through a coherently coupled double quantum dot device. Gate voltages on $G_L$ and $G_R$ are swept simultaneously to give an energy difference, $\varepsilon = E_R - E_L$, between the two levels. The different traces are taken at different microwave frequencies and offset such that the vertical axis gives the frequency. Typical peak current is 0.5 pA. The dashed lines show a linear dependence, $\varepsilon = hf$, expected for no coherent coupling, while the solid curve is a hyperbolic dependence with a covalent energy of 36 $\mu$eV. The left and right insets show schematic energy diagrams to show different current directions. At positive $\varepsilon$, the bonding and anti-bonding states are weighted respectively on the left dot and the right dot, and vice versa at negative $\varepsilon$. (b) The microwave frequency, $f$, dependence of the resonance condition, $\varepsilon$, taken from half the energy spacing between the positive and the negative peaks. Different symbols represent different gate voltage and magnetic field conditions. Solid lines show the hyperbolic dependence fitting to the data. [28]

($\varepsilon = 0$), and coupled by amplitude $T_c$. If one can place an electron $E_R$ on one level, for instance, at $t = 0$, the probability of finding an electron in the right dot, $P_R(t)$, should oscillate as $\cos^2(T_c t/h)$. The electron goes back and forth between the two levels with a period of $h/2T_c$. The gating of the oscillation controls the quantum state of superposition, which is the basis of a bit reversal operation in quantum logic gates. There are a number of ways to gate the coherent oscillation. Consider that the two levels are well separated in energy $\varepsilon \gg T_c$, to effectively "freeze" the oscillations into a steady state. If one can align the two levels $\varepsilon = 0$, for a short time $\tau$ necessary for gating and then misalign them again so $\varepsilon \gg T_c$, the probability $P_R(\tau)$ can be controlled by the time period. This technique has been demonstrated in superconducting small electron boxes by Nakamura *et al.* [30] Another way of control is via Rabi oscillations. The levels are separated, $\varepsilon \gg T_c$, as the steady state, and microwaves whose photon energy is matched to the energy spacing $\varepsilon = h\nu$ are irradiated for a short time $\tau$. The oscillating period is $h/2T_c J_1^2(\alpha)$, depending on the normalized ac amplitude $\alpha$ ($\approx eV/\hbar\omega$). [31]

### 2.6.4 Coupling to the environment

In the previous subsection, we argued that coherent tunneling coupling leads to coherent oscillations. The oscillations continue forever if the system is completely isolated from the environment, but some coupling to the environment is necessary for gating the oscillations, and for measuring the system. We now deal with the coupling of the double dot to the environment, which in general causes significant decoherence. It has been theoretically argued that how a two-level system behaves dynamically with coupling to the environment is a spin-Boson problem. [32] Depending on the spectral density function $J(\omega)$, which describes the energy-dependence of the coupling to the environment, the dynamics of a two-level system can show oscillations or localization. If the coupling is Ohmic, the energy dissipation is linear with energy, $J(\omega) \sim \omega$, so the dynamics behave in the same way as for a classical friction force. This gives localization (the system does not change), i.e. exponential decay without oscillation, or damped oscillations depending on the strength of the coupling to the environment. Super Ohmic dissipation, $J(\omega) \sim \omega^s$ and $s > 1$, leads to damped oscillations, while sub-Ohmic, $s < 1$, always leads to localization. Coupling to the environment is a crucial argument for long-lived coherence.

The double quantum dot system has tunable energy selectivity to the environment, since the system emits and absorbs quanta with energy equal to the energy spacing of the two levels. One can measure the energy-dependent coupling to the environment without any external spectrometer. Figure 2.25(e) shows the current spectrum measured by changing the gate voltages to sweep the configuration of the two levels from absorption (b), through resonance (c), to emission (d). [33] $V_{sd}$ is set sufficiently large ($V_{sd} = 140\ \mu eV$), so that the left

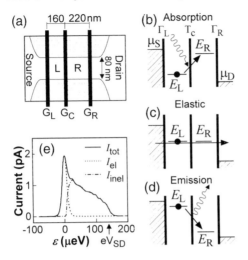

Fig. 2.25. (a) Schematic illustration of the double quantum dot device. (b)–(d) Diagrams showing energy (vertical axis) along the horizontal spatial axis through the dots for the situations: absorption, elastic tunneling and emission. Thick vertical lines denote tunnel barriers. The continuous-electron states in the leads are filled up to the Fermi energies, $\mu_s$ and $\mu_d$. The external voltage, $V_{sd}$, between the leads opens a transport window of size $eV_{sd} = \mu_s - \mu_d$. An elastic current can flow when $\varepsilon = 0$, otherwise a non-zero current requires absorption ($\varepsilon < 0$) or emission of energy ($\varepsilon > 0$). (e) Typical measurement of the current (solid) versus $\varepsilon$ at 23 mK. The measured current is decomposed into an elastic part (dashed) and an inelastic (dotted-dashed) part. [33]

level is occupied by an electron from the source, and the right level is kept empty as long as the two levels are in the transport window of $|\varepsilon| < 140\ \mu$eV. When the inelastic tunneling rate $\Gamma_{inel}$ is small compared to the tunneling rates through the left and the right barriers, $\Gamma_{inel}$ is directly determined from the measurement of current, $I = e\Gamma_{inel}$.

In general, emission contains a spontaneous and stimulated contribution, while the absorption contains only a stimulated contribution. Quantum optics predicts that the emission rate, $W_e$, and absorption rate, $W_a$, are related to the spontaneous emission spectrum, $A(\varepsilon)$ by $W_a(\varepsilon) = A(\varepsilon)$ and $W_e(\varepsilon) = (<n> + 1)A(\varepsilon)$. If one assumes Boson statistics for the environment, the average occupation number $<n>$ of the environment modes is given by the Bose-Einstein distribution function, $<n> = 1/(e^{\varepsilon/kT} - 1)$. The effect of a non-zero temperature on the current is shown in Fig. 2.26. [33] A higher temperature $T$ enhances the inelastic current on both the emission ($\varepsilon > 0$) and the absorption ($\varepsilon < 0$) sides. The inelastic current taken at the base temperature of 23 mK ($kT = 2\ \mu$eV) is the spontaneous emission contribution, since $<n> << 1$ for $\varepsilon > 2\ \mu$eV. The temperature dependencies of the emission rate and the absorption rate are plotted in Fig. 2.26(c). The experimental data of $W_a/A$ and $W_e/A$ versus

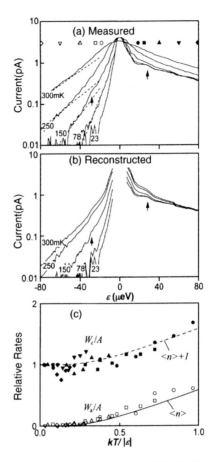

Fig. 2.26. (a) Measured current versus $\varepsilon$ for $T = 23$ to 300 mK. The current is measured for $eV_{sd} = 140$ meV while sweeping $V_{GR}$ and $V_{GL}$ simultaneously in opposite directions such that we only change the energy difference $\varepsilon$. Gate voltage is translated to energy $\varepsilon$ by a calibration that is better than 10% using photon-assisted tunneling measurements. Dashed lines indicate exponential dependence, $e^{\varepsilon/kT}$, for $|\varepsilon| \gg kT$. Arrows point to step-like structures on the emission side ($\varepsilon > 0$), and a shoulder on the absorption side ($\varepsilon < 0$). From fits to the elastic current part at 23 mK, we obtain $h\Gamma_R \sim hT_c \sim 1$ meV and $h\Gamma_L \sim 5$ meV for this data set. (b) Reconstructed current, $I_{tot}(\varepsilon) = I_{el}(23 \text{ mK}) + I_{inel}(\varepsilon, T)$ for different $T$. The spontaneous emission spectrum derived from the measured data at 23 mK and the Einstein's relation are used to reconstruct the full temperature and energy dependence. (c) The absorption rate $W_a$ (open symbols), and emission rate $W_e$ (closed symbols) normalized by the spontaneous emission rate A versus $kT/|\varepsilon|$. Circles, squares, upper- and lower-triangles, and diamonds are taken at $|\varepsilon| = 18, 24, 40, 60$, and $80$ meV, respectively (see also symbols in (a)). The solid line indicates the Bose-Einstein distribution, $\langle n \rangle$, while the dashed line shows $\langle n + 1 \rangle$. [33]

$kT/|\varepsilon|$ are well reproduced by the Bose-Einstein distribution function for $<n>$. This suggests that the inelastic tunneling observed in this experiment is caused by the coupling of the electronic systems to the Bosonic environment.

In semiconductor systems, acoustic phonons are the most likely Bosons to couple to a double dot. For a deformation-type coupling, the spontaneous emission rate is proportional to $\varepsilon$ for 3D phonons, and constant with $\varepsilon$ for 2D phonons. On the other hand, for a piezoelectric-type coupling, the energy-dependence is given by $1/\varepsilon$ for 3D phonons, and $1/\varepsilon^2$ for 2D phonons. The experimental spectrum of $A(\varepsilon)$ shows the energy dependence between $1/\varepsilon$ and $1/\varepsilon^2$, suggesting a piezoelectric interaction for 2D or 3D phonons. The wavelength of the phonon related to the energy resonance is comparable to the typical length scale of the sample geometry as defined by the Schottky gate metal. Step-like features indicated by the arrows in Fig. 2.25 could be due to these phonon resonances, and the phonon wavelength matches the system size, 380 nm in this case. [33]

## 2.7 Summary

In this chapter, the electronic properties of artificial semiconductor quantum dot atoms and molecules have been described. Atom-like properties of few-electron ground states such as shell filling and obeyance of Hund's first rule are all observed in a 2D circular dot or artificial atom. In the presence of a magnetic field, the atom-like ground states show a variety of transitions to maximize the total spin and the total angular momentum. The electronic phases between certain transitions are well defined in the plane of the magnetic field and the number of electrons ("B-N phase diagram"). The atom-like properties described are intimately linked to the symmetry of the dot geometry. In other words, breaking the symmetry can produce new electronic states. *In situ* control of the dot geometry is desirable for this, and as one of the candidates, a single electron transistor with multiple gates, is presented. For quantum dot molecules, the electronic properties of quantum mechanically strongly and weakly coupled two-dot systems are described. For vertically coupled double dot molecules, electron filling can be distinguished between these two different regimes: consecutive filling of delocalized symmetric states, reflecting quantum mechanical coherent coupling, is observed for the former, while alternate filling of two localized states, arising from the electrostatic coupling, is identified for the latter. On the other hand, laterally coupled double dot molecules are more useful for the *in situ* control of the coupling parameters. A microwave technique is a powerful tool for probing the energy spectrum of the coupled states, and can be employed to distinguish whether the states are delocalized or localized. The microwave photons provide the energy balance for transitions between the two coupled states. In the absence of microwaves, energy conservation is attained by coupling to the environment, i.e. phonon system in our case. This is well demonstrated from measurement of the temperature-dependence of inelastic tunneling between two localized states.

Quantum dot structures are ideally suited as probes to investigate interesting physics that have not yet been fully established, such as many-particle interactions, coherent or incoherent tunneling and coupling to other energy quanta, and also as a system in which quantum mechanics can be easily tested in the laboratory. Only a few examples are described in this chapter. Many more fundamental studies as well as applications for electrical and optical devices are underway and will be exploited in the future.

### [References]

1) For reviews, *Mesoscopic Phenomena in Solids*, ed. by B. L. Altshuler, P. A. Lee, and R. A. Webb, Elsevier (1991); U. Meirav and E. B. Foxman, *Semicond. Sci. Technol.*, **11**, 255 (1996); *Proc. of the NATO Advance Study Institute on Mesoscopic Electron Transport*, ed. by L. L. Sohn, L. P. Kouwenhoven, and G. Schoen, Kluwer Series E345 (1997).

2) M. Reed, *Scientific American*, **268**, 118 (1993); M. A. Kastner, *Physics Today*, **46**, 24 (1993); R. C. Ashoori, *Nature*, **379**, 413 (1996).

3) D. G. Austing, T. Honda, and S. Tarucha, *Semicond. Sci. Technol.*, **11**, 1995 (1995); D. G. Austing, T. Honda, Y. Tokura, and S. Tarucha, *Jpn. J. Appl. Phys.*, **34**, 1320 (1995)

4) S. Tarucha, D. G. Austing, T. Honda, R. J. van der Hage, and L. P. Kouwenhoven, *Phys. Rev. Lett.*, **77**, 3613 (1996).

5) M. Alonso and E. J. Finn, *Quantum and Statistical Physics*, Addison-Wesley (1968).

6) V. Fock, *Z. Phys.*, **47**, 446 (1928); C. G. Darwin, *Proc. Cambridge Philos, Soc.*, **27**, 86 (1930).

7) L. P. Kouwenhoven, T. H. Oosterkamp, M. W. S. Danoesastro, M. Eto, D. G. Austing, T. Honda, and S. Tarucha, *Science*, **278**, 1788 (1997).

8) M. Eto, *Jpn. J. Appl. Phys.*, **36**, 3924 (1997); S. Tarucha, T. Honda, D. G. Austing, Y. Tokura, K. Muraki, T. H. Oosterkamp, J. W. Janssen, and L. P. Kouwenhoven, *Physica E*, **3**, 112 (1998).

9) P. L. McEuen, E. B. Foxman, J. Kinaret, U. Meirav, M. A. Kastner, N. S. Wingreen, and S. J. Wind, *Phys. Rev. B*, **45**, 11419 (1992); P. L. McEuen, N. S. Wingreen, E. B. Foxman, J. Kinaret, U. Meirav, M. A. Kastner, Y. Meier, and S. J. Wind, *Physica B*, **189**, 70 (1993); see also for our disk-shaped dot, D. G. Ausing, Y. Tokura, T. Honda, S. Tarucha, M. W. S. Danoesastro, J. W. Janssen, T. H. Oosterkamp, and L. P. Kouwenhoven, *Jpn. J. Appl. Phys.*, **38**, 372 (1999).

10) A. H. MacDonald, S. R. E. Yang, and M. D. Johnson, *Aust. J. Phys.*, **46**, 345 (1993).

11) T. H. Oosterkamp, J. W. Janssen, L. P. Kouwenhoven, D. G. Austing, T. Honda, and S. Tarucha, *Phys. Rev. Lett.*, **82**, 2931 (1999).

12) S. Tarucha, D. G. Austing, T. Honda, R. J. van der Hage, and L. P. Kouwenhoven, *Jpn. J. Appl. Phys.* **36**, 3917 (1997).

13) D. G. Austing, T. Honda, and S. Tarucha, *Jpn. J. Appl. Phys.*, **36**, 4151 (1997); D. G. Austing, T. Honda, and S. Tarucha, *Semicond. Sci. Technol.*, **12**, 631 (1996); D. G. Austing, T. Honda, and S. Tarucha, *Physica E*, **2**, 583 (1998).

14) D. G. Austing, T. Honda, K. Muraki, Y. Tokura, and S. Tarucha, *Physica B*, **249-251**, 206 (1998).

15) F. R. Waugh, M. J. Berry, D. J. Mar, R. M. Westervelt, K. L. Campman, and A. C. Gossard, *Phys. Rev. Lett.*, **75**, 705 (1995).

16) T. Schmidt, R. J. Haug, K. V. Klitzing, A. Förster, and H. Lüth, *Phys. Rev. Lett.*, **78**, 1544 (1997).

17) G. Klimeck, G. Chen, and S. Datta, *Phys. Rev. B*, **50**, 2316 (1994).

18)  E. B. Foxman, P. L. McEuen, U. Meirav, N. S. Wingreen, Y. Meir, P. A. Belk, N. R. Belk, M. A. Kastner, and S. J. Wind, *Phys. Rev. B*, **50**, 14193 (1993).

19)  H. Pothier, P. Lafarge, P. F. Orfila, C. Urbina, D. Esteve, and M. H. Devoret, *Physica B*, **169**, 573 (1991).

20)  C. Livermore, C. H. Crouch, R. M. Westervelt, K. L. Campman, and A. C. Gossard, *Science*, **274**, 1332 (1996).

21)  N. C. van der Vaart, S. F. Godijn, Yu. V. Nazarov, C. J. P. M. Harmans, J. E. Mooij, L. W. Molenkamp, and C. T. Foxon, *Phys. Rev. Lett.*, **74**, 4702 (1995).

22)  Yu. V. Nazarov, *Physica B*, **189**, 57 (1993).

23)  P. K. Tien and J. R. Gordon, *Phys. Rev.*, **129**, 647 (1963).

24)  T. Fujisawa and S. Tarucha, *Superlattices and Microstructures*, **21**, 247 (1997).

25)  A. N. Korotokov, D. V. Averin, and K. K. Likharev, *Phys. Rev. B*, **49**, 7548 (1994).

26)  T. Fujisawa and S. Tarucha, *Jpn. J. Appl. Phys.*, **36**, 4000 (1998).

27)  T. H. Stoof and Yu. V. Nazarov, *Phys. Rev. B*, **53**, 1050 (1996).

28)  T. H. Oosterkamp, T. Fujisawa, W. G. van der Wiel, K. Ishibashi, R. V. Heiman, S. Tarucha, and L. P. Kouwenhoven, *Nature*, **395**, 873 (1998).

29)  T. Fujisawa, T. H. Oosterkamp, W. G. van der Wiel, S. Tarucha, and L. P. Kouwenhoven, *Inst. Phys. Conf. Ser. No. 162*, Chapter 9, 493 pp., IOP Publishing Ltd. (1999)

30)  Y. Nakamura, Yu. A. Pashkin, and J. S. Tsai, *Nature*, **398**, 786 (1999).

31)  C. A. Staford and N. S. Wingreen, *Phys. Rev. Lett.*, **76**, 1916 (1996).

32)  A. J. Legget, S. Chakravarty, A. T. Dorsey, M. P. A. Fisher, A. Garg, and W. Zwerger, *Rev. Mod. Phys.*, **59**, 1 (1987).

33)  T. Fujisawa, T. H. Oosterkamp, W. G. van der Wiel, B. W. Broer, R. Aguado, S. Tarucha, and L. P. Kouwenhoven, *Science*, **282**, 932 (1998).

# Chapter 3

# Electron Energy Modulation with Optical Evanescent Waves

## 3.1 Introduction

An electron beam is commonly used for generating and amplifying coherent electromagnetic waves over the wide-frequency spectrum of microwaves through to ultraviolet rays. Many kinds of electron beam devices, including the klystron, traveling wave tube (TWT), and backward wave oscillator (BWO), have been developed [1]. These beam devices have several advantages over semiconductor devices, such as a wide frequency tuning range and higher output power. These beam devices are mostly microwave or millimeter wave amplifiers or oscillators. Only free electron lasers (FEL) can oscillate in the visible region [2]. However, FELs use an electron accelerator as a beam source, so their applications are strongly limited. For general use, a new compact beam device similar to the semiconductor laser, is required.

In such compact beam devices, low energy electrons, i.e., a nonrelativistic electron beam, should be used for generating optical waves. However, there have been no reports on such beam devices except for the report by Schwarz and Hora (1969) [3]. They described that a low energy electron beam was modulated by an Ar-laser using a thin dielectric film as the interaction circuit. However, their experimental results have not yet been reproduced. Therefore, the first step of our study is to define the basic process in the electron-light interaction, including quantum effects such as electron-photon energy exchange and quantum modulation of an electron beam.

In this chapter, theoretical and experimental results are presented to show the feasibility of energy modulation of an electron beam with light using micro-gap circuits [4], particularly a metal micro-slit circuit. Recent progress in vacuum microelectronics [5] and micromachining techniques [6] have realized very compact electron emitters with dimensions in the order of several microns or less. The micro-structure of the interaction circuit could therefore be fabricated using these techniques.

## 3.2  Quantum Modulation of Electrons

For conventional beam devices using a cavity resonator such as klystrons, it has been thought that extending their operation to optical frequencies would be difficult. The reason for this is as follows. In order to obtain a signal gain in the beam devices, the electron beam must be density-modulated at a signal frequency. In the microwave region, this density modulation (bunching) is easily achieved, because the energy of the electrons changes continuously in proportion to the electric field of the signal applied to them, since signal quanta are much smaller than the energy changes of electrons. On the other hand, in the optical region, the electron energy changes by multiples of the energy of the signal photon. This effect would prevent smooth change of the beam density, diminishing the signal gain. Senitzky predicted theoretically that operation of the beam devices is limited at frequencies below the submillimeter wave [7].

In 1969, Schwarz and Hora, however, claimed that, based on their experimental results, the wave function of electrons could be modulated by a light wave. If this quantum modulation was realized, the bunching of electrons could be achieved even at the optical frequencies. Figure 3.1 shows a schematic drawing of their experimental system. In the experiment, a 50 keV electron beam with a dilute density and an Ar-laser (wavelength: $\lambda = 488$ nm) were used. The interaction circuit was a $SiO_2$ or an $Al_3O_2$ film with a thickness of less than 200 nm. The electron beam passed through the film illuminated with the 10W laser light and then hit a nonluminescent screen. They reported that light was emitted from the screen and that it had the same color as an incident laser light. They also described this effect as resulting from quantum modulation of

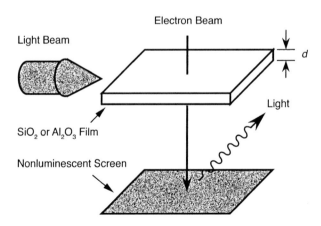

Fig. 3.1.  Schematic drawing of the Schwarz-Hora effect.

the wave function of coherent electron beams [8]. Though this "Schwarz-Hora effect" has not yet been observed by other people and its demodulation mechanism, in particular, is still not clear, their results have suggested the possibility of the realization of new electron beam devices utilizing the quantum effect.

### 3.3 Micro-Gap Interaction Circuits

Figure 3.2 shows three kinds of possible micro-gap circuits, (a) a dielectric film, (b) a metal film gap, and (c) a metal slit. The dielectric film is the interaction circuit used by Schwarz and Hora. The metal film gap is most similar to conventional circuits used in klystrons. The metal micro-slit circuit utilizes an optical near-filed on the slit to modulate an electron beam [9].

Though the "Schwarz-Hora effect" has still not been reproduced, we believe that the micro-gap circuits could be used as an interaction circuit for modulating electron energy with light (photon). This is because energy and momentum conservation is satisfied for the interaction. In the interaction

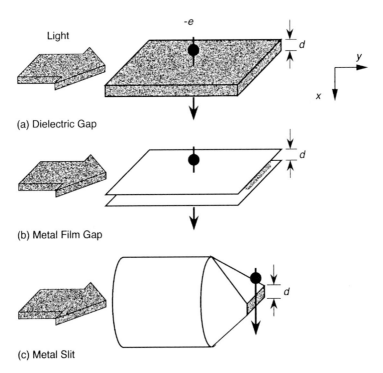

Fig. 3.2. Schematic drawing of three different gap circuits. The gap width $d$ is smaller than the wavelength of the incident laser light. The inset indicates the coordinates used in the text.

shown in Fig. 3.2, if an electron absorbs a photon, energy conservation requires a momentum difference $\Delta p$ between the electron and the photon [10],

$$\Delta p = p_f \cos\theta - p_i, \quad \text{for} \quad \theta = \sin^{-1}\left(\frac{\hbar\omega/c}{p_f}\right) \tag{3.1}$$

where $p_i$ and $p_f$ are the initial- and final-state electron momentum, respectively, $\theta$ is the angle of the scattered electron relative to the x-direction, i.e., the direction of incident electron motion, $\hbar = h/2\pi$, $h$ is Planck's constant, and $\omega$ is the angular frequency of light. If the momentum of the photon is much less than that of the electron, $\theta$ is approximately zero. In this case, $\Delta p$ in Eq. (3.1) is simplified to

$$\Delta p \sim \frac{\hbar\omega}{v} \tag{3.2}$$

where $v$ is the velocity of the electron. The interaction length in the gap circuits are limited so that an uncertainty exists in the momentum. This additional momentum $p_c$ from the circuit is given by

$$p_c \sim \frac{h}{d} \tag{3.3}$$

where $d$ is the gap width. If $p_c > \Delta p$, momentum conservation for the interaction can be satisfied. From Eqs. (3.2) and (3.3), it can be seen that the required condition for the gap width is

$$d \leq \beta\lambda \tag{3.4}$$

where $\lambda$ is the light wavelength, and $\beta = v/c$. For instance, for $\beta = 0.4$ (~50 keV electron energy), the gap width must be less than 200 nm for $\lambda = 488$ nm.

The same consideration can be made in a different way. A number of wave components with different wave numbers $k_{ev}$ are induced on, say, the slit shown in Fig. 3.2(c) with the incident light. On the basis of Fourier optics theory [11], the wave numbers $k_{ev}$ in the x-direction could extend from zero to infinity. From momentum conservation, $k_{ev}$ must be equal to $\Delta p/\hbar$. Using Eq. (3.2), the following relationship between $k_{ev}$ and $v$ is found:

$$k_{ev} = \frac{\omega}{v}. \tag{3.5}$$

Since $k_{ev}$ is larger than the light wave number $k_0 = \omega/c$ in free space, this wave is evanescent. This equation also represents phase matching between the electron and the optical evanescent wave with a phase velocity $v_p = \omega/k_{ev}$.

In contrast to the two circuits in Figs. 3.2(a) and (b), the electrons on the slit shown in (c) do not pass through the gap but interact with the evanescent wave traveling close to the slit surface. Energy exchanges between electrons and light can therefore be observed without interference such as electron scattering in the gap foils [12]. From this point of view, the metal slit is most suitable for investigating basic processes in the electron-light interaction.

### 3.4 Metal Film Gap and Dielectric Film Circuits

The electron transition rates in a metal film gap and a thin dielectric film were theoretically estimated. From the theoretical results, optimum gap widths giving the maximum transition rate to the electrons were determined.

The calculation model is shown in Fig. 3.3. The electron beam passes through the gap and travels in the +$x$-direction, and the laser light wave is propagated in the +$y$-direction. To simplify the calculation, we assumed that (i) all the electrons have the same initial velocity, (ii) propagation modes of the light wave in the gaps are restricted to the $TEM_{00}$ and the $TM_{01}$ modes for the metal film gap and the thin dielectric film, respectively, (iii) the electric field vector of the light is polarized to the $x$-direction, (iv) the gap materials do not have any propagation loss, and (v) the electrons exchange their energy only with the light. The assumption (ii) is valid for a gap width smaller than $\lambda/2$ and $\lambda/2\sqrt{n-1}$ for the metal and the dielectric gaps, respectively, where n is a refractive index of the dielectric film.

In accordance with the analyses by Marcuse [13], the transition rate for the electron energy in the metal film gap was calculated using the above assumptions. The calculated transition rate $w_m$ for an incident electron current $i$ is expressed as the following:

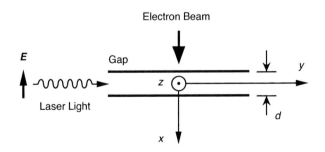

Fig. 3.3. The gap circuit configuration used for the theoretical analysis.

$$w_m = \frac{2qc\beta^2}{\varepsilon_0 \hbar^2 \omega^4} iP \sin^2(\omega d / 2v) \tag{3.6}$$

where $q$ is the electron charge, $c$ is the velocity of light, $\varepsilon_0$ is the dielectric constant of free space, and $P$ is the power density of the incident light.

The value of $w_m$ represents the rate at which an electron absorbs a photon with energy of $\hbar\omega$. Similarly, the transition rate to emit the photon can be calculated. The values of these two transition rates are almost the same. In the first instance, the energy changes of electrons have to be measured to find the corresponding interaction with the light. The electron transition rate related to the absorption of a photon is thus the major consideration.

Similarly, it is possible to calculate a transition rate $w_d$ for the dielectric film by quantizing the laser field of the fundamental $TM_{01}$ propagation mode. The expression giving $w_d$ is as follows:

$$w_d = \frac{qc}{\varepsilon_0 \hbar^2 \omega^4} iP \left| \frac{k_0 k_y d^2}{2n_i^2 \cos(k_{ix}d/2)} \right|^2$$

$$\times \frac{\left| 4 \dfrac{k_{ix}d \sin(k_{ix}d/2)\cos(\omega d/2v) - (\omega d/2v)\cos(k_{ix}d/2)\sin(\omega d/2v)}{(k_{ix}d)^2 - (\omega d/v)^2} \right|^2}{\left\{ \dfrac{k_0 d}{2n_i \cos(k_{ix}d/2)} \right\}^2 + \left[ \dfrac{\left(k_y^2 - k_{ix}^2\right)}{\left(n_i k_{ix}\right)^2} - \dfrac{n_i^2\left(k_y^2 - k_{ex}^2\right)}{\left(n_e k_{ex}\right)^2} \right] \dfrac{k_{ix}d}{2n_i^2} \tan(k_{ix}d/2)} \tag{3.7}$$

$$k_{ix}^2 + k_y^2 = \left(n_i k_0\right)^2, \quad k_{ex}^2 + k_y^2 = \left(n_e k_0\right)^2, \quad |k_{ex}| = \left(\frac{n_e}{n_i}\right)^2 k_{ix} \tan(k_{ix}d/2)$$

where $n_i$ and $n_e$ are the refractive indexes inside and outside the dielectric film, respectively, $k_0$ is the wave number of the laser light in free space, $k_y$ is the wave number in the $y$-direction, and $k_{ix}$ and $k_{ex}$ are the wave numbers inside and outside the film in the $x$-direction.

From Eqs. (3.6) and (3.7), the transition rates can be estimated as a function of the gap width. Figure 3.4 shows the calculated results used to find the optimum gap width. The parameters used in the calculation are $\beta = 0.5$, $\lambda = 780$ nm, $n_i = 1.45$ ($SiO_2$), and $n_e = 1$. As seen from Eq. (3.6), $w_m$ changes sinusoidally and has the peak at the gap width satisfying the equation

$$d = \beta\lambda\left(m + \frac{1}{2}\right) \tag{3.8}$$

where $m$ is an integer. Substituting $\beta = 0.5$ and $m = 0$ for the first peak, an optimum gap width of $\lambda/4$ is obtained. The variation of $w_m$ in Fig. 3.4 is different from the klystron theory in which the maximum value of $w_m$ is at $d \sim 0$. The difference between these comes from the different treatments of photon density in the gap. The klystron theory assumes that the total number of photons stored in the gap is constant, but in our treatment, it is assumed that the photon density in the gap is constant and determined by the incident laser power.

From Fig. 3.4, it can be seen that for the $SiO_2$ film, the first peak value of $w_d$ is 0.18 times that of $w_m$. The optimum film thickness of 0.38 is also longer than the optimum width of the metal film gap. In the dielectric film, the laser field is also distributed outside the film as an evanescent wave so that the number of photons inside the gap is smaller than that for the metal gap. The longer gap width increases the number of photons inside the dielectric film.

Figure 3.5 shows the calculated transition rates as a function of the light intensity for the metal film gap and the $SiO_2$ film. These transition rates represent the probability per unit time of one electron absorbing a photon. In the calculation, the optimum gap widths of $0.25\lambda$ and $0.38\lambda$ were used for the metal film gap and the $SiO_2$ film, respectively. Other calculation parameters are the same values as described previously. From Fig. 3.5, it can be seen that the transition rates are $1.1 \times 10^{-2}$ and $2 \times 10^{-3}$/sec for a power intensity of $10^6$

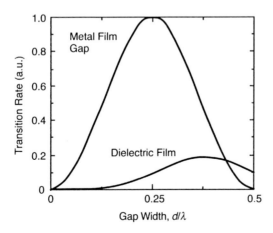

Fig. 3.4. Calculated transition rates of an electron with a velocity of $\beta = 0.5$ as a function of the gap width of the metal film gap and the $SiO_2$ dielectric film. The gap width is normalized to the laser wavelength.

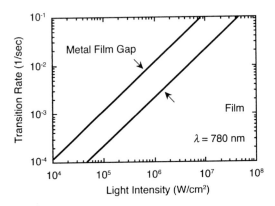

Fig. 3.5. Calculated transition rates as a function of the power intensity of the incident laser light for the metal film gap and the $SiO_2$ dielectric film at a wavelength of 780 nm.

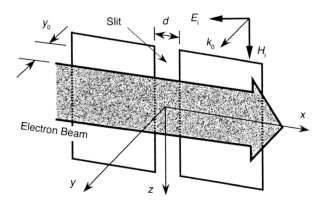

Fig. 3.6. Calculation model for the metal slit circuit.

W/cm² in the metal film gap and the $SiO_2$ film, respectively. The power intensity of the laser light corresponds to an output power of 790 mW focused onto a 10 $\mu$m diameter area.

Using Eq. (3.7), we can also estimate the electron transition rate for the $SiO_2$ film used by Schwarz and Hora. For their experimental parameters $-\lambda = 488$ nm and $v \sim 0.4c^-$ the optimum film thickness has been estimated to be about 170 nm which agrees with their estimation. In the $SiO_2$ film, the electron transition rate of about $2 \times 10^{-3}$/sec is calculated for a laser light with a power of 10 W and a beam diameter of 10 $\mu$m. They therefore probably obtained the signal electrons of more than 1,000 particles/sec for the electron beam used in the experiment.

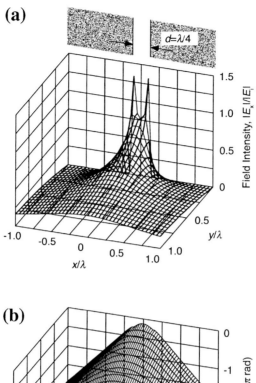

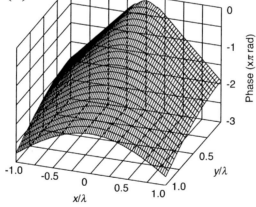

Fig. 3.7. Calculated (a) intensity and (b) phase distributions of the electric field component $E_x$ in the $x$-direction on the slit with a width of $d = \lambda/4$. $E_x$ is normalized to the incident field $E_i$. The positions $x$ and $y$ are also normalized to the wavelength.

## 3.5 Metal Micro-Slit

The field distributions of the evanescent waves on the metal slit shown in Fig. 3.2(c) were determined through classical electromagnetic analyses. Energy changes of electrons passing through the slit were also estimated with computer

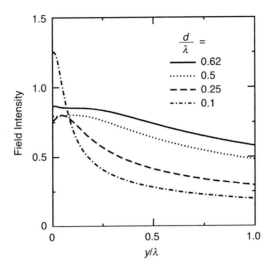

Fig. 3.8. Calculated field intensities of $E_x$ as a function of the distance $y/\lambda$ from the slit surface at $x = 0$ for different slit widths. The field intensity is normalized to $|E_i|$.

simulation using Lorentz force equations [14]. From the results, the optimum width of the slit was found. The calculated results were compared with those for the metal film gap.

### 3.5.1 Near-field distribution

The calculation model for a metal slit is shown in Fig. 3.6. In accordance with the analysis by Chou and Adams [15], electric field components $E_x$, $E_y$, and a magnetic one, $H_z$, in scattered waves from the slit were determined using the method of moments for an incident wave which has the field components $E_i$ and $H_i$. In this calculation, the following assumptions were made; (i) the metal slit consists of two semi-infinite plane screens with perfect conductance and a zero thickness, (ii) a plane wave polarized to the x-direction is normally incident upon the slit aperture.

Figure 3.7 shows the calculated (a) intensity and (b) phase distributions of $E_x$ on a slit with a width of $d = \lambda/4$. As seen from the result, the electric field in the proximity of the slit is well confined within the slit width, but its intensity and phase distributions are not uniform even at the slit surface. The field intensity decays and spreads quickly when y increases. The field intensity variations are plotted for various slit widths of $d/\lambda = 0.1$, 0.25, and 0.5 as a function of y at $x = 0$, and are shown in Fig. 3.8. For the slit width of $0.1\lambda$, the field intensity is 1.25 at $y \sim 0$, which is larger than that of the incident wave, and decreases to 0.5 at $y = 0.19\lambda$. It can be seen that the smaller the slit width

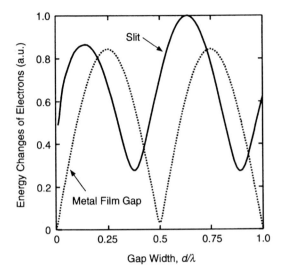

Fig. 3.9. Calculated maximum energy changes of electrons in the metal slit and the film gap circuits as a function of the gap (slit) width $d/\lambda$ for $v_0 = c/2$ at $y_0 = 0.01\lambda$.

is, the steeper the field decay is. These theoretical results show that the near-fields are localized at a distance within $y \sim d$ from the surface. These near-field distributions on the slit are similar to those on small apertures [16].

### 3.5.2 Optimum slit width

Energy changes of electrons passing through the near-field region on the slit were estimated with computer simulation. In Fig. 3.6, the electrons with an initial velocity $v_0$ move in the $x$-direction at an initial position $y_0$ from the surface, and are accelerated or decelerated by Lorentz force acting on them. The total energy change of the electron was numerically calculated by integrating small energy changes in small distances along the electron trajectory. The integration length of 20 times the operating wavelength was chosen to fully cover the near-field region on the slit. In the calculation, all the components, i.e., $E_x$, $E_y$, and $H_z$, in the near-field on the slit were taken into account.

Figure 3.9 shows the calculated maximum energy changes $\Delta W$ for the electrons with $v_0 = c/2$ as a function of the gap (slit) width $d$ at $y_0 = 0.01\lambda$. For comparison, the result calculated for the metal film gap using the same method is also indicated in the figure. $\Delta W$ is normalized to a maximum value for the metal slit at $d = 0.64\lambda$. In Fig. 3.9, the $\Delta W$-curve for the metal film gap has two peaks with the same values of 0.82 at $d = 0.25\lambda$ and $0.75\lambda$. These optimum gap widths are consistent with the ones in Eq. (3.8) and shown in Fig. 3.4. The

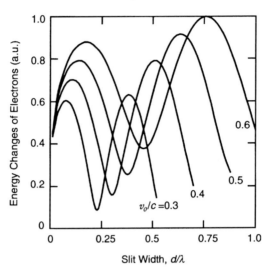

Fig. 3.10. Calculated maximum energy changes of electrons as a function of the slit width for different electron velocities from $0.3c$ to $0.6c$ at $y_0 = 0.01\lambda$. The calculated energy changes are normalized to the maximum value. The slit width is also normalized to the wavelength.

curve for the metal slit also has two peaks, but the optimum slit widths are smaller than those for the film gap.

Similar calculations were done for $v_0/c = 0.3$, 0.4, 0.5, and 0.6 and the results are shown in Fig. 3.10. The optimum slit width $d_m$ becomes wider as $v_0$ increases. From the calculation results, the following relationship between $d_m$ and $v_0$ was found:

$$d_m = \beta\lambda\left(m + \frac{1}{4}\right) \tag{3.9}$$

which is compared to Eq. (3.8) for the metal film gap. From Eq. (3.9), it can be seen that the optimum width in the slit is narrower than that in the film gap by approximately $\beta\lambda/4$. This could be due to the nonuniform distribution of the near-field. In Fig. 3.10, it should be noted that the second peaks are always larger than the first peaks. In the slit circuit, energy changes of electrons are given by roughly $qE_x L$, where $L$ is an effective interaction length. As seen from Fig. 3.8, the field intensity $|E_x|$ for the narrower slit width is larger than or approximately equal to the one for the wider slit width at $y = 0.01\lambda$. Thus, in the slit circuit, the increase of $L$ exceeds the decrease of the field intensity when the slit width increases.

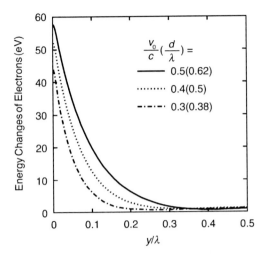

Fig. 3.11. Calculated maximum electron-energy changes as a function of $y/\lambda$ for different electron velocities $v_0/c = 0.5, 0.4$, and $0.3$. The slit widths $d/\lambda$ used in the calculation are indicated in the figure.

### 3.5.3 Interaction space

Figure 3.11 shows the calculated $\Delta W$ for different electron velocities as a function of $y$. Slits having the optimum widths $d_m/\lambda = (0.38, 0.5, 0.62)$ for $v_0/c = (0.3, 0.4, 0.5)$, respectively, were used for calculation. In the calculation, a $CO_2$ laser with $\lambda = 10.6\ \mu m$ and a power density of $10^8$ W/cm$^2$ was assumed as the incident wave. This power density corresponds to a 10 kW output power focused onto a 100 $\mu m$ diameter area.

In Fig. 3.11, when $y$ increases from zero to $0.5\lambda$, $\Delta W$ quickly decreases to near zero. These decay curves of $\Delta W$ represent effective field distributions of the laser wave interaction with the electrons, because $\Delta W$ is proportional to the field intensity at the point where the electrons pass the slit. The shapes of the $\Delta W$-curves are almost an exponential function of $y$. These results indicate that the electrons have interacted with evanescent waves in the near-field region on the slit. The decay constant $\alpha$ of an evanescent wave with a wave number $k_{ev}$ is given by [17]

$$\alpha = k_0 \sqrt{\left(\frac{k_{ev}}{k_0}\right)^2 - 1}. \tag{3.10}$$

Substituting Eq. (3.5) in Eq. (3.10), the decay constants for $v_0/c = (0.3, 0.4, 0.5)$ can be calculated and are $(3.2, 2.3, 1.7) \times k_0$, respectively. These decay constants agree with the ones estimated from the $\Delta W$-curves in Fig. 3.11. The computer simulation thus again supports the previous discussion in section 3.3 that the electron interacts with the evanescent wave satisfying Eq. (3.5).

An effective interaction distance from the slit can be defined as $y_i = 1/\alpha$, because the field intensity of the evanescent wave falls off by $e^{-1}$ at $y_i$. Using Eqs. (3.5) and (3.10), $y_i$ is expressed as

$$y_i = \frac{\lambda}{2\pi} \frac{\beta}{\sqrt{1-\beta^2}}. \tag{3.11}$$

From this equation, it is seen that the interaction space in the slit circuit is strongly limited, particularly for a lower energy electron beam. For $v_0 = 0.5c$ (~80 keV electron energy), the effective interaction space is less than about 1 $\mu$m for $\lambda = 10.6$ $\mu$m. Therefore, the electrons must be passed very close to the slit surface to obtain significant energy exchanges with light.

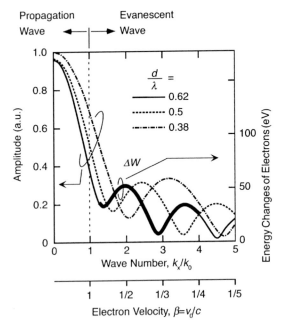

Fig. 3.12. Calculated wave-number spectra for the near-field distributions of $E_x$ on slits with different widths of $d/\lambda = 0.38, 0.5,$ and $0.62$. In the figure, $k_x$ is the wave number in the $x$-direction and $\beta$ is the initial electron velocity calculated substituting $k_x$ for $k_{ev}$ in Eq. (3.5). The thick curve represents the maximum electron-energy changes for different initial velocities from $\beta = 0.25$ to $0.8$ which correspond to an electron energy of 17 to 340 keV.

The field distribution of the evanescent wave is considerably different from the near-field distributions shown in Fig. 3.8 because the near-fields contain a large number of wave components with different wave numbers, including propagation waves. Electrons interact with only a single evanescent wave among them, as mentioned earlier. In order to see this in detail, a wave number ($k$) spectrum for the near-field distribution was calculated numerically using Fourier transform equations.

Figure 3.12 shows the calculated $k$-spectra of the near-fields $E_x$ at $y_0 = 0.01\lambda$ for the slits indicated in Fig. 3.11. The abscissa is the wave number $k_x$ normalized to $k_0$ and corresponding electron velocity calculated using Eq. (3.5). The ordinate is the amplitude of the wave which is also normalized to the maximum value at $k_x = 0$. The thick solid curve for the slit width of $d = 0.62\lambda$ represents the calculated $\Delta W$ for electrons with a velocity of $\beta = v_0/c$. As seen from Fig. 3.12, the evanescent waves are distributed in a wide range of $k_x$ from unity to above five times $k_0$. Therefore, electrons with any velocity can find the evanescent wave for interactions in the near-field region, and consequently $\Delta W$ for the electrons is proportional to the amplitude of the wave. These results suggest that an electron beam could be also used to measure the $k$-spectrum of an optical near-field appearing on the surface of an object with microscopic dimensions.

### 3.5.4 Estimation of signal electron numbers

As shown in Figs. 3.4 and 3.9, classical and quantum-mechanical treatments for the optimum gap width in the metal film gap have given the same results. From this fact, it would be expected that the theoretical predictions for the slit circuit in the classical treatment could be maintained in quantum-mechanical experimental conditions. Comparing the electron-energy changes for the metal slit and film gap in Fig. 3.9, the electron transition rate in the slit circuit is deduced to be comparable to that of the metal film gap, at least at the slit surface.

Based on the above assumptions, we estimated the number of electrons that have absorbed a laser photon in the metal slit, assuming a laser with $\lambda = 488$ nm and output power of 30 mW, an electron beam with a velocity $\beta = 0.5$ and a density of 1 mA/cm$^2$. Using Eqs. (3.9) and (3.11), the optimum gap length and the interaction distance with the slit were found to be 303 nm and 45 nm, respectively. If the metal slit is made at the end of a conventional single mode optical-fiber, the diameter of the propagating wave field is about 6 $\mu$m. Then electrons of $1.6 \times 10^7$ per second can pass through the interaction space of 45 nm × 6 $\mu$m. From Eq. (3.6), a transition rate of $1.7 \times 10^{-4}$ per second was obtained for the 30 mW laser power in the optical fiber. The number of signal electrons thus is about 2,700 particles per second. This number is enough to demonstrate experimentally the electron-energy modulation in the slit circuit.

## 3.6 Preliminary Experiment

### 3.6.1 The Inverse Smith-Purcell effect

In order to verify the theoretical predictions for the metal slit circuit, the Inverse Smith-Purcell (ISP) effect [18] was measured experimentally at submillimeter wavelengths [19]. The ISP effect makes use of a metal grating as an interaction circuit between electrons and light as shown in Fig. 3.13. The electrons pass over the grating surface perpendicularly to its grooves and interact with evanescent waves induced near the grating surface due to illumination by laser light.

Figure 3.14 compares idealized wave-number ($k$) spectra of the evanescent waves in the direction of an electron traveling through single slit and the multiple slit structures, which have a slit width $d$ and a period $D$. The multiple slit structure (i.e., grating) has periodic line $k$-spectra with a period $2\pi/D$. When a laser light with a wavelength $\lambda$ is illuminated at an incident angle of $\theta$, the $m$-th order evanescent wave has the wave number $k_m$

$$k_m = k_0 \cos\theta + \frac{2m\pi}{D}. \tag{3.12}$$

Substituting $k_m$ for $k_{ev}$ into Eq. (3.5), the phase matching condition for the ISP effect is obtained:

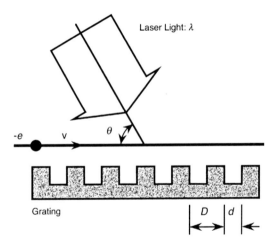

Fig. 3.13. Schematic drawing of the Inverse Smith-Purcell effect. $D$ is the period of the grating, $d$ is the gap width, $v$ is the electron velocity, and $\theta$ is the incident angle of the laser light.

$$m\lambda = D\left(\frac{1}{\beta} - \cos\theta\right).$$                    (3.13)

This condition indicates that only electrons with a velocity $v_0 = \beta c$ can interact with the laser light on the grating. In an actual experiment, an interaction length $L$ for the electrons is limited by a finite laser beam size so that the interaction condition in Eq. (3.13) for the electrons is relaxed by $\Delta v \sim v\Delta k/k_m$ as indicated in Fig. 3.14. The interaction strength with the laser light is proportional to the square root of the incident laser power, because it is determined by the amplitude of the evanescent wave, as mentioned previously.

In order to deduce the field distribution of an evanescent wave on a single slit from the experimental results for the ISP effect, a metal grating with a rectangular cross section was used. In addition, submillimeter wave lasers were chosen as a driving laser in order to measure the filed distribution accurately.

### 3.6.2 Experimental setup

Figure 3.15 shows a schematic diagram of interaction with the first order evanescent wave, because its amplitude is the largest among the wave components [20]. The experimental parameters used are listed in Table 3.1. Grating grooves with a rectangular cross section were cut into a Cu-alloy substrate, maintaining a tolerance of $\pm 2$ $\mu$m. The submillimeter wave $CH_3F$ and

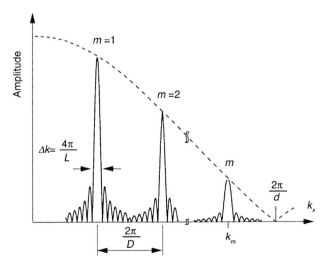

Fig. 3.14. Schematic drawing of the wave-number spectra for uniformly distributed fields at the gap in a single slit (dotted curve) and at the gaps in a multiple slit structure (grating: solid curves).

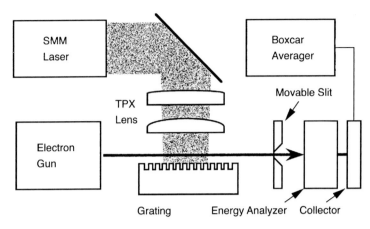

Fig. 3.15. Experimental setup for measuring the inverse Smith-Purcell effect. A SMM laser is a submillimeter wave laser and TPX is poly 4 methylpenten-1 which is a low-loss material in the submillimeter wave region.

Table 3.1. Experimental parameters.

| **LASER:** | | $CH_3F$ | $CH_2F_2$ | |
|---|---|---|---|---|
| Wavelength | : | 496 | 184.6 | $\mu m$ |
| Peak power | : | 1 ~ 86 | 1 ~ 56 | W |
| Pulse width | : | 30 ~ 100 | 25 ~ 35 | nsec |
| **GRATING** (fabricated, tolerance = ± 2 $\mu m$): | | | | |
| Period | : | 246 | 91 | $\mu m$ |
| Depth | : | 104 | 40 | |
| Width | : | 40 | 39 | |
| **ELECTRON BEAM:** | | | | |
| Initial energy | : | | 10 ~ 100 | keV |
| Current | : | | < 1 | $\mu A$ |
| Beam Diameter | : | | 240 | $\mu m$ |

$CH_2F_2$ lasers [21] were pumped by a current-pulsed Q-switched $CO_2$ laser [22]. The output pulse of the submillimeter wave laser had a single peak in the time domain for the fundamental transverse mode, which has a Gaussian profile. The typical output pulse shape in the $CH_3F$ laser is shown in Fig. 3.16. The peak power $P_i$ and the pulse width are 9 W and 80 nsec, respectively. Two lenses concentrated the laser beam at a normal incidence on the grating surface, i.e., $\theta = \pi/2$ in Fig. 3.13. The spot sizes on the surface were calculated to be 0.7 mm for the $CH_2F_2$ laser beam and 1.2 mm for the $CH_3F$ laser on the basis of Gaussian beam theory.

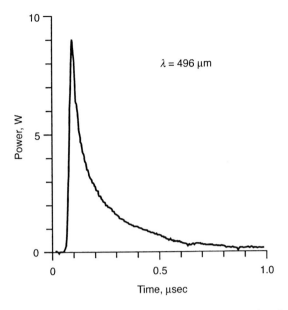

Fig. 3.16. Typical output pulse shape of the $CH_3F$ laser with a wavelength of 496 $\mu$m.

The electron energy was analyzed by using a retarding-potential technique [23]. Figure 3.17 shows the electron-energy spectrum measured without laser illumination for an electron beam with an initial energy $W_i$ of 80 keV and a current of 3 nA. This result shows that the energy analyzer has a resolution of 0.8 eV (full width at half maximum). This resolution includes the intrinsic thermal energy spread of the electron beam. A movable slit with a gap of 10 $\mu$m was placed at the end of the grating and was used to specify the position of electron beam above it. The position of the gap could be controlled to an accuracy of $\pm 3$ $\mu$m.

### 3.6.3 Electron energy spectrum

The pulsed laser output modulates the energy of the electron beam, so that the electron current through the analyzer changes during the pulse. The change in the collector current is measured by a box-car averager which is triggered by the laser pulse. Figure 3.18 shows the typical electron-energy spectrum (a) measured and (b) calculated with (solid line) and without (dotted line) laser illumination for $W_i = 80$ keV and $P_i = 12$ W (peak). The abscissa is a filter bias $W_f$ which corresponds to the retarding-potential of the energy analyzer. The ordinate is the change in electron current caused by the laser. The theoretical curve in Fig. 3.18(b) was calculated by a method similar to that described in

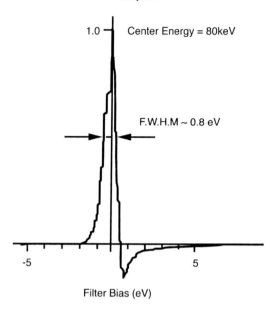

Fig. 3.17. Measured energy resolution of the electron-energy analyzer used in the experiment. The filter bias is a retarding potential for electrons with an initial energy of 80 keV.

section 3.5 when the shape of the laser pulse used here was taken into account [24].

Since the energy analyzer passes all the higher-energy electrons, it can be expected that for large bias voltages, the current change with laser illumination should be the same as the one without laser illumination. However Fig. 3.18(a) shows that the current change with laser illumination is slightly smaller than the one without laser illumination when $W_f > +30$ eV. This shows that the laser illumination deflects the electron beam, and consequently a part of the electron beam is clipped by the aperture before the collector. We calculate that 1.3% of the electrons are lost. We have also observed that this ratio becomes larger as the laser power increases, as theoretically expected. The measured maximum energy spread, $\Delta W = 64$ eV, is about a quarter of the theoretical spread. There might be two reasons for this discrepancy. Firstly, the amplitude of the evanescent wave may not be the same as the theoretical one because of fabrication errors in the groove dimensions. Secondly, noise in the electron-energy spectra may contribute to making an error.

### 3.6.4 Phase matching condition

Figure 3.19 shows experimental results for the phase matching condition between the electron velocity and the phase velocity of the evanescent wave.

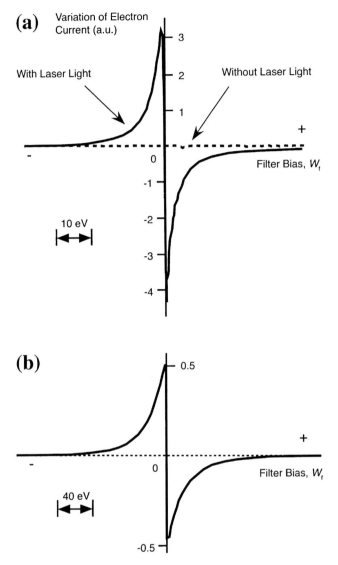

Fig. 3.18. (a) Measured and (b) calculated electron-energy spectra for the electron energy of 80 keV and a peak laser power of 12 W.

The ordinate is electron energy spread normalized to its maximum value. The laser power was between 8.3 and 12 W. The abscissa is the initial electron energy $W_i$. As the interaction length between electrons and waves is finite, an effective interaction can occur for electrons with a certain range as pointed out

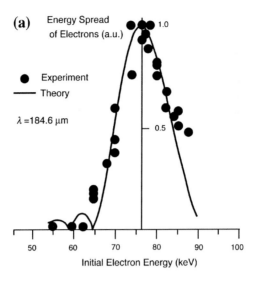

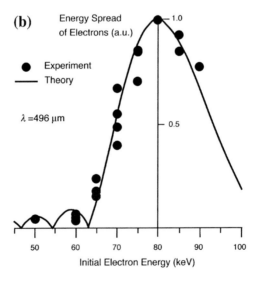

Fig. 3.19. The changes in electron-energy spread versus the initial energy of electrons for (a) the CH$_2$F$_2$ laser ($\lambda = 184.6$ $\mu$m) and (b) the CH$_3$F laser ($\lambda = 496$ $\mu$m). The points represent the experimental values normalized to the maximum values. The solid lines are the theoretically predicted change of the energy spread for an interaction length (a) 1.6 mm and (b) 3 mm, respectively.

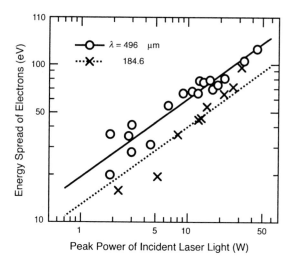

Fig. 3.20. The energy spread of electrons as a function of the peak power of the incident laser light.

in section 3.6.1. The largest energy spreads are produced at $W_i = 76.3$ keV at $\lambda_1 = 184.6\,\mu$m and 80 keV at $\lambda_2 = 496\,\mu$m, respectively. From Eq. (3.13), theory would predict $76.3 \pm 4$ keV at $\lambda_1$ and $77.5 \pm 1.6$ keV at $\lambda_2$. By curve fitting, we can deduce that the effective interaction lengths are 1.6 mm for $\lambda_1$ and 3 mm for $\lambda_2$. Theoretical plots for the lengths are given by the solid curves in Fig. 3.19. These experimental results show that the theoretical considerations for the electron-light interaction in the multiple slit structure is valid allowing for experimental errors. It should be noted that the solid curves represent the wave-number spectra for the first-order evanescent waves on the grating.

### 3.6.5 Laser power dependence

Figure 3.20 shows the measured maximum energy spread of electrons as a function of the laser power. The solid and dotted lines indicate theoretical changes which have a slope of 0.5 as described in section 3.6.1. The agreement between the theoretical and experimental results shows that the electron-energy changes are proportional to the field intensity of the incident wave, i.e., the square root of the laser power, as predicted by the theory. The deviation from the theoretical lines is somewhat large at a laser power of less than 5 W. These would be due to a reading error of the maximum energy change from the measured energy spectrum for electrons with small energy changes.

### 3.6.6 Field distributions

As described earlier, the field intensity of the evanescent wave on the grating is proportional to $\exp(-\alpha y)$, where $y$ is a distance from the grating surface. The decay constant $\alpha$ is given by Eq. (3.10). Figure 3.21 shows experimental results that show the field decay characteristics of the first-order evanescent wave interacting with the electron beam. In Fig. 3.21, the abscissa is an electron position which is the position of the movable slit mentioned earlier. The ordinate is the energy spread of the electron beam passing through the slit. The initial electron energy is the center energy shown in Fig. 3.19, i.e., 76.3 keV at $\lambda = 184.6$ $\mu$m and 80 keV at $\lambda = 496$ $\mu$m. The solid lines indicate the theoretically predicted changes, i.e., $\exp(-0.022y)$ and $\exp(-0.06y)$ for the $CH_3F$ and the $CH_2F_2$ laser drives, respectively. The experimental results are in good agreement with the theory, except for the points near the grating surface at $\lambda = 184.6$ $\mu$m. These values of the electron-energy spread are small in comparison with the theoretical ones, and the difference becomes larger as the beam is moved closer to the grating surface. Therefore, the gap of 10 $\mu$m in the movable slit could be too large to measure rapid decay of the evanescent wave at 184.6 $\mu$m. The experimental results are a direct verification of the evanescent wave theory for the inverse Smith-Purcell effect, and thus for the metal slit interaction circuit.

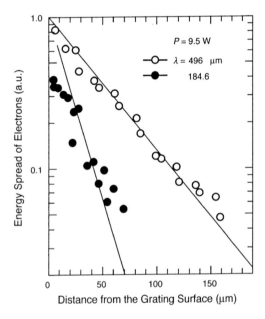

Fig. 3.21. The change of energy spread of the electron beam interacting with a first-order evanescent wave as a function of the distance from the grating surface. The solid lines are the theoretically predicted changes and $P$ is the incident laser power.

## 3.7 Fabrication of the Micro-Slit

As described in section 3.1, the first goal of our study is to demonstrate experimentally the quantum effect which appears in the electron-light interaction using a metal slit circuit. Figure 3.22 shows a conceptual drawing of the experimental system, where an electron-energy analyzer is used to detect electron-energy changes with laser light. Since the electron-energy analyzer can resolve an energy change of 1 eV or less for electrons with 80 keV energy, many kinds of lasers with a wavelength of less than 1.2 $\mu$m can be used to measure the quantum effect in the experiment. A metal micro-slit is fabricated at the end of an optical fiber so that the laser beam is guided to a slit with microscopic dimensions without any precise adjustment.

In the experiment, a key device is the metal micro-slit with a sub-micron width. The micro-slit is fabricated on the top of the ridge, as shown in Fig. 3.22. This ridge structure is needed to decrease the image force acting on the electrons passing very close to the metal slit surface and to avoid a collision of the electrons with the slit. In order to fabricate such a slit circuit, a chemical etching technique can be utilized [25]. The fabrication process is shown in Fig. 3.23. This fabrication method is almost the same as that for conventional aperture probes used in scanning near-filed optical microscopy [26], except for the photolithographic process. In Fig. 3.23(d), the slit width must be controlled by adjusting the etching time because the metal thickness at the top of the ridge is thinner than it is at other places.

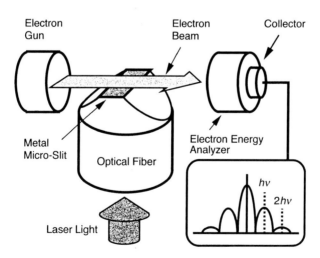

Fig. 3.22. Conceptual drawing of the experimental setup to demonstrate energy modulation of electrons with laser light using a metal micro-slit as the interaction circuit.

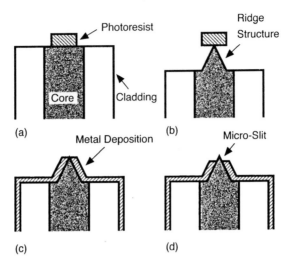

Fig. 3.23. Process sequence for the fabrication of a metal micro-slit at the end of an optical fiber: (a) after patterning photoresist with dimensions of 30 $\mu$m $\times$ 10 $\mu$m, (b) after etching the optical fiber in HF:NH$_4$F (=1:6.5), (c) after evaporating metal at a right angle to the desired slit surface, (d) after wet chemical etching of the metal coated ridge structure to open a slit aperture.

Figure 3.24 shows the ridge structure fabricated at the center of the core of an optical fiber with a 125 $\mu$m diameter (a), and the same ridge after etching a 270 nm-width metal slit (b). The ridge structure has a length of 5.5 $\mu$m for a core diameter of 8.8 $\mu$m, a height of 7.2 $\mu$m, and a taper angle of 80°. Both the flatness along the length of ridge and the radius of curvature at the top of the ridge are less than 30 nm. The radius of curvature is comparable to that of conventional aperture probes with hole diameters of several tens of nanometers. In Fig. 3.24(b), the coating metal is aluminum and its thickness is about 300 nm. Another probe with a 180 nm-width slit was fabricated using gold instead of aluminum as a coating metal, using the same fabrication process.

In order to fabricate slit probes with a narrower width, a smaller taper angle than 80° for the ridge structure is required to make precise fabrication of the slit through chemical etching. A ridge structure with a smaller taper angle of about 20° could be fabricated by choosing an optical fiber with a higher doped core and the appropriate etching solution, in a way similar to how conventional probes with nanometric apertures are fabricated [27].

### 3.8 Summary

Electron-light energy exchanges in metal gap circuits were theoretically analyzed. The optimum dimensions for three gap circuits such as a metal film gap, a dielectric film, and a metal micro-slit were determined. The metal

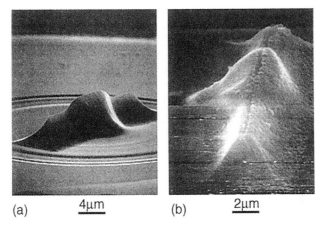

(a)        4µm                (b)        2µm

Fig. 3.24. Metal micro-slit fabricated at the center of the end of a single-mode optical fiber: (a) ridge structure with a taper angle of 80°, (b) metal slit with a width of 270 nm fabricated on the ridge structure.

micro-slit is most suitable for a study of basic processes in the interaction because electrons can exchange their energy with only the incident light without any interference. Electrons interact with an optical evanescent wave on the slit. Some of the theoretically obtained results were verified experimentally at submillimeter wavelengths. Theoretical and preliminary experimental results showed that experimental observation of the electron-energy modulation of optical laser light could be feasible. The results also indicated that electrons passing through an optical near-field region on the slit interact with only a single component of the evanescent waves, the phase velocity of which component is equal to the electron velocity. This specific feature in the interaction could be utilized for electron-beam microscopy that detects a wave-number spectrum for the near-filed distribution which appears on the surface of an object with nano-structures.

[References]

1)    R. G. E. Hutter, *Beam and Wave Electronics in Microwave Tubes*, edited by H. J. Reich, D. Van Nostrand Co. Inc., Tronto (1960).

2)    *Free-Electron Lasers, Advances in Electronics and Electron Phys.*, Suppl. 22, edited by C. A. Brau, Academic, New York (1990).

3)    H. Schwarz and H. Hora, *Appl. Phys. Lett.*, **15**, 349 (1969).

4)    K. Mizuno and S. Ono, *IEEE Proc.*, **63**, 1075 (1975).

5)    C. A. Spindt, I. Brodie, L. Humphrey, and E. R. Westerberg, *J. Appl. Phys.*, **47**, 5248 (1976).

6)    J. Mohr, C. Burbaum, P. Bley, W. Menz, and U. Wallrabe, *Micro System Technologies*, edited by H. Reichl, Springer-Verlag, Berlin (1990).

7)    I. R. Senitzky, *Phys. Rev.*, **95**, 904 (1954).

8)   H. Hora and P. H. Hndel, in *Advances in Electronics and Electron Physics*, Vol. 69, edited P. W. Hawkes, Academic, New York (1987).

9)   J. Bae, S. Okuyama, T. Akizuki, and K. Mizuno, *Nucl. Instrum. & Methods in Phys. Research A*, **331**, 509 (1993).

10)  R. H. Pantell, in *AIP Conf. Proc.*, No. 87, p. 863, edited by R. A. Carigan, F. R. Huson, and M. Moath, AIP, New York (1982).

11)  G. A. Massey, *Appl. Opt.*, **23**, 658 (1984).

12)  P. N. Denbigh and C. W. B. Grigson, *J. Sct. Instrum.*, **42**, 305 (1965).

13)  D. Marcuse, in *Engineering Quantum Electrodynamics*, edited by G. Wade, Academic, New York (1970).

14)  R. Ishikawa, J. Bae, and K. Mizuno, in *Tech. Digest of 5th Int. Conf. on Near Field Opt. and Related Tech.*, pp. 173–174, Shirahama, Japan (1998).

15)  T. Y. Chou and A. T. Adams, *IEEE Trans. Electro Magnetic Compatibility*, **EMC-19**, 65 (1977).

16)  Y. Leviatan, *J. Appl. Phys.*, **60**, 1577 (1986).

17)  D. P. Tsai, H. E. Jackson, R. C. Reddick, S. H. Sharp, and R. J. Warmack, *Appl. Phys. Lett.*, **56**, 1515 (1990).

18)  K. Mizuno, S. Ono, and O. Shimoe, *Nature*, **253**, 184 (1975).

19)  K. Mizuno, J. Bae, T. Nozokido, and K. Furuya, *Nature*, **328**, 45 (1987).

20)  J. Bae, H. Shirai, T. Nishida, T. Nozokido, K. Furuya, and K. Mizuno, *Appl. Phys. Lett.*, **61**, 870 (1992).

21)  M. S. Tobin, *Proc. IEEE*, **73**, 61 (1985).

22)  J. Bae, T. Nozokido, H. Shirai, H. Kondo, and K. Mizuno, *IEEE J. Quantum Electron.*, **30**, 887 (1994).

23)  J. F. Graczyk and S. C. Moss, *Rev. Sci. Instrum.*, **40**, 424 (1969).

24)  J. Bae, K. Furuya, H. Shirai, T. Nozokido, and K. Mizuno, *Jap. J. Appl. Phys.*, **27**, 408 (1988).

25)  J. Bae, T. Nozokido, T. Okamoto, T. Fujii, and K. Mizuno, *Appl. Phys. Lett.*, **71**, 3581 (1997).

26)  T. Pangaribuan, S. Jiang, and M. Ohtsu, *Electron. Lett.*, **29**, 1978 (1993).

27)  M. Ohtsu, *J. Lightwave Tech.*, **13**, 1200 (1995).

# Chapter 4

# Interactions of Electrons and Electromagnetic Fields in a Single Molecule

## *4.1 Single Electron Tunneling and Photon-Assisted Tunneling*

### *4.1.1 Introduction*

Single electron charging effect was established theoretically by Likharev in the 1980s [1], following the pioneering work on electron tunneling through granular metallic islands that was done in the 1960s [2]. It is surprising that single electron charging can be observed even when the dimension of double tunnel junction electrodes is macroscopic, namely, involving ~$10^{23}$ electrons. The charge $Q$ induced at the surface of the electrode increases continuously as bias voltage increases, in line with $Q = CV$, where $C$ is the capacitance of the double tunnel junction [1]. Nevertheless, under appropriate conditions, electron tunneling at discrete amounts of elemental charge $e$ can be detected on a one-by-one basis.

Coulomb blockade is a phenomenon in which electrons are unable to tunnel through a barrier until the battery power provided to the tunnel junctions is higher than Coulomb energy. The Coulomb energy (or charging energy) associated with this tunnel junction is $e^2/C$, where $e$ is the electron charge and $C$ is the double tunnel junction's capacitance. This theory is applicable to macroscopic electrodes, but not to nanostructure electrodes. The meaning of capacitance on the nanometer scale must be re-considered.

Tunneling through nanostructures has been studied by scanning tunneling microscopy and also in the field of semiconductor superlattice structures, but without taking into account electron correlations. There are two reasons for this. In STM, well-defined double tunnel junctions have not been realized, since in the tip-molecule-substrate configurations examined previously, molecules are adsorbed firmly to the substrate, and in such cases, tunneling electrons are included with substrate electrons. In the field of semiconductor superlattices, the two-dimensional structure (perpendicular to the superlattice layer) is large enough to treat electron tunneling in a one-electron approximation.

The reason why macroscopic electrodes are still used to study single electron charging effects is because there are serious technical restrictions on realizing nanostructured electrodes. Although the state-of-the-art electron lithography can narrow the structure dimension down to a submicrometer level, it is still in a regime where the macroscopic concept of capacitance is valid.

Shrinking the structure of electrodes that form tunnel junctions requires a consideration of the effective size of electron confinement. Small structures causing single electron charging have been realized using constriction gates fabricated on a semiconductor substrate [3–8]. The reason semiconductors can be used as the medium for studying single electron charging is because of their Fermi wavelengths being as large as 1 $\mu$m.

However, Fermi wavelengths in metals are limited to a few nanometers. This is the reason quantum effects have not yet been observed with metals [9]. It should be noted that the single electron charging effect observed is not necessarily a quantum effect. From the energy viewpoint, if Coulomb energy is large enough, this effect can be observed in metal electrodes at room temperature. Schonenberger and others observed this effect in metal clusters of 5 nm diameter at room temperature [10]. This fits with the orthodox theory, since Coulomb energy is $e^2/C$ under macroscopic capacitance. Such phenomena have been observed using nanometer size metal electrodes, but the effects characterized by the Fermi wavelength in metals have not.

Fermi wavelength is not definable for molecules. The extent of electron wavefunction in a molecule can be estimated from the Bohr radius (0.053 nm). Because electron confinement in molecules has not been studied, single electron charging in a molecule can be used as an appropriate tool for this purpose.

Our purpose here is to try to understand how single electron charging occurs in a nanostructure, e.g., in a liquid crystal (LC) molecule. Although single electron charging was also observed using molecules, it is still not clear how molecular orbitals play a role in this phenomenon [11]. One proposed theory is based on a model. But at present, determining Coulomb energy (correlation energy) quantitatively is impossible. Since the Hartree-Fock calculation does not match experimental results [12], it is difficult to know if single electron charging can occur in an area as small as a molecule. By detecting the interaction of molecules with electromagnetic fields, we show that the molecule does play a role in single electron charging.

After the above introduction, we present in the following experimental procedures and results. We discuss an experiment in which we irradiate infrared (IR) light on a nanostructure double tunnel junction. A theory of single electron charging for nanostructures is proposed to apply quantum mechanics to the specific field of scanning-tunneling microscopy (STM). We also evaluate whether IR light irradiation really suppresses current.

### 4.1.2 Experimental

#### 4.1.2.1 I-V curves

Our experiment used the following molecule/substrate combinations: 4'-cyano-7-alkylbiphenyl (7CB) LC molecule on Pt(111). A Pt(111) substrate was polished to 0.1 micrometer roughness, chemically etched in a solution of HCl and HNO$_3$, rinsed with deionized water, and finally dried in air. A Peltier element was used to raise the temperature from the LC crystalline to the nematic phase transition point. The *I-V* curves were measured at room temperature. A commercial scanning tunneling microscope (Nanoscope II) was used to make STM imaging and to measure *I-V* curves, and a chemically etched PtIr wire was used as the tip. The tip was held above the molecule (Fig. 4.1) so that tip, molecule, and substrate form a double tunnel junction (Fig. 4.2). During *I-V* measurements, tunneling currents were measured while the bias voltage was scanned over the predesigned range.

The bias voltage was scanned with tip height held constant. The acquisition cycle includes two parts: running the feedback for a period of time to establish the set-point current, then turning the feedback off and ramping the bias

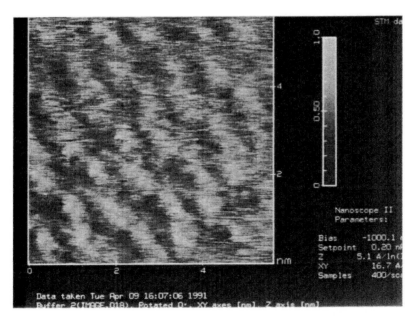

Fig. 4.1. 5.6 × 5.6 nm² STM image of a 7CB LC molecule on graphite. The alkyl moiety of the molecules appears darker than the cyanobiphenyl moiety. The image of LC molecules on graphite is shown because of the difficulty in getting high resolution molecular images on Pt(111) substrates. We think the characteristics of liquid crystal cause the same mosaic to form on Pt substrates [16].

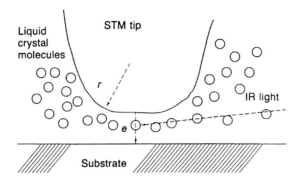

Fig. 4.2. Schematic view of a LC molecule under the apex of an STM tip. The molecule is irradiated by IR light. Light reaches the molecule under the tip apex even though its wavelength is shorter than the tip radius [17].

voltage. With 2000 points along the ramp, bias modulation was added to the voltage and the current was measured. This process of modulating bias voltage was repeated 16 times. The curve was obtained by repeating the cycle 40 times.

The bias was scanned from +1 V to −1 V for one second so that the current caused by the polarization of molecules surrounding the tip did not dominate the current through the double barrier. Since voltage scanning takes a long time, the instrument's stability may be insufficient and the gap may not be constant. The dominant current term comes from these effects in each scanning cycle; namely, the characteristic feature is embedded in a low frequency component of a few Hz. But since these phases do not correlate, averaging makes the low frequency component zero. Repeating 40 cycles should make the noise component $1/\sqrt{40}$ of each cycle.

Measurements were carried out in ambient pressure. Since the tip was immersed into a LC droplet, each molecule under the tip was surrounded by many other molecules. We assume that such environment around the double tunnel junction is free from contamination during measurement.

Figure 4.3(a) shows the $I$-$V$ curve of 7CB on Pt(111) and Fig. 4.3(b) shows its differential conductance at room temperature. The bias voltage is −1 V and the set point current is 0.2 nA. The data clearly show a Coulomb staircase [13]. The different stepwidths (indicated by arrows) show that this system cannot be explained by conventional Coulomb blockade theory.

### 4.1.2.2 IR light irradiation

In the experimental set-up in Fig. 4.2, the sample was irradiated by a monochromatic light of 2642 nm (0.47 eV) wavelength at a precision within ±0.6 nm. This was the longest monochromator wavelength used. Since inverse dispersion is 17.1 nm/mm and slit width is 0.5 mm, spectral bandwidth is 8.5

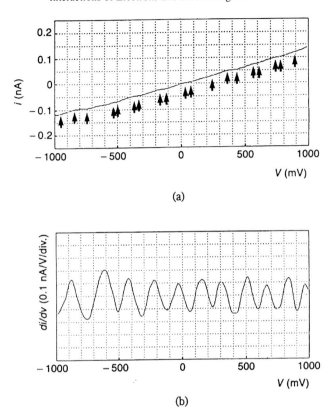

(a)

(b)

Fig. 4.3. (a) Current as a function of bias voltage applied to a 7CB sample on Pt(111) shows a staircase effect. (b) The differential conductance of Fig. 3(a). The different step width indicates that this system cannot be explained by conventional Coulomb blockade theory. The bias voltage is –1 V (sample negative) and the set point current is 0.2 nA.

nm (full width half maximum). The IR source light was introduced to a monochromator of 0.5 mm exit slit width and 5 mm height. The monochromatic light impinged on the sample.

To confirm that light at these wavelengths does not excite the molecule's vibrational modes, IR spectroscopy was measured (Fig. 4.4). Wavelengths shorter than 3100 nm (0.4 eV) will not excite the vibrational mode so that the phenomenon of the above IR irradiation is caused by electronic excitation of the molecules. Hartree-Fock calculations of the 7CB molecule's vibrational mode further confirmed that wavelengths beyond the experimental range do not excite the vibrational mode.

Figure 4.5(a) shows the *I-V* curve with light irradiation (0.47) and without irradiation (0). The curve without irradiation is the normal staircase. When the

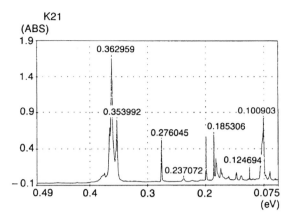

Fig. 4.4. The IR absorption spectra of the 7CB LC molecule. The absorption amplitude (the vertical axis) was measured from 0.075 to 0.49 eV (the horizontal axis).

sample is irradiated, the first observable step edge in the positive bias range is above 0.47 V. Below 0.47 V, a monotonic current increase is observed. In the negative bias region, the current continues to increase monotonically even beyond –0.47 V. These features are more clearly revealed in the derivative conductance diagrams of Fig. 4.5(b). Furthermore, Fig. 4.5(b) shows the dependency of $dI/dV$ curves upon the photon energy of excitation light. When the excitation photon energy is increased from 0.47 eV to 0.6, 0.7, 0.8 and 0.9 eV, the onset position for the first evident step edge also shifts to higher bias. That is, the higher the photon energy is, the higher the positive bias at which the first step edge appears.

### 4.1.3 Discussion

Single electron charging is characterized by Coulomb energy $Uc$ when an extra electron is localized in the central electrode of a double tunnel junction. This section introduces a quantum mechanical analysis employed in the STM field. The concept behind this analysis is that electrons tunnel through the states of the molecule that are added by the amount of Coulomb energy caused by the additional electron entering the molecule [13]. The substrate, STM tip and molecular states are expressed with the Anderson model:

$$H = \sum_{i,j,\sigma} H_{ji} n_{ij}^{\sigma} + U \sum_{a} n_{aa}^{\uparrow} n_{aa}^{\downarrow} \qquad (4.1)$$

where $a$ is the molecular state, $\sigma$ is the spin state, $H_{ji}$ is a coefficient based on the one-electron approximation between electrons, $n_{ij}$ is the electron number

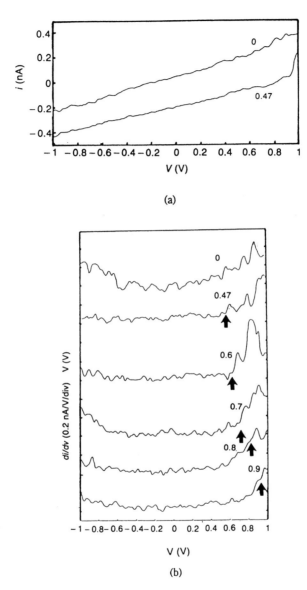

(a)

(b)

Fig. 4.5. (a) The *I-V* curve with (0.47) and without (0) monochromated IR of 0.47 eV irradiation. The first step edge shifts toward the lager voltage by an amount equivalent to the photon energy. (b) Derivative *dI/dV* curves with and without monochromated light irradiation acquired from 0.47 to 0.9 eV with various photon energies, which shows that electron charging effects are suppressed when the magnitude of the applied bias is less than the photon energy. The photon energy for each curve is indicated on the curves.

operator and $U$ is the Coulomb repulsion between electrons in the molecule. The tunneling current is calculated to be the number of electrons per unit time from tip to substrate. The current is the expected value of the product of the electron number operator between different spins. We will next attempt to evaluate whether light intensity is enough to cause the effect observed above. The tunnel current was suppressed at the energy of irradiated light, 0.47 eV.

The result of IR irradiation described in section 4.2.2 can be viewed as an electronic excitation phenomenon of molecules caused by photon radiation. Within this framework, the tunnel current through the excited state can be compared in the following way with Eq. (4.1):

$$H_p = \hbar\omega + H_0, \qquad (4.2)$$

where $\hbar$ is the Plank's constant and $H_0$ is the Hamiltonian constant described in Eq. (4.1).

The power of 1 micrometer of light through the monochromator with the above 3000 K color temperature was 30 $\mu$W [14]. The 2.64 $\mu$m intensity we used was calculated to be 3 $\mu$W through the monochromator. When this light impinged on the sample, it exited from a 2.5 mm$^2$ slit. In the free space approximation, 2.64-$\mu$m photon flux density through the slit is $2.64 \times 10^{14}$ photons/sec. If we assume that the absorption coefficient is 0.01, then the number of photons in the molecule under the STM tip is $2.64 \times 10^{12}$ photons/sec.

The decay time of excited states is estimated from fluorescence decay time. Fluorescence decay time was measured using UV light from a restriction of the light source. It consists of two components: $1.27 \times 10^{-9}$ sec. and $11.4 \times 10^{-9}$ sec. (Fig. 4.6). Since photons reach the molecule under the tip every $10^{-12}$ sec., which is much shorter than decay time, the molecule keeps its excited state. Since tunneling current was 0.1 nA, the number of electrons entering the molecule is $10^{10}$/sec. When the tunneling electron enters the molecule at an average $10^{-10}$ sec. time interval, it is always affected by the excited state.

But the molecules reside in the very small 1 nm space under the tip, and it is not clear how much photon flux density is in the molecules. When light is confined in a small tip-substrate area shorter than the wavelength, the (conventional) photon concept collapses and it is appropriate to consider the phenomenon from the viewpoint of a high frequency electrical field which is modulated by the light at the tip and the substrate. The electrical field 1 $\mu$m from the molecule under the tip apex is calculated to be $1.4 \times 10^{-1}$ V/m using the relation $I = \varepsilon_0 c |E|$, where $I$ is light intensity, $\varepsilon_0$ is the dielectric constant in vacuum and $E$ is the high frequency electric field.

If the transition is caused by electron excitation of photons, the transition probability $P$ is

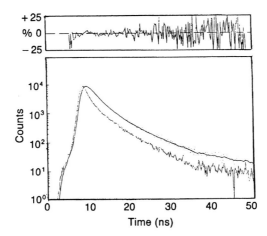

Fig. 4.6.   The photon's time evolution counts fluorescence from the 7CB molecule. The lowest line is the source light's time evolution. The upper dotted line shows detected fluorescence light. The upper solid line is calculated to fit the detected light. The wavelength of detected light was set to 430 nm.

$$P = \frac{2\pi}{h} \rho(f)\left|\langle f|pE|i\rangle\right|^2, \tag{4.3}$$

where $\langle f|pE|i\rangle$ is the electron transfer matrix element from the initial to the final state due to perturbation (transition moment), $p$ is dipole moment, $E$ is electric field and $\rho(f)$ is the density of the final state. The transition moment is the electrode dipole moment of the molecule multiplied by the high frequency electric field strength.

The light reached at the molecule under the tip should not be treated as a free wave but as an evanescent wave. Since the electric field is not much in the free space, it is not clear how much it is in the 1 nm space. If the amplification is $10^3$ and is due to tip apex geometry [15], then the electric field at the molecule is $1.1 \times 10^2$ V/m. The molecule's electric dipole moment that we used is calculated to be 4.8 D ($3.3 \times 10^{-30}$ Cm). Here, the electric dipole moment is along the cyano-group parallel to the C-N bond. The transition moment is then calculated to be $10^{-8}$ eV. It is not known whether this transition moment is sufficient for molecule excitation.

If we suppose this transition moment to be sufficient, then the electron tunneling resonates with the states of the excited molecule, accompanied by the additional amount of electron correlation, as described in Eq. (4.2). It is assumed that interaction between surrounding molecules and substrate provides adequate molecular states that correspond to each amount of photon energy. Figure 4.7 shows a possible energy diagram associated with this effect. The

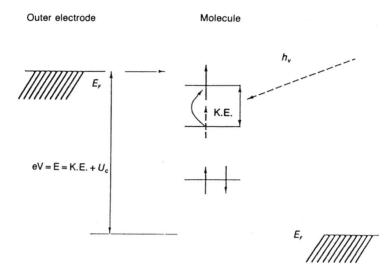

Fig. 4.7. Energy diagram of a molecule irradiated by light. When the molecule is irradiated, electrons tunnel to higher kinetic energy levels, keeping the number of electrons in the molecule constant. This is why Eq. (4.1) is applicable to this irradiated case. The state's electrostatic charging energy is the same as the state without irradiation, and it is included in Eq. (4.1) here. The bias voltage must be raised to overcome the energy increased by photon energy when the sample is irradiated.

central energy levels stand for the excited molecular states after irradiation. The incoming electron strikes the potential barrier described by Eq. (4.2), and the battery connected to the double tunnel junction has to provide the same amount of potential to make the electrons tunnel through. Here, the potential of the molecule in the excited state is higher than in the ground state by an amount equivalent to photon energy.

## 4.2 STM-Induced Photon Emission from Single Molecules on Cu(100)

### 4.2.1 Introduction

The atomic resolution of a scanning tunneling microscope (STM) arises from the small tail overlap between the wavefunctions of both tip and substrate across the tunneling barrier. What happens optically when a single molecule is inserted into this sub-nanometer gap and the molecule junction is slightly biased? This is the issue we will address in the present paper.

The appeal of single-molecule measurements is the unraveling of normal ensemble averaging that occurs when a large number of molecules are probed simultaneously, so that one can directly observe the true distribution and time-resolved information of a given property for a particular molecule or many single molecules but on a one-by-one basis [16]. Inelastic processes excited by tunneling electrons are known to carry chemically specific information on molecular systems, but are very difficult to detect by conventional scanning tunneling spectroscopy due to their small contribution to the total tunneling current despite the recent elegant example of acetylene on Cu(100) [17]. The combination of STM with optical techniques are particularly attractive for single-molecule studies, because one can not only image individual molecules on a surface, but also through probing photons emitted via inelastic processes, gain insights into the tunneling region: the specific characteristics of diverse excitations of molecules and substrates as well as modified optical properties due to quantum confinements [18–20]. (A single molecule is itself a simple quantum system for studying electron-photon interaction and other fundamental quantum-mechanical effects.) Tunneling of electrons or holes from an STM tip provides an extremely localized source of low-energy carriers (only a few eV), which offers nanometer resolution in photon emission mapping and thus gives access to the tip-sample electromagnetic interaction and its variations even on the atomic scale.

Photon emission stimulated by tunneling electrons can be traced back to the research by Lambe and McCarthy in 1976 on metal-oxide-metal junctions [21]. Theoretically, photon emission from metal surfaces was attributed to radiative decay of plasmons. These plasmons were excited either by inelastic electron tunneling [22,23] in which a tunneling electron gives away a fraction of its energy to a plasmon *during tunneling*, which is consecutively converted to a photon; or by a hot-electron mechanism [24] in which an electron tunnels elastically and surface plasmons are then excited by the hot electrons *injected into* the surface. The first experimental evidence of STM-induced light emission via inelastic tunneling was reported by Gimzewski and coworkers from Ag films [25]. The application of similar approaches to the $C_{60}$ molecules on Au(110) gave the first experimental demonstration of spatially resolved photon emission of individual molecules on a surface [26]. It is also noteworthy that the use of STM as an electron source for generating cathodoluminescence (CL) was also reported for GaAs/AlGaAs quantum wells [27,28] and there are also reports of use of a conductive and transparent indium-tin-oxide (ITO) tip for both carrier injection and photon collection [29]. An ITO substrate was also used to suppress the quenching of molecular emission, but the $10^{-5}\%$ quantum efficiency (number of photons produced per electron injected) reported there is still very low [30].

However, the detailed mechanism of STM-induced photon emission from single molecules is still not well understood, in particular with regard to how exactly a molecule gets excited and couples with the local electromagnetic field to give out light. This work is conducted in an effort to gain an understanding of the mechanism and to improve the quantum efficiency of electroluminescence (EL) via molecules for future molecular electronics. We report here a new and highly controllable technique—*electromechanical resonance pumping*—to produce intense photon emission from single molecules, and furthermore, present direct evidence of wavelength-resolved spectra of STM-induced luminescence that registers molecular vibrational excitations. The emission mechanism and blinking of molecules will also be briefly discussed.

### 4.2.2 Experiment

The experiments were performed in an ultrahigh vacuum system (UHV) with facilities for sample heating and cooling, molecule deposition, and characterization by a scanning tunneling microscope (JEOL JSTM-4500XT). The base pressure of the system is ~2 × $10^{-10}$ Torr. The experiments were

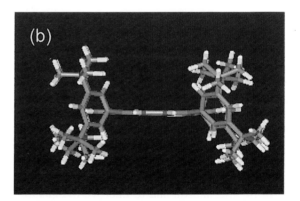

Fig. 4.8. (a) Structure and (b) conformation of Cu-TBP porphyrin molecules.

conducted at 80 and 295 K, but no significant differences were observed in terms of STM images and photon emission, so only data at 295 K are reported. Cu(100) substrates ($3 \times 7 \times 0.5$ mm$^3$) were mechanically polished using 0.3-$\mu$m alumina paste followed by electrochemical polishing in a solution of 66% H$_3$PO$_4$ and 34% methanol (20 mA, 2.8 V, 10 min, 4°C). These were thoroughly rinsed in methanol before being loaded into the UHV chamber. Atomically clean Cu(100) surfaces were obtained after cycles of sputtering (Ar$^+$, 1 keV, ~1 $\mu$A, ~800°C, 30 min) and annealing (~500°C, 30 min). Cu-tetra-(3,5 di-tertiary-butyl-phenyl)-porphyrin (Cu-TBPP) molecules (Fig. 4.8) were deposited onto Cu(100) by sublimation from a Knudsen cell at ~300°C or higher. The sample was then annealed at ~200°C for 5 min. All images were taken in the constant current topographic (CCT) mode with the sample biased. Tips used for imaging and photon emission were prepared by electrochemical etching of a W wire in NaOH followed by *in situ* cleaning in vacuum by Ar$^+$ ion sputtering or heating the tip apex up to ~1000°C.

Photons emitted from the tunneling gap were transmitted through a viewport and collected by a condenser lens located outside the UHV at an angle of ~20° with respect to the sample surface, with a detection solid angle of about 0.03 sterad. These photons were refocused onto an optical fiber ($\phi = 400 \ \mu$m) connected to a spectrophotometer (Hamamatsu PMA-100). This device is capable of measuring optical spectra and counting photons simultaneously. It includes a grating spectrometer (C5094) and an intensified CCD with Peltier-cooled multialkali (S-20) and multichannel plate (MCP), which operates in a pulse counting mode at a dark count rate of ~20 counts per second (cps) at –15°C for the wavelength range of 350–850 nm. To ensure that emitted photons were counted properly, the intensity was also measured by, respectively, a photomultiplier (PMT) (Hamamatsu H6180-01, dark counts ~10 cps at 25°C, 300–650 nm), and a cooled digital CCD camera (Hamamatsu C4880-40, average dark counts ~0.02 electron/pixel/sec at –50°C, 300–1100 nm).

### 4.2.3 Results and discussion

#### 4.2.3.1 Why Cu-TBPP molecules?

Figure 4.8 shows the structure and conformation of a Cu-TBPP molecule. The steric repulsion among the bulky substituents (legs) on the phenyl drives the phenyl rings roughly perpendicular to the porphyrin ring. STM images of a Cu-TBPP molecule on Cu(100) consist of four bright lobes, as shown in Fig. 4.9, (a) for a very low coverage and (b) for a coverage close to one monolayer. The distance between the trans-lobes is ~1.8 nm, slightly less than the molecular dimension of ~2 nm across. The bright lobes are thus assigned to the di-*t*-butylphenyl (DTP) side-groups, in accordance with previous reports [31,32] and theoretical simulation [33]. The characteristic four-lobe pattern is a registry of Cu-TBPP and is used for subsequent molecular recognition and controlled tip positioning above a molecule. Porphyrin molecules are chosen

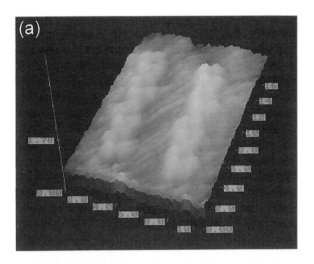

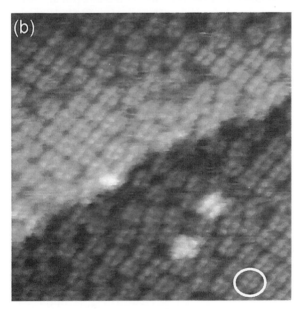

Fig. 4.9. STM images of Cu-TBPP on Cu(100) at 295 K, (a) for a small coverage, $9 \times 9$ nm$^2$, +0.5 V, 0.2 nA, no annealing was applied; (b) for a monlayer coverage, $25 \times 25$ nm$^2$, +1.5 V. 2 nA, the sample was annealed at ~200°C for 10 min.

because they are fluorescent molecules with aromaticity and are known to have high photoluminescence efficiency at converting $\pi \rightarrow \pi^*$ electronic excitations into visible and near UV lights in their free state or in solution [34]. However, in close proximity to metallic surfaces nonradiative damping of molecular excited states by metal surfaces is thought to drastically quench fluorescence [35]. Our experiments on the planar TPP-porphyrin (with the C(CH$_3$) replaced by H) molecules appear to support the above claim. In order to suppress the fast electron energy dissipation to metal surfaces, a new architecture of porphyrin molecules is designed to be somehow electronically decoupled from the substrate. The use of bulky *t*-butyl groups makes the phenyl rings rotate out of the plane of the porphyrin ring, which dramatically weakens the $\pi$-electron delocalization over the porphyrin-phenyl bond and renders it a predominantly single $\sigma$-bond. The electronic states of the porphyrin core are thus electronically decoupled from the surface since it hangs ~0.5 nm above it. Strong STM-induced photon emission is indeed observed from this molecule, as illustrated below. In addition, the crown-like cap configuration of Cu-TBPP together with its weak chemisorbed nature on Cu(100) might also serve as a squeezable tunneling junction and enable the molecule to sustain a relatively large current for the photon emission.

### 4.2.3.2 Molecular origin of photon emissions

The implication of light emission from single molecules comes from our photon emission experiments performed on the surface in Fig. 4.9(a) with a very low coverage of Cu-TBPP molecules. Strong photon emission was observed readily by CCD and even by the naked eye when the tip was positioned above a single molecule, but no significant light was observed when the tip is located above the bare Cu surface under the same excitation condition. This observation is in agreement with the report of C$_{60}$ on Au(110) [26], in which the detected photon intensity is greatest when the tip is placed above an individual C$_{60}$ molecule. A molecular origin for the emitted light was also claimed previously by Flaxer *et al.* [30] on the basis of the coverage dependence of the emission efficiency.

### 4.2.3.3 Technical approaches of STM-induced photon emission

Photon emission from the molecules starts to be detected by CCD when the bias voltage is above ~2 V in both polarities, becomes strong and stable above ~2.5 V, and appears to reach a local maximum around 3.5–4 V. Further increase of the bias voltage above ~7 V often leads to short but incandescent emission ("bleaching") of the molecule, which suggests that the molecule is probably modified or damaged [32]. There are three ways we used to produce photon emission from single molecules when the tip is positioned above it and a moderate bias voltage (e.g., 4 V) is designated. The first is to apply a duration-controllable voltage pulse with the feedback loop "off". As this setup may cause an abrupt change of tip-sample bias voltages, the photon emission so produced is probabilistic and the longest one lasted for only a few ten seconds but sometimes could be very strong and visible to the naked eye. The second is to keep the feedback loop "on" to maintain a constant current at a few nanoamperes (nA) *without tip-oscillation*. It is noteworthy that the overall current remains stable under this setup but there are still fine current fluctuations during the emission process. The emission event in this case is quite reproducible and stable but generally weak unless the tunneling current is increased up to a few ten nA (e.g., 50 nA). In other words, the emission intensity increases with increasing currents. The third is to keep the feedback loop "on", but allow the *tip to oscillate* at a frequency of a few kHz through the control of the feedback low-pass filter frequency by making use of the mechanical resonant vibration of the experimental device (mainly the STM head). The tunneling current was usually set around 1 nA, low enough to avoid tip-crashing into the metal surface. The value of set-currents, however, appears not so critical in this case as long as the tip does not crash into the metal surface since the tip-oscillation causes current fluctuation simultaneously. The tunneling current, unlike case two, is not well-defined due to the fluctuation. The averaged current was found to be around a few hundred nA but the peak value could be as high as a few microamperes depending on the oscillation magnitude. This approach (we call it the *electromechanical resonance pumping* in a sense that the piezomechanical oscillation of the tip is regulated by an electronic signal), is very effective in producing intense and reproducible light emission from molecules, and the photon emission can last almost as long as desired. Further measurements on the controlled tip oscillation through a well-defined voltage modulation rather than the mechanical resonances are underway. In both cases two and three, one can switch the light emission "on" or "off" simply through control of the feedback loop. The emitted light could be so strong that it is visible (orange-colored) to the naked eye when the environment is relatively dark. However, the intensity of photon emission in all the above three cases does not appear to be constant; in other words, the light blinks, shining unsteadily or going on and off rapidly. Blinking is thought to be the general fluorescent behavior of single molecules and nanostructures [16,36–38] and will be briefly discussed below in respect of the present system. It is worthwhile to point out an important observation about the relationship between the photon emission and tunneling current behavior: light is emitted only when the current fluctuates, although a quantitative relationship is still hard to discern. We speculate that for intense light emission from molecules the STM tip is very close or even somehow in contact with the molecule, rendering the molecular junction more-or-less symmetric (i.e., $R_{tip\text{-}molecule} \sim R_{molecule\text{-}substrate}$).

### 4.2.3.4 Intensity and quantum efficiency

Photon intensity data were measured on a surface with a monolayer coverage of Cu-TBPP as shown in Fig. 4.9(b). Although the intensity of photons emitted from an STM is generally low, the signal-to-background ratio is actually quite high, above 50 in the present setup. Count rates of up to $10^4$ cps per nA were recorded even for the small detection solid-angle of 0.03 sterad. The intensity measurements from different techniques (PMT, ICCD, or digital CCD) gave rise to a consistent value of ~$10^9$ cps for the total intensity after calibrations of the PMT quantum efficiency, solid angle, and attenuation or loss through transmission. Since the tunneling current during photon emission is in the order of a hundred nA, the quantum efficiency is estimated to be in the order of $10^{-3}$ photons/electron, around 0.1%. This value is at least ten times larger in comparison with the photon emission from clean metal surfaces [39–41]. The molecular origin of the photon emission is evident and the enhancement is presumably related to the quantum confinement of electrons in the molecule [42] and associated resonant tunneling process. It is noting worthy that the above intensity data were usually integrated/averaged over a few minutes, the actual intensity for the precise moment of molecular light emission should be higher because the integration also counts the dark moment due to the blinking aspect of molecules or other factors. Our preliminary measurements for such exact moment of molecular light emission gave a quantum efficiency up to 1%, but further measurements with better time resolution on current and photon emission are needed to confirm and quantify this number.

### 4.2.3.5 Optical spectra

Figure 4.10 shows an optical spectrum acquired over 200 s from nearly monolayered Cu-TBPP molecules on Cu(100) at a preset excitation condition of +4 V and 2 nA with tip oscillation. The actual average current, according to what we measured for similar measurements, was about 100 nA. Note that the photon intensity in Fig. 4.10 is not corrected for the device response. For bias voltages less than 2 V, no meaningful peaks were observed on the spectra even for a long integration time, suggesting a threshold around 2 V for the light emission. Speaking in general terms, without looking into fine structures, the two broad peaks are observed at 662 and 715 nm. The peak at ~715 nm is almost invariant with respect to the excitation voltage and sample location, suggesting an intrinsic feature associated with the Cu surface, likely related to its red-shifted fluorescence (or tip-induced surface plasmon) [43,44]. However, the shape and intensity for the peaks from 600 to 660 nm are dramatically different from that of the pristine Cu(100) surface, especially through its unique three shoulder-peaks approximately at 600, 630, and 660 nm. This additional spectral feature is attributed to the fluorescence from the molecules. Moreover, the relative intensity of three shoulder peaks depends on the particular sample site measured, changing from place to place, which implies that the fine spectra are related to the local molecular arrangements. One other important observation

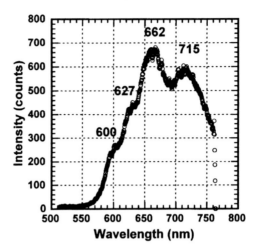

Fig. 4.10. Optical Spectra of Cu-TBPP on Cu(100) at 295 K at a preset excitation condition of +4 V and 2 nA integrated over 200 s. The intensity data is not corrected for the device response.

is that the variation of the bias voltage from 2 to 10 V appears not influential to the peak positions. The cutoff at 550 nm can be attributed to the electronic interband transition in the copper [45,46].

One amazing aspect of Fig. 4.10 is the presence of three shoulder-peaks superimposed with the broad peak around ~630 nm (~2 eV). The peak spacing of the fine structure is ~0.1 eV $\approx$ 800 cm$^{-1}$. This energy falls in the range of molecular vibrational excitations and is a strong indication that the photon emission arises from the coupling of molecular electronic states with the local electromagnetic field between the tip and substrate. What this vibrational peak can be attributed to has not yet been clarified. However, in reference to the range of vibrational frequencies reported for the chemical bonds of organic molecules, there are two possibilities. One is the C-C stretching or bending (700–1250 cm$^{-1}$) between the phenyl-porphyrin single bonds, the other is the C-H deformation vibration of ring hydrogens on the phenyl groups or the ring deformation vibration itself (700–900 cm$^{-1}$). We speculate that the observed frequency of ~800 cm$^{-1}$ is related to one of the above vibrational modes with the ring C-H deformation vibration the most likely one. The tip oscillation or current fluctuation is likely to cause molecular deformation and symmetry lowering due to mechanical pumping or field fluctuation. A Franck-Condon-like mechanism is possibly involved during this process: vertical electronic transition occurs rapidly, which is followed by the relatively slow relaxation of molecules.

It is noteworthy that our preliminary experiments on both electrically and optically excited spectra appear to show interesting similarities and differences in terms of energy absorption and emission. In principle, these spectra should be related to the electronic feature of the free-base- or Cu-TBP porphyrin, particularly to the 18-$\pi$-electron aromatic system delocalized along the inner 16-membered ring of the molecule. Indeed, the absorption spectrum of porphyrins can be successfully rationalized by Gouterman's four-orbital model (two HOMO $\pi$-orbitals and two LUMO $\pi^*$ orbitals), which gives rise to a relatively weak (pseudoparity-forbidden) Q-band in the visible region and an intense (optically allowed) Soret or B-band in the near-UV [34]. Our absorption spectra for Cu-TBPP reproduce nicely such spectral character with the small Q-band at 539.6 nm (2.3 eV) and the sharp B-band at 417.0 nm (3.0 eV). However, while the fluorescence spectra of the free-base $H_2$-TBP porphyrin give a normal peak at 650 nm in solution, metal-based Cu-TBP porphyrin molecules show no fluorescence either in solution or on a glass substrate deposited with molecules. The IR spectra in solution do show a peak around 800 cm$^{-1}$ although there are many more other and stronger peaks. All these observations tend to suggest that different excitation mechanisms are involved in the excitation process by electrons and photons, and might lead to different electronically excited states.

### 4.2.3.6 Threshold

The threshold at $V_b \sim 2$ V is in agreement with the HOMO-LUMO gap of Cu-TBPP molecules ($E_g \sim 2$ eV). Visible-light emission from molecules can occur only when $V_b \geq E_g$ no matter what kind of mechanism is involved, although the energy gap may be slightly reduced upon the adsorption of molecules on the surface and insertion into the tunneling junction. Adsorption of Cu-TBPP on Cu(100) results in broadening of HOMO and LUMO states, in particular the LUMO [32]. Accordingly, the tip can also contribute to the enhancement of the tunneling current by pressure-induced distortion of the molecular orbitals and resultant broadening when in contact with the molecule. The local maxima between 3.0–4.0 V may be related to the molecules' higher absorption efficiency of electron energy in this range (in rough parallel with optically allowed B-band transition versus forbidden Q-band transition), or to the field emission resonance inside the tunneling gap, or both. Further experiments to clarify this issue are underway.

### 4.2.3.7 Mechanism

The STM-induced photon emission on clean metal surfaces (Cu, Ag, Au) is reported to arise from the formation of coupled and localized surface plasmons of the tip and sample, the so-called tip-induced plasmon (TIP) modes [44]. The close proximity of the tip to a metal surface induces localized plasmon modes upon tunneling which are characterized by a strongly enhanced electric field in the cavity formed by the tip and sample. This sort of resonance mechanism produces photon emission through inelastic tunneling processes and radiative decay of localized dipolar plasmons. However, the TIP modes alone cannot explain the enhanced emission above the molecules. The presence of a molecule would lead to an increase in the distance between the tip and metal surface and thus to a smaller field enhancement (or TIP). This in turn would result in intensity minima above molecules, contradictory to the experimental observations. Conversely, the molecule itself is the primary photon source and plays an active role in the photon emission process. Enhanced photon emission above molecules suggests a strong coupling between the molecules and the electromagnetic modes of the cavity between the tip and metal surface. Coupling between excited molecules or atoms and the plasmons of a metal surface is a well-known phenomenon [45]. A tunneling electron or an absorbed molecule can excite a surface plasmon in the metal substrate, which causes the enhancement of local electromagnetic field and in turn helps to excite the molecules. Viewing the tip-sample geometry as a cavity, the fluorescence of molecules may be enhanced in a way that is similar to those invoked in the classical explanation of surface-enhanced Raman scattering [46]. The molecular fluorescence stimulated by inelastic tunneling was also used previously to explain STM-induced light emission from dichoroanthracene layers [30]. However, a detailed mechanism of how molecules get excited and then decay to the ground state still awaits further investigation.

### 4.2.3.8 The role of tip and molecule

Surface plasmon on a perfectly smooth metal surface cannot radiate since such a process would not be able to simultaneously conserve energy and momentum. However, when a tip is brought in the proximity to the surface or when the surface is roughened, the translational invariance along the surface is broken and photon emission can occur. Although it has not yet been well understood how a molecule couples with the local electromagnetic field and enhances the photon emission, it is reasonable to consider the molecule as a quantum system and tunneling electrons as being more-or-less confined in the molecule during their transport path. The confinement-induced enhancement in the radiative rate and reduction in the nonradiative rate would result in a higher efficiency of luminescence from the semiconducting molecule [42]. On the other hand, each molecule on the surface can, in a sense, be viewed as a protrusion, acting as a scattering center and facilitating the conversion of plasmons into photons [47]. The unique crown-like conformation with bulky legs also helps to suppress the otherwise fast nonradiative decay [48] by weakening the interaction of the molecule with the substrate, the so-called electronic decoupling.

### 4.2.3.9 Molecular blinking

When an STM tip or the substrate is used to inject carriers (e.g., electrons), the inelastic tunneling channels contribute only a small fraction to the total tunneling current. In other words, for a given tunneling current associated with a particular sample site, an overwhelming proportion of the total current is carried by elastic tunneling. The inelastic current that gives out photon emission is reported to be less than 0.1% and is affected by factors such as local dielectric properties, surface geometry and density of states for inelastic processes [39,41]. These factors are subjected to variation during photon emission since the molecule inside the tunneling gap and its surrounding arrangement are likely to be modified, e.g., yielding quenching defects. The resultant fluctuation in the inelastic current component might be partly responsible for the blinking of molecules.

### 4.2.3.10  Field fluctuation and tip oscillation

A model developed by Johansson and coworkers [49] takes the role of the tip into account by approximating the tunneling from a metallic sphere to a planar metal surface. The intensity of the emitted light from the tunneling electrons was found to split into two separate factors. One is the strength of the electromagnetic field in the vicinity of the tip giving us a measure of the strength of the coupling between tunneling electrons and the electromagnetic field. The other is the fluctuations in the tunneling current that act as a source for the radiation. The presence of the tip close to the sample surface creates a localized interface plasmon built up by surface charges of opposite polarity on the tip and the sample surface, respectively. This plasmon resonance leads to a considerable enhancement of the light emission from the tunneling electrons with a quantum efficiency of the light emission in the order of $10^{-4}$. Similar arguments may apply to the photon emission for a surface with absorbed molecules if the electronically excited molecule can be viewed as an oscillating charge distribution or a dynamic dipole moment surrounded by an oscillating electric field. If surface plasmons are excited by fluctuations of the tunneling current (or field), specifically by its high frequency components [41,49,50], a moderate tip oscillation (without tip-crashing) would offer a dynamic electric field, fluctuations in tunneling currents, and as a result, a dynamic dipole moment for the effective excitation of plasmon resonances. Our experimental results appeared to indicate that the intensity of light emission parallels the magnitude of current fluctuation. When the tunneling current is absolutely stable, i.e., no fluctuation at all, no photon emission is observed. In the case of tip oscillation through the control of the feedback loop filter frequency, a higher frequency implies a larger magnitude of tip oscillation and resultant current fluctuation, and thus generally gives a stronger light emission (which is also of better control and stability). These observations are roughly in agreement with the model proposed above. In other words, the vertical tip displacement will modify the electromagnetic modes and thus the coupling of tunneling electrons and molecules with the electromagnetic field and eventually the photon emission.

### 4.2.4 Summary

We have developed a new and powerful technique—electromechanical resonance pumping—to produce strong photon emission from single molecules of Cu-TBPP on Cu(100) via tunneling electron excitation. The general physical process of STM-induced molecular light emission may be pictured as follows. When an STM is positioned above a molecule and a bias voltage (e.g., +3.5 V) is applied across the tunneling gap, a net tunneling electron current flows from the tip to the substrate via the absorbed molecule. The molecule is excited by the resonant coupling of molecular electronic states with the localized electromagnetic modes (enhanced by the surface dipolar plasmons) and then decays to the ground states with emission of light. Further investigation into the details of this electroluminescent process—energy supply or absorption, energy transfer or excitation, and energy decay or emission for the tip-molecule-substrate structures—is underway. On the other hand, the combination of STM with optical techniques promises electronic and vibrational spectroscopy at the single-molecule level, which will facilitate the identification of individual molecules ("molecular recognition") and luminescent defects.

### Acknowledgments

We thank Dr. J. K. Gimzewski, Dr. R. Berndt, Dr. V. J. Langlais, and Dr. H. Isago for helpful discussions.

[References]

1)   D. V. Averin and K. K. Likharev, *Mesoscopic Phenomena in Solids*, Chap. 6, edited by B. L. Altshuler, P. A. Lee, and R. A. Webb, North Holland (1991).

2)   C. A. Neugebauer and M. B. Webb, *J. Appl. Phys.*, **33**, 74 (1962).

3)   T. A. Fulton and G. L. Dolan, *Phys. Rev. Lett.*, **59**, 109 (1987).

4)   P. J. M. van Bentum, R. T. Smorkers, and H. van Kempen, *Phys. Rev. Lett.*, **60**, 2543 (1988).

5)   L. J. Geeligs, M. Peters, L. E. M. de Groot, A. Verbruggen, and J. E. Mooij, *Phys. Rev. Lett.*, **63**, 326 (1989).

6)   M. H. Devoret, D. Esteve, H. Grabert, G. L. Ingold, H. Pothier, and C. Urbina, *Phys. Rev. Lett.*, **64**, 1565 (1990).

7)   B. Su, V. J. Goldman, and J. E. Cunningham, *Science*, **255**, 313 (1992).

8)   A. Sakai (private communication) (1994).

9)   C. Schonenberger, H. van Houten, H. C. Donkersloot, A. M. T. van der Putten, and L. J. G. Fokkink, *Phys. Scr.*, **T45**, 289 (1992).

10)  C. M. Fischer, M. Burghard, S. Roth, and K. v. Klizing, *Europhys. Lett.*, **28**, 129 (1994).

11)  H. Nejo and M. Aono, *Mod. Phys. Lett. B*, **6**, 187 (1992).

12)  N. Shima, *JRDC Pre-research Program*, p. 17, Abstract (1994).

13)  H. Nejo, *Nature*, **353**, 640 (1991).

14)  N. Iwai (private communication) (1994).

15)  *Scanning Tunneling Microscopy II*, edited by R. Wiesendanger and H.-J. Guntherodt, Springer Verlag (1995).

16)  W. E. Moerner and M. Orrit, *Science*, **283**, 1670 (1999).

17)  B. C. Stipe, M. A. Rezaei, and W. Ho, *Science* **280**, 1732 (1998).

18) J. K. Gimzewski, *Photons and Local Probes*, edited by O. Marii and R. Möller, p. 189, IBM, Netherlands (1995).

19) R. Berndt, *Scanning Probe Microscopy*, Chap. 5, p. 97, edited by R. Wiesendanger , Springer, Berlin (1998).

20) J. K. Gimzewski and C. Joachim, *Science*, **283**, 1683 (1999).

21) J. Lambe and S. L. McCarthy, *Phys. Rev. Lett.*, **37**, 923 (1976).

22) D. Hone, B. Mühlschlegel, and D. J. Scalapino, *Appl. Phys. Lett.*, **33**, 203 (1978).

23) R. W. Rendell, D. J. Scalapino, and B. Mühlschlegel, *Phys. Rev. Lett.*, **41**, 1746 (1978).

24) J. R. Kirtley, T. N. Theis, J. C. Tsang, D. J. MiMaria, *Phys. Rev. B*, **27**, 4601 (1983).

25) J. K. Gimzewski, J. K. Sass, R. R. Schlittler, and J. Schott, *Europhys. Lett.*, **8**, 435 (1989).

26) R. Berndt, R. Gaisch, J. K. Gimzewski, B. Reihl, R. R. Schlittler, W. D. Schneider, and M. Tschudy, *Science*, **262**, 1425 (1993).

27) D. L. Abraham, A. Veider, Ch. Schönenberger, H. P. Meier, D. J. Arent, and S. F. Alvarado, *Appl. Phys. Lett.*, **56**, 1564 (1990).

28) T. Tsuruoka, Y. Ohizumi, S. Ushioda, Y. Hono, and H. Hono, *Appl. Phys. Lett.*, **73**, 1544 (1998).

29) S. Sasaki and T. Murashita, *Jpn. J. Appl. Phys.*, **38**, L4 (1999).

30) E. Flaxer, O. Sneh, and O. Cheshnovsky, *Science*, **262**, 2012 (1993).

31) T. A. Jung, R. R. Schlittler, and J. K. Gimzewski, *Nature*, **386**, 696 (1997).

32) J. K. Gimzewski, T. A. Jung, M. T. Cuberes, and R. R. Schlittler, *Surf. Sci.*, **386**, 101 (1997).

33) P. Sautet and C. Joachim, *Chem. Phys. Lett.*, **185**, 23 (1991).

34) I. B. Berlam, *Handbook of Fluorescence Spectra of Aromatic Molecules*, Academic Press, New York (1965); M. Gouterman, in *The Porphyrins*, Part A, pp. 1–165, edited by D. Dolphin, Academic Press, New York (1978).

35) D. H. Waldeck, A. P. Alivisatos, and C. B. Harris, *Surf. Sci.*, **158**, 103 (1985).

36) R. M. Dickson, A. B. Cubitt, R. Y. Tsien, and W. E. Moerner, *Nature*, **388**, 355 (1997).

37) M. Nirmal, B. O. Dabbousi, M. G. Bawendi, J. J. Macklin, J. K. Trautman, T. D. Harris, and L. E. Brus, *Nature*, **383**, 802 (1996).

38) Z. Wu, T. Nakayama, S. Qiao, and M. Aono, *Appl. Phys. Lett.*, **73**, 2269 (1998).

39) R. Berndt and J. K. Gimzewski, *Phys. Rev. B*, **48**, 4746 (1993).

40) R. Berndt, R. R. Schlittler, and J. K. Gimzewski, *J. Vac. Sci. Technol.*, **B9**, 573 (1991).

41) B. N. J. Persson and A. Baratoff, *Phys. Rev. Lett.*, **68**, 3224 (1992).

42) A. P. Alivisatos, *Science*, **271**, 933 (1996).

43) M. Welkowsky and R. Braunstein, *Solid State Commun.*, **9**, 2139 (1971).

44) R. Berndt, J. K. Gimzewski, and P. Johansson, *Phys. Rev. Lett.*, **67**, 3796 (1991).

45) A. Adams, J. Moreland, and P. K. Hansma, *Surf. Sci.*, **111**, 351 (1981).

46) A. Otto, I. Mrozek, H. Grabhorn, and W. Akemann, *J. Phys. Condens. Matter*, **4**, 1143 (1992).

47) N. Kroo, J. P. Thost, M. Völker, W. Krieger, and H. Walther, *Europhys. Lett.*, **15**, 289 (1991).

48) P. Avouris and B. N. J. Persson, *J. Phys. Chem.*, **88**, 837 (1984).

49) P. Johansson, R. Monreal, and P. Apell, *Phys. Rev. B*, **42**, 9210 (1990).

50) M. Tsukada, T. Shimizu, and K. Kobayashi, *Ultramicroscopy*, **42/44**, 360 (1992).

# Theory of Electronic and Atomic Processes in Scanning Probe Microscopy

## 5.1 Introduction

A new epoch of surface science has been opened up by the remarkable progress in scanning probe microscopy [1]. Moreover novel research area of materials science, i.e., the science of nano-scale materials has been brought about thanks to the development of SPM. In spite of the rapid development in SPM as an experimental tool, however, a firm theoretical basis for it has not in general been fully established. It is only for the fundamental aspects of scanning tunneling microscopy (STM), that present theoretical understanding of its mechanism seems to be satisfactory [2]. There still, however, remain various fundamental problems related to subtle tunneling processes in STM, though compared to these, the theoretical understanding of atomic force microscopy (AFM) has lagged far behind [3]. Although the recent rapid development of non-contact mode atomic force microscopy (nc-AFM) has achieved real atomic resolution [4–6], the mechanism ensuring this is unclear. One still does neither know the reason assuring atomic resolution of nc-AFM, nor the nature of the interaction force in the intermediate regime of the van der Waals and chemical region.

In this article, we will describe the theoretical background of STM, frictional force microscopy (FFM), and dynamic mode AFM mainly on the basis of theoretical simulations. By copmaring various kinds of scanning probe microscopy (SPM), we will find the characteristic feature of each type of microscopy and related spectroscopy more transparently. Roughly speaking, the real atomic resolution of STM is achieved by a remarkable concentration of the tunneling current onto a single apex atom at the mini-tip [2]. It might be supposed that the force contributing nc-AFM images also concentrates on the tip apex atom, but how can this be possible [7]? This is a problem related to the nature of the long and short range forces between the tip and the surface and their distribution over atoms. Theoretical understanding is also needed of how the actual cantilever dynamics is affected by the interaction force field [8] and how they provide atomistic images [7].

We first introduce the theory of STM based on the first principles density functional approach [2]. Bardeen's perturbation theory [9] enables us to calculate the tunnel current from first principles for an atomic mini-tip model placed on any position on the sample surface. Theoretically simulated images are obtained by the tunnel current distribution that is thus calculated. For most of the cases, they are more or less similar to a map of the local density of states (LDOS) of the surface for smaller bias voltages. The reason for it is explained in the next section in details. For certain cases the tunnel current map shows a remarkable deviation from the LDOS map, the reason of which is ascribed to the tip effect [10]. From theoretical simulations using various models of the tip, one sees clearly the effect of the atomic structure and the shape of the tip on the STM/STS. Various strange but interesting features of STM/STS such as the negative differential tunnel conductance [11,12] and abnormal images [13] can only be explained using the simulations with atomic tip models [2]. The mechanism whereby atomic resolution by the tip is achieved can only be understood by such theoretical analyses. We might conclude that the mechanism for atomic resolution by the STM is concentration of the tunneling current predominantly on the single apex atom of the tip. In section 5.2, we will describe the theoretical framework of the simulation, and from sections 5.3 through 5.5, some remarkable features of STM/STS will be presented based on numerical calculations using the theory.

STM can be used as a tool for manipulating the atomic objects on the surface, and to investigate the electronic properties of the nano-structures on the surface. To study these phenomena using a theoretical approach, it is necessary to extend the local density functional (DFT) method to be applicable to open non-equilibrium systems. In section 5.6, a method for the purpose called the recursion-transfer matrix method is introduced [14,15]. The atomic nature of the formation of the quantum point contact [16] and quantum transport through atom bridges using this method are described in section 5.6, and section 5.7 [17,18], respectively.

Information about the surface observed by the atomic force microscopy (AFM) is quite different from that by STM/STS. This can be easily understood by considering that the tip-surface interaction force is contributed by the total valence electrons and their response to the external substance, rather than those of only the electrons near the Fermi level, which solely contribute to the tunnel current in the STM. The interaction force might have a long attractive tail continuing to the van der Waals regime. However this part of the force dose not bare the atomic sensitivity, in particular to the lateral direction. Then the true atomic resolution of the AFM images might be attained by the short range attractive or repulsive force which is a tiny portion of the net force and felt only by a single or a few atoms on the tip apex.

How can such atomically sensitive force be separated and detected by the dynamic mode AFM? This is not a trivial problem. Recently the nc-AFM has been successfully demonstrated to achieve a real atom scale resolution [4,5,6]. However the mechanism ensuring this is still puzzling and it is still not known what quantities actually form the nc-AFM images. To establish a firm basis for an understanding of the nc-AFM images, the first step would be to describe the relationship between the cantilever dynamics and the tip-surface interaction [7,8,19,20]. Next, the nature of the tip-surface interaction force and its distribution over the tip surface should be clarified from theoretical approach in order to create realistic models.

In section 5.8 we shall discuss some topics relating to the friction force microscopy (FFM) based on the results of the theoretical simulations [21]. The theory reproduces the experimental images fairly well, and we will clarify how the Tomlinson's stick-slip model [22] manifests itself in the FFM images. We then consider the fundamental problems of the dynamic mode AFM [8]. Aspects of non-linear dynamics of the cantilever will be carefully studied in sections 5.9 and 5.10. Various types of remarkable behaviors, such as the dynamic touching mode and hysteresis due to the bimodal nature will be discussed [8], and how the interaction potential will reflect on the frequency shift curve of the large amplitude mode of the nc-AFM will be explained in section 5.10 [19,20].

## 5.2 Tunneling Current in STM

In the usual experimental conditions for STM, the distance between the tip and the sample surface is rather long, in the order of 10 Å. Perturbation treatment of the tunneling current is therefore valid, if the applied bias voltage is not very strong. Within the framework of the perturbation theory, it is possible to carry out a detailed analysis of the tunneling current [2]. This method also affords a clear understanding of the tip effect and the relationship between the current map and the LDOS (local density of states) map.

Starting from Bardeen's perturbation treatment [9], we can express the tunneling current in STM as follows:

$$I = \frac{2\pi e}{\hbar} \int \{f(E) - f(E + eV)\} A(R, E, E + eV) dE \qquad (5.1)$$

$$A(R, E, E') = \int dr dr' V_T(r) V_T(r') G^T(r, r', E') G^S(r + R, r' + R, E) \qquad (5.2)$$

where $A$ is the energy density of the tunnel current, $f$ is the Fermi distribution function, $V_T(r)$ is the potential in the tip and $V$ is the tip bias voltage. $G^T$ and $G^S$ are the imaginary part of the Green's function of the surface and the tip, respectively. The integral of Eq. (5.1) ranges over the Fermi level offset

Chapter 5

between the two electrodes. From these relations, an important fact is found, namely, if there is a very small protrusion close to the sample surface, the tunnel current is almost entirely concentrated on this region. This is confirmed by expanding the smooth part of the surface Green's function, i.e., the Surface Green's function divided by the natural decay factor

$$\gamma(z; E) = \exp\left(-z\sqrt{2m|E|} \, / \, \hbar\right) \tag{5.3}$$

around this protrusion. This leads to a moment expansion of the tunnel current [2], in which each term corresponds to the respective order of the derivative of the surface Green's function (smooth part) multiplied by the geometrical moment of the weight function,

$$\omega(r, r') = V_T(r)V_T(r')\sum_v \psi_v(r)\psi_v^*(r')\gamma(z; E)\gamma(z'; E)\delta(E' - E_v). \tag{5.4}$$

In the above, $\psi_v$ is the tip eigen function. If we take only the first term of the expansion, the naive relationship [23],

$$I(R) \propto V\rho(R, E_F) \tag{5.5}$$

results for small $V$, where $\rho$ is the LDOS (local density of states) of the surface at the protrusion site $R$ of the mini-tip. Indications of significant deviations from the naive relationship have often been noticed and they are mostly related to the microscopic states of the tip. For example, much finer structures of STM

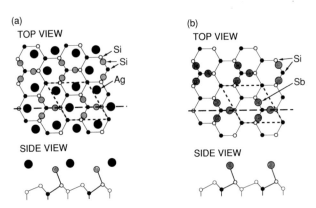

Fig. 5.1. (a) Modified HCT model for the structure of the Si(111) $\sqrt{3} \times \sqrt{3}$ -Ag surface, and (b) milkstool model for the Si(111) $\sqrt{3} \times \sqrt{3}$ -Sb surface.

images are often observed than expected merely by the surface LDOS. This can only be explained by there being a contribution from the higher order derivatives of the surface wave function, by the higher order terms of the moment expansion. If several atoms in the tip apex are located on the same level, abnormal STM images are often observed [13]. This is due to the interference of the tunnel current components flowing through different atoms. To perform a quantitative simulation of the STM images including the tip effects, the first-principles calculation of the tunnel current based on Eqs. (5.1) and (5.2) would, in any case, be necessary [2].

## 5.3 *The STM Images of Si(111)* $\sqrt{3} \times \sqrt{3}$ *-Ag and -Sb Surfaces*

For the positive surface bias, the observed STM image of Si(111) $\sqrt{3} \times \sqrt{3}$ -Ag surface [24] showed a distribution of the bright spots in a honeycomb lattice arrangement. On the other hand, the HCT (Honeycomb-Chained-Trimer) model (see Fig. 5.1(a)) proposed by Takahashi on the basis of the X-ray standing wave method [25] and by Katayama and others by coaxial impact collision ion scattering spectroscopy (CAICISS) [26] is consistent with many experimental facts, but not including STM. The energy band structure of the HCT model calculated by Watanabe and others [27] revealed a semiconductor character, as seen in Fig. 5.2(a). Moreover, the overall density of states (DOS) was consistent with the UPS experiment [28]. A serious question thus arises; how can the HCT model be reconciled with the observed honeycomb pattern of STM?

Watanabe, Aono and Tsukada [27] performed a theoretical simulation of the STM image of this surface with a tip model $W_{10}[111]$ consisting of ten W atoms. The theoretical STM image showed an excellent agreement with the

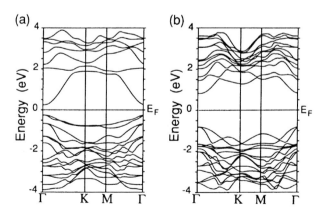

Fig. 5.2. Calculated energy-band dispersion of (a) the Si(111) $\sqrt{3} \times \sqrt{3}$ -Ag, and (b) the Si(111) $\sqrt{3} \times \sqrt{3}$ -Sb surface.

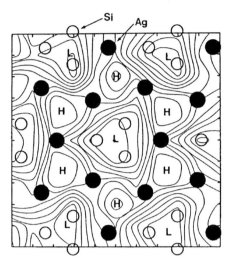

Fig. 5.3. Contour map of the tunneling current for HCT model of Si(111) $\sqrt{3} \times \sqrt{3}$ -Ag surface calculated by the tip model $W_{10}[111]$. Surface bias voltage is 1.0 V and the tip height is 3.7 A.

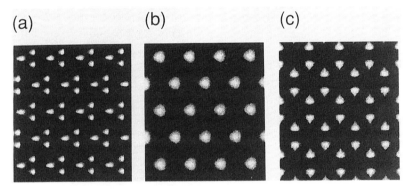

Fig. 5.4. Simulated STM images of the Si(111) $\sqrt{3} \times \sqrt{3}$ -Sb surface. Surface bias voltages are (a) –2.0 V, (b) 2.0 V, and (c) 1.0 V.

observed image [24]. The reason the HCT model gives a honeycomb pattern can be understood by Fig. 5.3; namely, the bright spots do not correspond to any of the surface atoms, but are located at the inner region of the Ag trimer. This can be explained by the fact that the lowest unoccupied surface states of Si(111) $\sqrt{3} \times \sqrt{3}$ -Ag has amplitude large in the center region of the triangle made of Ag atoms. This is an interesting example of how naive interpretation of STM images sometimes fails.

$W_{10}[111]$

$W_{14}[1\overline{1}0]$

$W_{13}[1\overline{1}0]$:Without Apex

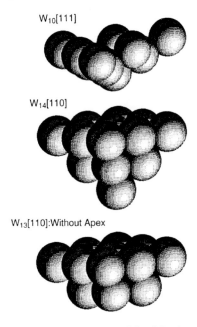

Fig. 5.5.  Some cluster models of the tip.

The Sb chemisorbed Si(111) $\sqrt{3} \times \sqrt{3}$ surface has a milk-stool structure as shown in Fig. 5.1(b). This structure is related to the HCT model of the Si(111) $\sqrt{3} \times \sqrt{3}$-Ag surface (Fig. 5.1(a)); the milk-stool structure, aside from the detailed bond lengths, is obtained from the HCT structure by removing the top Ag layers and replacing the Si atoms in the next layer by Sb atoms. We performed a simulation of the STM images [29] for this surface. Figure 5.4 shows the theoretical STM images for the surface bias voltages, $Vs = -2.0$ V, $+2.0$ V, and $+1.0$ V, respectively. For the case of $Vs = -2.0$ V, it can be observed that each Sb atom in the milk-stool appears as a distinct bright spot forming a trimer. For the case of $Vs = 2.0$ V, the three atoms of the trimer cannot be seen separately, but form a single large bright area centered at the milk-stool. Interestingly enough, almost the same honeycomb pattern appears for the case of $Vs = 1.0$ V (as shown in Fig. 5.4(c)) as that of the HCT model. The spatial position where the bright spots are located in the honeycomb of the Sb chemisorbed surface is outside the milk-stool and thus at exactly the same location as in the case of the Si(111) $\sqrt{3} \times \sqrt{3}$ Ag surface. Figs. 5.2(a) and 5.2(b) show the energy band structure of Si(111) $\sqrt{3} \times \sqrt{3}$-Ag (HCT) and that of the Si(111) $\sqrt{3} \times \sqrt{3}$-Sb (milk stool) surface, respectively [29]. The surface energy band which contributes to the honeycomb spots in the STM image was

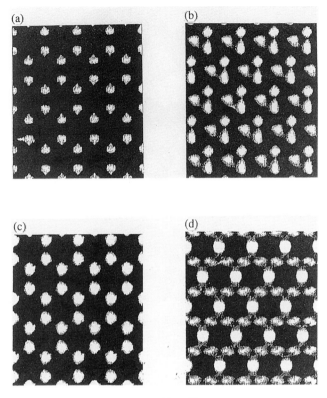

Fig. 5.6. Theoretical STM images of Si(111) $\sqrt{3} \times \sqrt{3}$ -Ag by different models of the tip for (a) $W_{10}[111]$, (b) $W_9[111]$, (c) $W_{14}[110]$, (d) $W_{13}[110]$.

found to be the lowest unoccupied bands for both the surfaces. We can see from Fig. 5.2 that the lowest surface state band in each surface has a remarkably similar dispersion although that in the Si(111) $\sqrt{3} \times \sqrt{3}$ -Sb (milk stool) has shifted to a somewhat higher energy position and merged into the higher conduction bands. As indicated by the STM images, the wavefunction of the lowest unoccupied surface state relative to the substrate is almost the same for both the surfaces.

## 5.4 The Effect of a Microscopic Structure at the Tip

According to a number of theoretical simulations, it seems that normal STM images can be produced by a tip with a single atom on its apex [13]. However, if there are a few atoms on the top plane of the tip, i.e., at the same height from the sample surface, abnormal images result. This is due to the spreading of the tunnel current over these atoms on the tip.

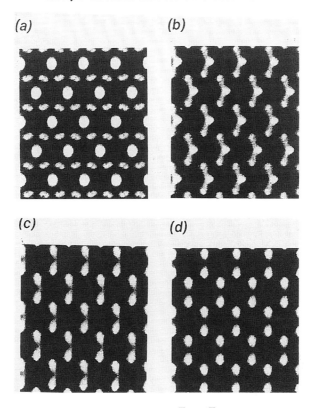

Fig. 5.7.  Theoretical STM image of the Si(111) $\sqrt{3} \times \sqrt{3}$-Ag surface (HCT model) by the $W_{13}[110]$ tip. The tip-surface distance is 3.7 A and the surface bias voltage is 0 V. (a), (b), (c) and (d) correspond to the tilting angles 0, 10, 20, 30, degrees, respectively.

Let us look at this feature using the example of the Si(111) $\sqrt{3} \times \sqrt{3}$-Ag (HCT) surface. With the tip models of $W_{10}[111]$ or $W_{14}[110]$ (Fig. 5.5), normal images of the unoccupied state, which has been observed in experiments, are reproduced quite well. If we remove the apex atom from these tips, the calculated images are very different from those obtained by the tip models of $W_{10}[111]$ or $W_{14}[110]$ as shown in Fig. 5.6. These abnormal images are formed because the tunnel current is divided into four atoms for the case of the $W_{13}[110]$ tip, or three atoms for the case of the $W_{9}[111]$ tip (Fig. 5.5). The abnormal images change drastically with the rotation of the tip around its symmetry axis, though the normal STM images obtained by the tip $W_{10}[111]$ or by the tip $W_{14}[110]$ hardly change with such rotation [30].

If the spreading of the tunnel current into several atoms is due to the presence of more than one atom on the same height of the tip, a slight tilting of the tip axis from the surface normal direction will recover the correct STM

(a)                                          (b)

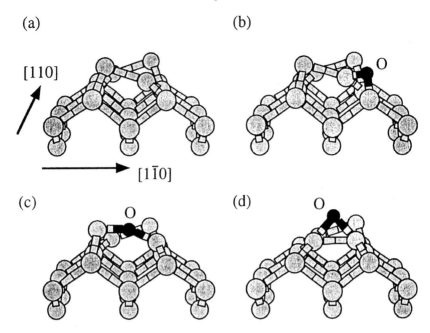

[110]

[1$\bar{1}$0]

(c)                                          (d)

Fig. 5.8. Structures of the clean Si(001) surface (a), and the oxygen chemisored Si(001) surfaces at the back bond of the down dimer site (b), the middle of the dimer bond (c), and above the dimer (d).

image. To confirm this, we attempted to make a theoretical simulation of STM image with a tilted tip geometry [31]. Figures 5.7(a) through 5.7(d) show the simulated results of STM images obtained by the tip $W_{13}$[110] for the tilting angle $\theta = 0$, 10, 20, and 30 degrees from the surface normal, respectively. With an increase of the tilting angle $\theta$, the simulated images gradually recover to the normal image. For $\theta = 20$ degrees, the images are almost the same as the normal one. For the case of the $W_9$[111] tip, similar results are obtained for the recovering of the abnormal images; in this case only small tilting angle $\theta = 10$ degrees is required for recovering the image to the normal one. The features obtained by the simulation confirm the assumption that the tunnel current concentrates almost totally on the single apex atom when the normal images are obtained [2].

## 5.5 STM Images of Oxygen Chemisorbed Si(001) Surfaces

STM images are most sensitive to the surface electronic states near the Fermi level, and they do not directly reflect the surface atomic structure. The latter is observed by STM indirectly through its influence on the surface electronic states. Examples of this fact have already been seen in section 5.3,

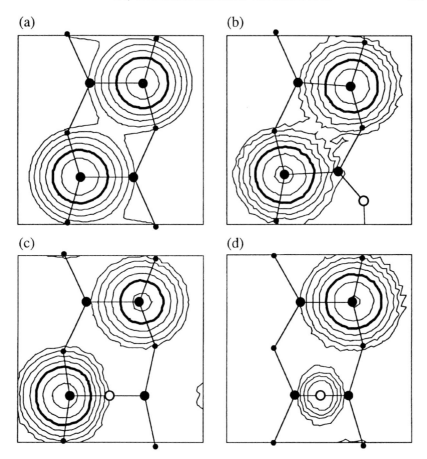

Fig. 5.9. Theoretical STM images of the four structures shown in Fig. 5.8. The surface bias voltage is −2V, and the tip height is 2.4 A. The labels (a) through (d) of the images correspond to those for the structures in Fig. 5.8.

but in this section, we will discuss the case of the oxygen chemisorbed Si(001) surface further [32]. Figure 5.8 shows several stable and quasi-stable structures of the oxygen chemisorbed Si(001) surface as well as the clean Si(001) surface. In the lowest energy structure, the oxygen atom is inserted in the back bond of the down dimer atom (Fig. 5.8(b)). In the second lowest structure the oxygen atom is inserted in the dimer bond as seen in Fig. 5.8(c). On the other hand the oxygen atom is located slightly above the dimer bond at the next higher energy structure (Fig. 5.8(d)). Figure 5.9 shows theoretically simulated STM images corresponding respectively to the structures shown in Fig. 5.8 [32]. The bias voltage is surface negative (−2.0 V), but the qualitative features of the STM images for the oxygen chemisorption structure do not change so much depending on the bias polarity, if the absolute value is not so large.

For the clean surface, a bright region appears around each up dimer atom. This is a well established fact, and the strong tunnel current originates from the highest occupied dangling bond located on this Si site. A remarkable feature of the STM images in Figs. 5.9(b) and (c) is that the oxygen atom inserted in the bonds cannot be seen at all in the STM image. This is strange, since the oxygen ionic radius should be very large and we might expect a large electron density near the oxygen region. The oxygen atom thus behaves like a transparent object in the STM image, though it can be seen slightly if it is located in a high position, as in the geometry of Fig. 5.9(d). This is an interesting example of the STM image only detecting an electron distribution close to the Fermi level.

## 5.6 *Formation of Atomic Point Contact and Electron Transmission through Atom Bridges*

When the tip approaches very close to the surface, some part of the current between the tip and the surface starts to flow in a ballistic way through tiny holes in the tunnel barrier. This is the atomic picture of the point contact formation, and various important features of the quantum transport are expected to appear.

To discuss quantum transport with any finite bias voltages, as well as various atomic processes under the non-equilibrium open condition, the first principles recursion transfer matrix method (RTM) provides a powerful theoretical approach [14,15]. In this method, every wave function is solved in the form of the scattering waves, which are linear combinations of the 2-dimensional plane waves in the surface parallel direction multiplied by the unknown coefficient functions in the surface normal direction. Since the scattering waves are rigorously divided into the right and the left electrode waves, we can assign the two different Fermi levels of the two electrodes to the respective kinds of wave functions. The unknown coefficients are solved in the form of the transfer matrix in a collective way, which satisfies a three-term matrix difference equation transformed from the Schroedinger equation [14,15]. The matrix difference equation is solved numerically with the use of the recursive relation as well as the exact boundary conditions satisfied in the two boundary planes assumed deep inside the electrodes.

Using this method, we investigated the atomic nature of the point contact formation between the Al mini-tip and the Si(111) surface [16]. The top panel of Fig. 5.10 shows a reduction in the tunnel barrier height in front of the tip apex atom when it approaches to surface. The barrier is also lowered by the bias voltage i.e., the barrier height for the surface bias 2 V is lowered compared with that for the bias 0 V. It is remarkable that the current increases almost in a simple exponential way over a wide range where the barrier height is steeply reduced. This result shows that the apparent work function obtained by a

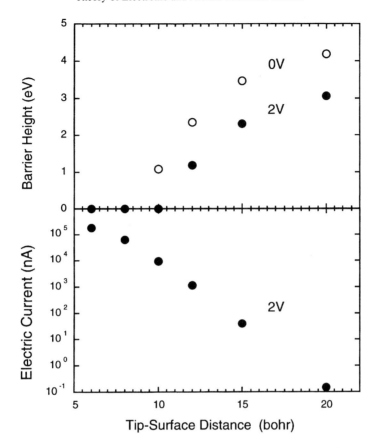

Fig. 5.10. The tunnel barrier height at the apex of the Al tip placed on an ideal Si(111) surface, and the current as the function of the tip height.

derivative of the logarithm of the current does not mean the real tunnel barrier, but includes the complicated effect of a change in the microscopic distribution of the electron potential. When the tip height approaches 5 a.u., which is the distance corresponding to the sum of the atomic radii, the current approaches a certain saturated value. The spatial distribution of the tunnel current is shown in the top panel of Fig. 5.11 for the tip height, 12, 10, and 8 a.u., as well as the potential barrier above the Fermi level. For the cases of tip height 10 and 8 au, a tiny hollow appears in the potential barrier in front of the apex atom, which is enlarged with the approach of the tip to the surface. The saturation of the current is caused, since the tunneling regime changes to the ballistic regime. The value of the conductance estimated at the lower bias voltages turned out be close to unity in the unit of the quantized value, $G_0 = 2e^2/h$.

## Potential Barrier and Electric Current

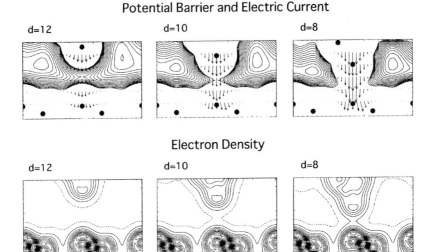

Fig. 5.11. The potential and the current distribution in the space between the Al tip and Si(111) surface. The surface baias is +2V. The bottom shows the valence electron distribution.

### 5.7 Quantum Transport through Atom Bridges

Electron transport properties through the atom-wire bridges (hereafter called atom bridges) have attracted much attention recently. They are formed by experimental techniques such as scanning tunneling microscopy (STM) and mechanically controllable break junction (MCB). Their remarkable features have been experimentally revealed [33].

All relevant quantities in the scattering problem can be resolved in terms of the so-called eigenchannels [34], which are independent paths of the current, i.e., there are no scattering events among these due to the nano-structure itself. They are mixtures of scattering states which diagonalize $t^{\dagger}t$, $t$ being the transmission amplitude matrix,

$$Ut^{+}tU = \mathrm{diag}\{T_i\}. \tag{5.6}$$

Using the Landauer formula [34,35] and this basis, the total conductance, $G$, is expressed by the sum of the individual channel transmissions, $T_i(E_F)$ (in the quantized unit, $2e^2/h$),

$$G = \frac{2e^2}{h} tr(t^{+}t)(E_F) = \frac{2e^2}{h} \sum_i T_i(E_F), \tag{5.7}$$

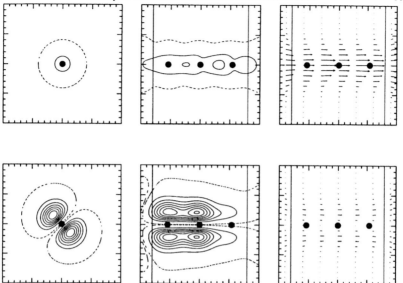

Fig. 5.12. Channel LDOS (left and center) and channel current density (right) of the Al 3-atom-bridge at the Fermi level. The top and the bottom panels correspond to the first and the second channel, respectively.

where $E_F$ is the Fermi energy.

We will investigate a model of the Al atom bridge consisted of three aluminum atoms with a bond length of 5.4 a.u. between two semi-infinite jellium electrodes. The distance between the jellium electrodes and the Al atom attached to it is chosen to be 2.6 a.u. The conductance of the wire is calculated to be 1.9 $(2e^2/h)$, which is decomposed into three eigen channels; the corresponding conductance is 0.99, 0.47 and 0.47 for the first, the second and the third channel, respectively [17,18]. The existence of the three channels is also confirmed by recent experiments [36]. Figure 5.12 shows the channel LDOS and the channel current density [37] at the Fermi level. The top and bottom panels show the first and the second channel of the wave incident from the left electrode, respectively. The third channel is the same as the second channel except for a $\pi/2$-rotation around the wire-axis. Figure 5.13 shows the channel DOS around the center atom. The thick solid (broken) line shows the first (second) channel DOS and the thin solid (broken) line show the integrated DOS. The long tail in the high energy side associated with the peak corresponding to the channel onset, can be understood as the feature of the 1D energy band. The channel transmission also supports this view, since it saturates at unity for higher energy. From this, it is understood why the lowest channel with much smaller DOS contributes a larger amount to the conductance than the higher

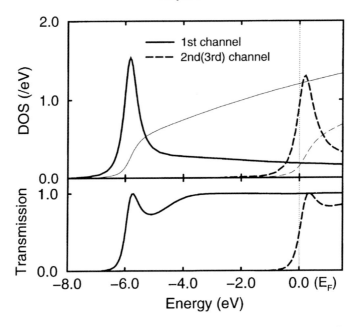

Fig. 5.13. The Channel DOS of the center atom (top) and the channel conductance (bottom) for the Al 3-atom-bridge as the function of energy.

channels at the Fermi level. It is remarkable that even for very short chain of 3 atoms already 1D band like feature appears in the local electronic state and to the electron flow. The on-set energy of the second and third channel is close to the Fermi level, which is consistent with the band-structure of the infinite atomic chain of Al.

Next we shall consider the effect of bending of the Al and Na atom wires on the channel properties [17,38]. The inset in Fig. 5.14 shows the model of the bent system. The middle atom is laterally displaced with length of $d$. The splitting of the degeneracy for the second and the third channels is induced due to the lowering of the symmetry. Figure 5.15 shows the channel DOS and the channel transmission as a function of the energy measured from the Fermi level for the displacement of $d = 2$ a.u. The channel DOS's of the first and the second channel show a significant change from the corresponding ones in the straight bridge due to the interactions of the states in the channels, in particular at the onset of the second channel. Namely the first channel DOS has an additional peak at the onset of the second channel, and the peak of the second channel DOS decreases drastically because of the interaction with the first channel.

Figure 5.16 shows the channel LDOS and the channel current for the bent

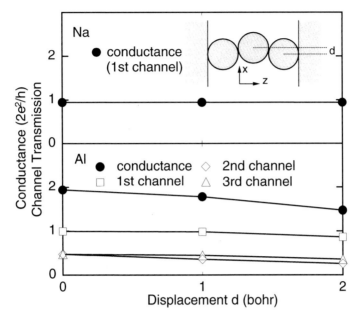

Fig. 5.14. Conductance and channel transmissions of Na (upper) and Al (lower) atom wire as a function of displacement of $d$. Upper inset: Schematic representation of the model used. Three Na or Al atoms are sandwiched between jellium electrodes.

bridge of $d = 2$ au. Figures from the top to bottom panel correspond to the first through the third channel, respectively. It is remarkable that a local loop current circulating the central atom is induced for the first and the second channel. These current components would sensitively interact with a local magnetic moment of this atom. Figure 5.14 shows the conductance and the channel transmissions for Al atom bridges for various values of $d$. With increase of the displacement $d$, each channel transmission decreases which results in the reduction of the total conductance. The transmission values of the second and the third channel are not the same except for the straight bridge. The transmissions of these channels decrease significantly with bending. In the Al wire, the Fermi level cut the onset of the second and the third channel, which induces the sensitivity of the channel conductance with the shape of the atom bridge. This situation is rather different from that of the Na atom bridges, where we confirmed the conductance dose not change from the value of unity in the quantization unit by the bending [38]. This is because the Fermi level of the alkali atom bridge located enough above the lowest channel and below the onset of the higher channels.

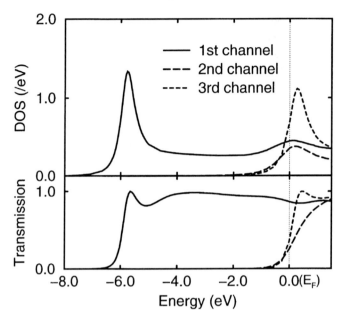

Fig. 5.15. The channel DOS of the center atom of the bending three-aluminum-atom wire (top) and the channel transmission.

## 5.8 Frictional-Force Microscopy

Atomic Force Microscopy (AFM) is a powerful experimental method for detecting force operating between a single microscopic tip apex and a surface. When lateral force is detected by the tip during its scanning of the surface, it is called frictional-force microscopy (FFM) or lateral-force microscopy (LFM) [39]. Observed two-dimensional FFM images include information of the microscopic stick-slip motion of the tip [40–42], which is closely related to the origin of the friction.

In the theoretical simulations, the system is mimicked by a single-atom tip connected to a three-dimensional spring and a monolayer graphite surface [43,44,45]. If the graphite surface is assumed to be rigid, the total potential energy consists of the harmonic elastic energy of the cantilever, and the microscopic tip-surface interaction energy. The effect of the surface deformation can be put into the effective cantilever spring constant. Minimization of the total energy is performed for each of the cantilever basal positions, which gives the equilibrium position of the tip atom and the lateral force acting on the cantilever.

As a result of the simulation, the following stick-slip model proposed by Tomlinson [22] for the friction mechanism is confirmed. First of all, with an

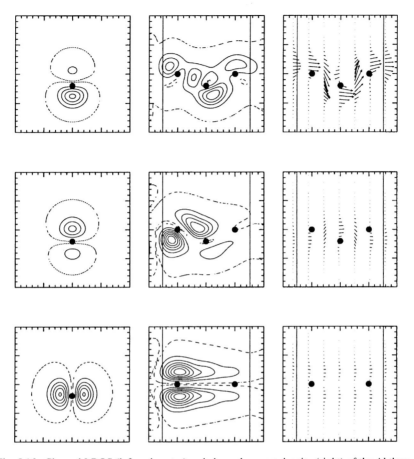

Fig. 5.16. Channel LDOS (left and center) and channel current density (right) of the Al three-atom wire at Fermi level. The top, the middle and the bottom panels are the first, the second and the third channel of the states incident from the left electrode, respectively. The left panels are contours of the channel LDOS on the plane including the center atom and parallel to the electrode surface. The center and right panels are contours of the channel LDOS and channel current density, respectively, on the plane including the three atoms (top and the middle) or the plane perpendicular to it (bottom). An area with a side of 20 bohrs is displayed.

increase in the load, the corrugation amplitude of the tip-surface interaction increases. For a low load, the total potential energy surface is nearly parabolic and only a single minimum appears, as shown in Fig. 5.17(a). However for a high load, the total potential energy surface has several quasi-stable points corresponding to the local minima, as shown in Fig. 5.17(b). Therefore for the low load, when the cantilever is scanned, the tip atom moves continuously following the single minimum of the potential energy surface. However, for

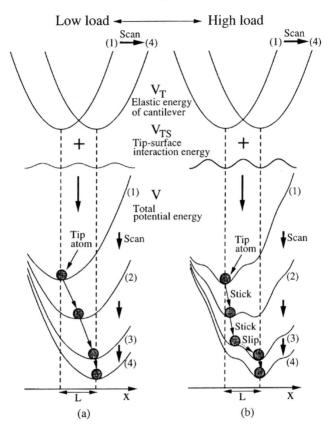

Fig. 5.17. Schematic model of Tomlinson mechanism.

the high load, this movement is restricted only for a certain period of the scan (1)–(3), and it makes a sudden discrete jump from one minimum to another deeper minimum when the energy barrier between the two minima disappears. The elastic energy stored in the cantilever then dissipates with the multiple generation of phonons.

The images obtained by the theoretical simulation and experiment are compared in Figs. 5.18(a)–(c). Here, the lateral force is represented by $F/k$, the deflection of the cantilever. $F_x/k_x$ and $F_y/k_y$, which are the quantities expressed by the gray scale images in Fig. 5.18, denote components parallel and vertical to the scan direction, respectively. Details of FFM experiments are described in Refs. 40)–42). As shown in Fig. 5.18, the simulated and experimental images of $F_x/k_x$ show an excellent agreement with each other. However, the magnitude of the load for the theory is smaller by about two orders than that by experiment. This might be due to the fact that a single-atom tip model is

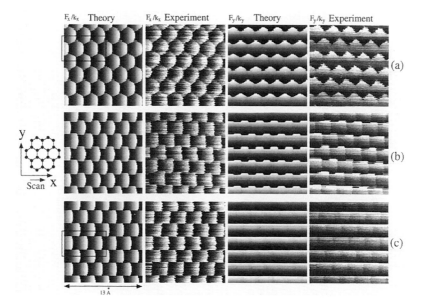

Fig. 5.18. Frictional-force images of $F_x/K_x$ and $F_y/K_y$ obtained by theoretical simulation and experiment under the constant-height mode. Average vertical load $<F_z>$ is (a) theory: 0.675 nN experiment: 44 nN, (b) theory: 1.10 nN, experiment: 122 nN, and (c) theory: 1.40 nN, experiment: 327 nN.

used in the simulation, though in a real system the background force felt by many atoms on the tip contributes to the force. In the images of the force vertical to the scan direction (right side of Fig. 5.18(a)), we find that the zigzag pattern corresponding to the C-C bonds of the graphite lattice appears for the smaller load (Fig. 5.18(a)), but with the increase of the load, it gradually vanishes to become a simple straight pattern parallel to the scan direction.

In the hatched regions B and C in Figs. 5.19(a-1) and (b-1), the tip atoms can be located stably. Such regions are therefore called hereafter sticking domains. The sticking domains are discontinuously located around the hollow sites and on the lattice sites for the low load (a-1). However when the load increases, as shown in Fig. 5.19(b-1), sticking domains with a reduced size appear only around the hollow sites. The shape of the sticking domain is mapped into the corresponding pattern for the basal position of the cantilever, and this pattern is observed as the FFM image. The sharp boundaries in the FFM image dividing the bright and the dark area are due to a discrete jump of the tip at the thick lines shown in Figs. 5.19(a-2) and (b-2). The patterns of these lines are in excellent agreement with those of the frictional force images in Figs. 5.18(a) and (c). Change in the FFM image depending on the load follows that of the shape of the stable domain of the cantilever basal position.

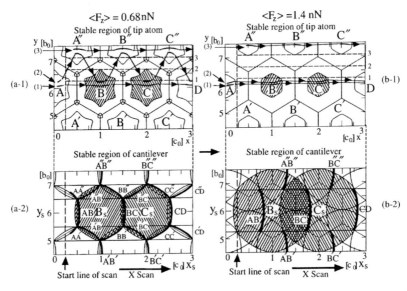

Fig. 5.19. (a-1) and (b-1) denote sticking domains of the tip atom in $(x, y)$ plane. (a-2) and (b-2) denote stable regions of the cantilever basal positions in $(xs, ys)$ plane. $<F_z>$ = (a) 0.675 nN and (b) 1.40 nN. Regions represented in (a) and (b) correspond to those within rectangular boxes in Figs. 5.18(a) and 5.18(c), respectively. bo = 1.42 A, and co = 2.46 A.


## 5.9 Tapping-Mode Atomic Force Microscopy

In this subsection, we discuss the dynamics of the cantilever motion in dynamic-mode atomic force microscopy (AFM), and reveal various important features relevant to the image formation. Though many of the features are common for the dynamic modes AFM, i.e., the non-contact mode AFM and the tapping-mode AFM, we focus in this subsection on the tapping mode AFM and topics for the non-contact mode AFM will be presented in the next subsection. For the large amplitude dynamic-mode AFM, the amplitude of the cantilever oscillation is of the order of several hundreds of Angstroms, though the tip-surface interaction zone does not extend over several Angstroms above the surface. If we mimic the cantilever part with the tip as a point mass attached to a harmonic spring (Fig. 5.20), its dynamics can be described by the trajectory in the phase space, as seen in Fig. 5.21. In the large area outside the interaction zone, the trajectory shows an elliptic form because the motion is the same as that of the harmonic oscillator. But at the turning point where the tip enters the interaction zone, the trajectory deviates from an elliptic one due to the tip-surface interaction. For example, for the hard wall model, the elliptic form is truncated by the almost vertical line parallel to the momentum axis due to the sharp repulsive potential of the surface.

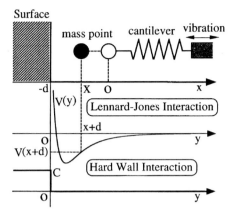

Fig. 5.20. Harmonic oscillator model used in the calculation.

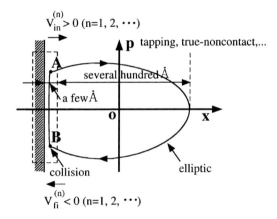

Fig. 5.21. Dynamics of the cantilever described by the trajectory in the phase space.


For the strong repulsive interaction with the surface, the traverse time $\tau$ from the lower edge B to the upper edge A of the nearly vertical section of the trajectory is the duration time of the tip-surface collision. The order of this time $\tau$ is estimated to be $10{-}10^2$ n sec, which is much shorter than the oscillation period, i.e., about $1{-}10$ $\mu$sec for most cases. However, the collision time $\tau$ is much larger than the oscillation period of each atom $10{-}10^2$ f sec. Therefore, for the dynamics of individual atoms, the tip motion in the interaction zone can be treated as quasi-static on the whole. On the other hand, the same tip motion in the contact region can be regarded as very rapid, almost instantaneous, compared with the macroscopic oscillation of the cantilever.

Thus two issues are involved in theoretical analyses of dynamic-mode AFM images: the first is to determine the collision time $\tau$, turning point, and dissipation energy in a single collision event from the atomic dynamics of the contact region. The second issue is to determine the cantilever motion with the thus obtained collision condition. From the cantilever motion, the quantity actually observed as the AFM image would be clarified. We will focus below only on the latter issue and study the cantilever dynamics with a simple harmonic spring model.

The motion of the cantilever we are considering is extremely nonlinear, and therefore various unexpected features can emerge. If the equilibrium position of the tip is rather far from the surface, it is quite seldom that the tip

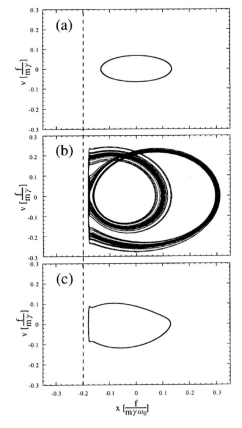

Fig. 5.22. Traces of the tip in the phase space for the Lennard-Jones interaction potential for $w = $ (a) $0.5\omega_0$, (b) $0.6\omega_0$, and (c) $0.7\omega_0$.

comes into the interaction zone. One might imagine that even after an accidental tip-surface collision event, the tip motion will decay into that of the forced harmonic oscillator, the amplitude of which is generally quite small except very close to the resonance. However, this is not necessarily the case; the tip can collide repeatedly with the surface without decaying into the small forced harmonic oscillator motion. The kinetic energy is sustained by a favorable phase-matching to the external force [8]. Some of the unexpected features of the cantilever dynamics can be seen in numerical analyses of the equation of motion. For example, Fig. 5.22 shows several examples of the tip trajectory in the phase space for the Lennard-Jones interaction potential.

For the case where the tip approaches close to the surface from a far position, the energy of the cantilever motion changes with the tip height, as shown in Fig. 5.23. If the tip is far from the surface, the tip motion shows a non-touching mode of behavior. The energy of the tip is thus almost constant as a function of the tip height. When the tip approaches the surface so close that it collides with the surface, the oscillation changes to the touching mode and the cantilever energy starts to decrease remarkably with the tip height. On the other hand, when retracting the tip from the surface in the touching regime, the oscillation energy increases. The energy increases beyond the initial value at the infinite tip height due to the dynamic touching mode. But when the tip is retracted further beyond a certain critical distance from the surface, the cantilever energy jumps discontinuously from the higher value to the initial value at the far away distance. This is nothing more than a transition from the dynamic touching mode to the non-touching mode. The spectrum thus shows a hysteresis between the approaching and the retracting processes.

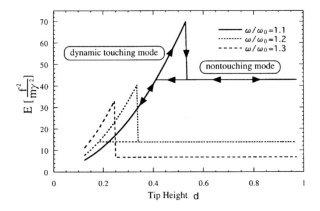

Fig. 5.23. Energy of the cantilever as a function of the tip height $d$ for various frequencies.

## 5.10 Theory of Noncontact-Mode Atomic-Force Microscopy

In this section, we present an analytical perturbation theory in order to clarify detailed relationships between the cantilever-forced vibration and the tip-surface interaction potential. As a test case, the theory will be applied to a system with a van der Waals tip-surface interaction potential, although it is applicable to any form of the potential. We find important aspects of the resonance curve. One of these gives a clue to the question "What information of the tip-surface potential can be obtained from the frequency shift?" The other important feature is the dynamic bi-stability which exerts a marked influence on observation of the resonance peak in the regime where the turning point of the tip enters the repulsive regime. This aspect has been already found in the previous subsection for the low $Q$-values, but we will discuss now this in detail, in particular for the high $Q$-value regions. The results from the analytical perturbation method agree fairly well with those obtained by a numerical integration of the equation of motion.

In the nc-AFM model, a tip with a vertical length $h$ is assumed to be located at the end of the cantilever, as shown in Fig. 5.24. If the cantilever basal position is forced to vibrate with amplitude $l$ around $z = u_0$ with a driving frequency $\omega$ i.e., $z = u_0 + l\cos(\omega t)$, the equation of motion of the cantilever becomes

$$m\frac{d^2}{dt^2} + \frac{m\omega_0}{Q}\frac{dz}{dt} + k(z - u_0 + h) + \frac{dV(z)}{dz}$$
$$= l\cos(\omega t), \tag{5.8}$$

where $m$ is the effective mass of the cantilever and tip, $Q$ is the $Q$-value of the cantilever due to the friction term, $k$ is the spring constant of the cantilever, $\omega_0 = \sqrt{k/m}$ is the resonance frequency of the free cantilever, and $V(z)$

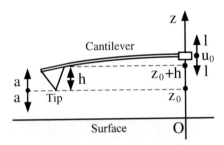

Fig. 5.24. The model of nc-AFM. A tip with a height $h$ is forced to vibrate by vibrating the cantilever basal position externally.

represents the tip-surface interaction potential, where $z$ indicates the distance between the tip apex and the rigid substrate surface. Equation (5.8) is transformed into the following form:

$$\ddot{u} + u = \varepsilon f(u, \dot{u}) + \varepsilon A \cos(\Omega t) \tag{5.9}$$

with

$$f(u, \dot{u}) = -2\mu\dot{u} - C_0 - C_1 \frac{dV(u + z_0(u_0))}{dz}, \tag{5.10}$$

$$\mu = \frac{1}{2Q\varepsilon}, \quad C_0 = \frac{z_0(u_0) - u_0 + h}{\varepsilon}, \quad C_1 = \frac{1}{\varepsilon k}, \tag{5.11}$$

$$A = \frac{l}{\varepsilon}, \quad \Omega = \frac{\omega}{\omega_0}. \tag{5.12}$$

Solution of the differential equation (5.9) can be analyzed using the perturbation theory of the Krilov-Bogoliubov-Mitropolsky method [46]. If the solution $u(\tau)$ of Eq. (5.9) is assumed to be $u = a\cos(\Omega\tau - \Phi) + O(\varepsilon)$ with $\dot{a} = O(\varepsilon)$ and $\dot{\Phi} = O(\varepsilon)$, and $\dot{a}$ and $\dot{\Phi}$ approximated up to the first order $\varepsilon$ can be obtained from the solution of the following equations:

$$\dot{a} = -\varepsilon\mu a + \frac{\varepsilon}{2} A \sin\Phi \equiv F(a, \Phi), \tag{5.13}$$

$$\dot{\Phi} = \Omega - 1 + \varepsilon g(a, u_0) + \frac{\varepsilon}{2a} A \cos\Phi \equiv G(a, \Phi). \tag{5.14}$$

In the above $g(a, u_0)$ is given by

$$g(a, u_0) = -\frac{C_1}{2\pi a} \int_0^{2\pi} \frac{dV(z_0(u_0) + a\cos\psi)}{dz} \cos\psi \, d\psi. \tag{5.15}$$

For a certain class of solutions, the motion of the cantilever approaches a stable simple sinusoidal oscillation with fixed values of $a$ and $\Phi$. It can be seen that the fixed point is at the critical points of Eqs. (5.13) and (5.14), which are given by $F(a, \Phi) = G(a, \Phi) = 0$. By deleting $\Phi$ at the critical point, the following frequency-amplitude relationship:

$$a = \frac{Ql}{\sqrt{1 + 4Q^2 \left(\Omega - 1 + \varepsilon g(a, u_0)\right)^2}} \tag{5.16}$$

is obtained. This gives an analytical expression of the resonance curve shown in Fig. 5.25. From Eq. (5.15) the maximum amplitude is realized at the frequency,

$$\Omega = 1 - \varepsilon g(a, u_0). \tag{5.17}$$

This therefore gives the following frequency shift of the peak value of the resonance curve:

$$\Delta v = \frac{\omega_0}{2\pi} \left(\Omega(a_{max}) - 1\right) = -\varepsilon v_0 g(a_{max}, u_0) \tag{5.18}$$

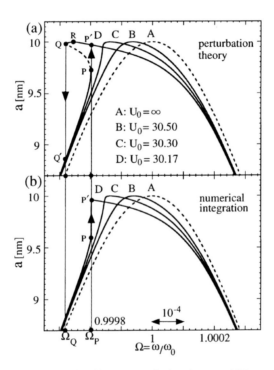

Fig. 5.25. Resonance curves obtained by (a) perturbation theory and (b) numerical integration of equation of motion, for $u_0$ = A: $\infty$, B: 30.5, C: 30.3 and D: 30.17 nm. R is a resonance peak of the curve D.

with $v_0 = \omega_0/2\pi$. The most important conclusion of the present theory is that the frequency shift is in proportion to the quantity $g(a, u_0)$ given by Eq. (5.15), which is a weighted average of the tip-surface interaction force around the turning point of the tip.

Figure 5.25 shows a comparison between the resonance curves for different tip heights ($u_0$), obtained by (a) the perturbation theory mentioned above and (b) numerical integration of Eq. (5.9). The resonance curve A for $u_0 = \infty$ is shown for the reference. As the cantilever approaches the surface, the resonance peak shifts towards a lower frequency, i.e., from A to D. This behavior of the resonance curve, in particular, the peak shift in Fig. 5.25(a) is excellently reproduced by the perturbation theory as seen in Fig. 5.25(b). With the approach of the turning point towards the surface, the attractive force tip feels becomes more significant, and this causes the shift of the resonance frequency to the lower side. However, when the attractive force becomes very steep and strong, the tip dynamics shows a bistable nature as is seen from the analytical theory. In this regime the resonance curve shows a hysteresis, i.e., the motion of the tip follows the different mode when the frequency is changed from the higher side or from the lower side. The curve of the frequency shift as a function of the tip height, which is called as the force spectrum, is obtained by this relation (5.17), (5.18). Figure 5.26 shows an example of the force spectrum curve for case of the tip-surface interaction potential of a Lennard-Jones type.

Since ncAFM makes images by plotting the frequency shift as a function of the lateral tip position, we can make a theoretical simulation of the ncAFM images with the relation Eqs. (5.15) and (5.18). The necessary information for this is the tip-surface interaction force in a wide area covering the long tail of

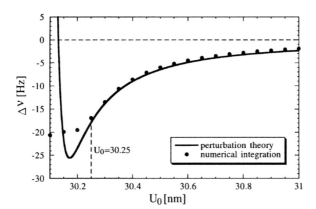

Fig. 5.26. Frequency shift $\Delta v$ as a function of the cantilever height $u_0$. Solid curve is obtained by perturbation theory and dots by numerical integration.

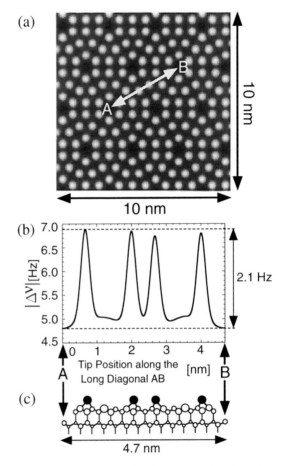

Fig. 5.27. (a) A simulated nc-AFM image and (b) its cross section along the long diagonal AB of the unit cell of the Si(111) 7 × 7 surface (c) the side view of the DAS model.

the attractive force region. It is rather difficult to calculate this from the first-principles, which is only reliable around the range of the chemical interaction, i.e., up to about 3 Å from the surface. Therefore some conventional methods should be utilized to extrapolate the force from the first-principles results near the surface to the region far distant from the surface.

We have used this idea to carry out a theoretical simulation of the ncAFM images of the Si(111) surface. The force between the surface and a Si cluster tip calculated by Perez and others [47] is mimicked by the sum of a pairwise Morse potential over atoms on the tip and the surface. In the tip model, we attached the apex cluster to the base part of the tip with a suitable cone shape with a spherical top. The base part contributes predominantly to the strong van der Waals type force, which is not sensitive to the lateral tip position. The sensitivity of the frequency shift to the short range part of the force explains the atomic resolution of the ncAFM. Figure 5.27 shows simulated images of the Si(111) $7 \times 7$ DAS model for various average tip heights. The corrugation of the image becomes more enhanced with a reduction in the tip height. The simulated image compares fairly well with the experimental ones, and demonstrates the validity of the simulation method [48].

## 5.11 Summary

In the present article, theoretical bases for the STM, FFM, tapping mode AFM and ncAFM are explained at the hand of several case studies. For the case of STM, behavior at the atomic point contact and quantum transport through atom bridges is also discussed.

Why STM images are so sensitive to surface electronic states is explained and how atomic resolution is attained for the usual cases. This is attributed to the extreme concentration of the tunnel current on a single apex atom on the mini-tip. Depending on the tip shape, this cannot always be realised. In such cases, abnormal images which are sensitive to detailed tip geometries and kind of tip atom result. Therefore we can say that except some delicate quantitative aspects, the mechanism of the atomic resolution of the STM images seem to be understood well. Although we have not discussed this here, scanning tunneling spectroscopy (STS) often reflects the electronic states of the tip apex more strongly. Negative differential tunnel resistance, for example, can be only understood by taking the tip effect into account [11,12].

There are many examples of interesting behavior in surface electronic states which can only be observed by STM. In this article, we mentioned as a typical case the spatial character of the surface states of the Si(111) $\sqrt{3} \times \sqrt{3}$-Ag, and its similarity to the surface state of Si(111) $\sqrt{3} \times \sqrt{3}$-Sb. Recently STM observation of the interference patterns of the surface states has revealed remarkable properties extending over several nano-meters. Theoretical analyses for these patterns would afford important information for the surface science of nano-structures. Furthermore, theoretical aspects of BEEM(ballistic electron emission microscopy) raise interesting problems. These topics cannot be treated in the present article, because of the limitation of the volume.

We discussed aspects of quantum transport through quantum channels formed between the tip and the surface at the hand of examples of point contact and atom bridges. The first-principles transfer matrix method provides an eigen channel analysis approach, which is effective for studying microscopic transport and its relationship to the atomic or molecular orbitals in the bridge region. How the concept of conductance quantization should be modified in atomic scale systems, and how the quantum tunneling features continuously change to those of ballistic flows through channels is elucidated.

Finally, we provided some theoretical bases for various modes of AFMs, i.e., FFM, tapping mode AFM and ncAFM. Compared to STM, the present understanding of AFM is not satisfactory. One reason for this is that there is no established first-principles theory for calculating the interaction force over a wide spatial range from the chemical to the van der Waals regime. This is a pity, since the enormous potentiality of the ncAFM, including atomic resolution, has already been demonstrated experimentally. However, as we have shown in this article, experimental observables such as the frequency shift forming ncAFM image or the force spectra can be derived from the microscopic force distribution in the 3-dimensional space outside the surface. If the force distribution is acutally calculated, it would be used as the input data for the simulation procedure proposed here. In the present article we conventionally extended the force in the chemical range calculated by the density functional theory to a wider region by fitting it to the phenomenological potential. This is a rather conventional approach, but it provides a practical method for ncAFM image simulation and force spectra. Further case studies which include detailed comparison of experimental results are necessary.

## [References]

1)  G. Binning and H. Rohrer, *Review of Modern Phys.* **59** (1987) 615;
    P. Hansma and J. Tersoff, *J. Appl. Phys.* **61** (1987) 615;
    *Scanning Tunneling Microscopy I, II, III*, ed. by H. J. Guntheroht and R. Wiesendanger, Springer (1992)
2)  M. Tsukada, K. Kobayashi, N. Isshiki and H. Kageshima, *Surface Sci. Reports* **13** (1991) 265;
    M. Tsukada, K. Kobayashi, N. Isshiki, S. Watanabe, H. Kageshima and T. Schimizu, in *Scanning Tunneling Microscopy III*, Springer (1993)
3)  G. Binnig, C. F. Quate and Ch. Gerber, *Phys. Rev. Lett.* **56** (1986) 930
4)  S. Kitamura and M. Iwatsuki, *Jpn. J. Appl. Phys.* **34** (1995) L145
5)  R. Erlandsson, L. Olsson and P. Martensson, *Phys. Rev. B* **54** (1996) R8309
6)  T. Uchihashi, M. Ohta, Y. Sugawara, Y. Yanase, T. Sigematsu, M. Suzuki and S. Morita, *J. Vac. Sci. & Technol. B* **15** (1997) 1543
7)  N. Sasaki, Dr. Thesis, Univ. of Tokyo (1996)
8)  M. Tsukada, N. Sasaki, R. Tamura, N. Sato and K. Abe, *Surf. Sci.* **401/3** (1998) 355
9)  J. Bardeen, *Phys. Rev. Lett.* **15** (1961) 57
10) S. Watanabe, M. Aono and M. Tsukada, *Ultramicroscopy* **42–44** (1992) 105; *Appl. Surface Sci.* **60/61** (1992) 437
11) P. Bedrossian, D. M. Chen, K. Mortensen and J. A. Golovchenko, *Nature* **342** (1989) 258
12) M. Tsukada, K. Kobayashi and N. Isshiki, *Surf. Sci.* **242** (1991) 12; *J. Vac. Sci. Technol. B* **9** (1991) 492
13) N. Isshiki, K. Kobayashi and M. Tsukada, *Surf. Sci.* **238** (1990) L439
14) K. Hirose and M. Tsukada, *Phys. Rev. Lett.* **73** (1994) 150
15) K. Hirose and M. Tsukada, *Phys. Rev. B* **51** (1995) 5278
16) N. Kobayashi, K. Hirose and M. Tsukada, *Jpn. J. Appl. Phys.* **35** (1996) 3710
17) M. Tsukada, N. Kobayashi and M. Brandbyge, *Prog. in Surface Sci.* **59** (1998) 245
18) N. Kobayashi, M. Brandbyge and M. Tsukada, *Jpn. J. Appl. Phys.* **38** (1999) 336–338
19) N. Sasaki and M. Tsukada, *Jpn. J. Appl. Phys.* **38** (1999) 192–194
20) N. Sasaki and M. Tsukada, *Appl. Surf. Sci.* **137/3-4** (1999) (in press)

21)  N. Sasaki, M. Tsukada, S. Fujisawa, Y. Sugawara, S. Morita and K. Kobayashi, *J. Vac. Sci. Technol. B* **15** (1997) 1479

22)  G. A. Tomlinson, *Philos. Mag.* **7** (1929) 905

23)  J. Tersoff and D. R. Hamann, *Phys. Rev. Lett.* **50** (1983) 1998; *Phys. Rev. B* **31** (1985) 805

24)  E. J. van Loenen, J. E. Demuth, R. M. Tromp and R. J. Hamers, *Phys. Rev. Lett.* **58** (1987) 373

25)  T. Takahashi, S. Nakatani, N. Okamoto, T. Ichikawa and S. Kikuta, *Jpn. J. Appl. Phys.* **27** (1983) L753

26)  M. Katayama, R. S. Williams, M. Kato, E. Nomura and M. Aono, *Phys. Rev. Lett.* **66** (1991) 2762

27)  S. Watanabe, M. Aono and M. Tsukada, *Phys. Rev. B* **44** (1991) 8330

28)  T. Yokotsuka, S. Kono, S. Suzuki and T. Sagawa, *Surface Sci.*, **127** (1983) 35

29)  S. Watanabe, M. Aono and M. Tsukada, *Surface Sci.* **287/288** (1993) 1036

30)  K. Kobayashi and M. Tsukada, *J. Vac. Sci. Technol. A* **8** (1990) 170

31)  S. Watanabe, M. Aono and M. Tsukada, *Jpn. J. Appl. Phys.* **32** (1993) 2911

32)  T. Uchiyama and M. Tsukada, *Phys. Rev. B* **53** (1996) 7917

33)  in *Nanowires*, ed. by P. A. Serena and N. Garcia, Kluwer Academic Pub., Dordrech (1997)

34)  M. Buttiker, *IBM J. Res. Dev.* **32** (1988) 63

35)  M. Buttiker, Y. Imry, R. Landauer and S. Pinhas, *Phys. Rev. B* **31** (1985) 6207

36)  N. Kobayashi and M. Tsukada, *Jpn. J. Appl. Phys.* (in press)

37)  M. Tsukada, N. Kobayashi and M. Brandbyge, *Prog. Surface Sci.*, **59** (1998) 245

38)  N. Kobayashi, M. Brandbyge and M. Tsukada, *Surface Sci.* (in press)

39)  C. M. Mate, G. M. McClelland, R. Erlandsson and S. Chiang, *Phys. Rev. Lett.* **59** (1987) 1942

40)  S. Fujisawa, E. Kishi, Y. Sugawara and S. Morita, *Tribo. Lett.* **1** (1995) 121; *Phys. Rev. B* **52** (1995) 5302

41)  S. Fujisawa, E. Kishi, Y. Sugawara and S. Morita, *Phys. Rev. B* **58** (1998) 4909

42)  S. Morita, S. Fujisawa and Y. Sugawara, *Surf. Sci. Rep.* **23** (1996) 1

43)  N. Sasaki, K. Kobayashi and M. Tsukada, *Phys. Rev. B* **54** (1996) 2138; *Surf. Sci.* **92** (1996) 357–358

44)  N. Sasaki, K. Kobayashi and M. Tsukada, *Jpn. J. Appl. Phys.* **35** (1996) 3700; *NATO-ASI Series, Micro/Nanotribology and Its Applications*, ed. by B. Bhushan, p. 355, Kluwer Academic Publishers, Dordrecht (1997)

45)  N. Sasaki, M. Tsukada, S. Fujisawa, Y. Sugawara, S. Morita and K. Kobayashi, *Phys. Rev. B* **57** (1998) 3785

46)  N. Bogoliubov and Y. A. Mitropolsky, in *Asmptotic Methods in the Theory of Nonlinear Oscillations*, Gordan and Beach, New York (1961)

47)  R. Perez, M. C. Payne, I. Stich and K. Terakura, *Phys. Rev. Lett.* **78** (1997) 678

48)  N. Sasaki and M. Tsukada, *Appl. Surf. Sci.*, **137** (1999) 399

*Chapter 6*

# Tunneling-Electron Luminescence Microscopy for Multifunctional and Real-Space Characterization of Semiconductor Nanostructures

## 6.1 Introduction

Mesoscopic optical and electronic properties that result from quantum effects appear in semiconductor nanometer-sized structures, or nanostructures, and these properties are very different from those in macroscopic structures. [1] Progress in crystal growth and micro-process technology has enabled the atomically controlled fabrication of artificial semiconductor nanostructures and devices. These have been actively studied for the purpose of achieving lower power consumption, faster operation, and advanced functions through mesoscopic effects. To improve the performance of the nanostructures and devices, their individual optical and electronic properties in local regions must be characterized, even where many nanostructures with a high density are integrated. Furthermore, it is important to characterize both buried structures and semiconductor surfaces with single-digit nanometer-level (<10 nm) spatial resolution. The intensities and emission spectra of luminescence due to the radiative recombination of electron-hole pairs confined in the nanostructures sensitively reflect atomic-scale variations in structures as well as the quality of materials in nanometer-sized regions. Therefore, luminescence measurements in local regions, or luminescence microscopy, are effective for characterizing nanostructures and have been widely used for this purpose. In luminescence microscopy, photon-induced luminescence (photoluminescence: PL) [2] and high-energy electron induced luminescence (cathodoluminescence: CL) [3] have been widely used. However, these conventional methods cannot easily produce nanometer-level lateral spatial resolution. Thus, experimental results include statistical fluctuations that result from the convolution of the number of nanostructures, so individual nanostructures are difficult to evaluate. If the nanostructure density is low enough for there to be only one structure in an excitation region, this nanostructure can be characterized. This method, however, is not generally applied. A new concept is therefore needed for luminescence microscopy with nanometer-level spatial resolution that will enable the characterization of the optical and electronic properties of individual nanostructures even if they are fabricated with a high density in local regions.

Tunneling-electron luminescence (TL) microscopy using tunneling electrons emitted from an STM tip for the luminescence excitation has recently been used for real-space characterization of the optical and electronic properties of local regions with nanometer-level spatial resolution. [4–7] In this chapter, we explain the limitations of conventional luminescence microscopy, then describe the setup and performance of a TL microscope used for real-space characterization of the optical and electronic properties with single-digit nanometer-level spatial resolution.

## 6.2 Limitations of Conventional Luminescence Microscopy

In this section, we will discuss conventional luminescence microscopy that uses PL or CL mainly in terms of the factors that limit the spatial resolution and the excitation-energy tunability when this form of microscopy is used for real-space characterization of semiconductor nanostructures.

### 6.2.1 Lower limits of spatial resolution

In PL microscopy, electron-hole pairs are generated by light excitation. The light is usually a laser light from a source far from the sample and it is focused by lenses or mirrors. In this case, the width of the excitation region cannot be less than about half of the wavelength of the focused excitation light (submicron level) because of the diffraction limit.

In CL microscopy, on the other hand, high-energy electrons (cathode rays) emitted from a scanning electron microscope (SEM) are used for excitation. The SEM has a very high spatial resolution for the characterization of surface morphologies because of its fine (nanometer level) beam diameter and easy two-dimensional scanning operation. Despite the nanometer-level beam diameter, though, the cathode rays generate electron-hole pairs in large volumes because the injection electrons, whose energies are as high as the kilo-electron volt order, scatter widely in the sample. [8]

Moreover, the carriers diffuse away from the generation volume before radiative recombination occurs, so the size of the photon emission source expands to the sum of the carrier generation and carrier diffusion in size. In PL and CL microscopy, a lens or mirror collects luminescence, and these collect light from the entire source because of the long distance between the source and the collection optics. Their spatial resolution cannot therefore be smaller than the source size, which is at least a submicron-meter level. The carrier diffusion length can be reduced by changing the carrier movement from a straight line to the cyclotron path by applying vertical magnetic fields to the samples. [9] However, the applicable magnetic field intensities have an upper limit so that the spatial resolution, even CL microscopy, is still limited to the submicron level despite the fine beam that would otherwise enable nanometer-level spatial resolution. Thus, conventional PL and CL methods are still inadequate for real-space characterization with nanometer-level spatial resolution because of the diffraction limit of the light, the wide scattering of cathode rays, and the long carrier diffusion.

### 6.2.2 Excitation-energy tunability

Radiative recombination that emits luminescence usually occurs between the lowest energy level in direct-gap semiconductors. However, much higher energy states, however, exist in materials and nanostructures. Moreover, hot electron phenomena appear in much higher energy ranges. It is also important for the characterization that these energy states be investigated. This can be done by measuring the dependence of the luminescence intensity on the excitation energy. Almost all energy states of materials exist in the energy range up to 10 eV, so excitation methods should allow tuning or scanning the excitation energies from the band gap up to 10 eV. However, the PL and CL methods are not capable of this. For example, although the highly monochromatic laser light used in the PL methods can be finely tuned to the desired energy states, the energy range available from an individual laser medium can be as narrow as several milli-electron volts. It is also difficult to produce tunable light with an energy above several electron volts.

The cathode rays from SEM have high injection energies on the kilo-electron volt level, so CL is difficult to use for directly investigating energy states except for a radiative-recombination state around the band gap.

A new form of luminescence microscopy is therefore needed for real-space characterization of nanometer-scale regions over a wide energy range.

Luminescence microscopy can be categorized according to the excitation and luminescence collection methods used (Table 6.1). A main factor restricting

Table 6.1. Luminescence microscopy categorized by the excitation method.

| | Excitation | | | |
|---|---|---|---|---|
| | Near field | | Far field | |
| | electron (tunneling) | photon (evanescent) | electron | photon |
| Luminescence detection | | | | |
| N. F. | TL (tip/tip) | NOM (tip/tip) | CL (–/tip) | NOM (lens/tip) |
| F. F. | TL (tip/lens) | NOM (tip/tip) | CL | PL |

NOM: near-field optical microscopy.

improvement of the spatial resolution of the luminescence microscopy to the nanometer level is use of far-field effects in both the excitation and collection of luminescence. Near-field effects should therefore be used to enable multifunctional characterization with nanometer-level spatial resolution. The optical near-field effect is an evanescent wave [10] and the electronic near-field effect is an electron tunneling. [11]

## 6.3 Tunneling Electron Luminescence (TL) Microscopy

Tunneling electrons emitted from the tip of a scanning tunneling microscope (STM) are a superior source of excitation in luminescence microscopy since it overcomes the limits of conventional luminescence microscopy.

### (a) Excitation method: Electronic near-field (tunneling electron)

When a conductive tip is close to a sample, within the electron-wavelength order in distance, electrons can transport through a potential barrier between the tip and the sample even though their energies are lower than the potential. The tunneling electrons can induce photon emission through certain processes that depend on the materials involved. At metal surfaces, photon emission is due to surface plasmons. In direct-gap semiconductors, it is due to a radiative recombination of the tunneling electrons with majority carriers and secondarily generated electron-hole pairs; this is called tunneling-electron luminescence (TL). In this chapter, we limit our discussion to the TL from semiconductors and consider the application of the TL microscope to the characterization of semiconductor nanostructures. Use of the tunneling electrons provides a suitable excitation method for luminescence microscopy to characterize the optical and electronic properties in nanometer-level local regions of materials because of the following advantages:

(1) Nanometer-size generation volumes can be formed because the beam diameter is typically at the angstrom level and the scattering length is less than several tens of nanometers.

(2) Injection energies can be continuously tuned over a wide energy range from 0 to over 10 eV; the energy states of most materials and nanostructures lie within this range.

(3) Changing positive and negative polarities in the tip biases can produce both electron injection and hole injection (the pulling out of electrons).

(4) High excitation power can be effectively injected into nanometer-size local regions.

### (b) Collection method: optical near-field (tip collection)

Conventional "lens collection" TL microscopes combine a metal for current injection and a lens or mirror for luminescence collection. Since the tip is aligned perpendicular to the sample, the lens is angled off to the side and is positioned far from the photoemission source.

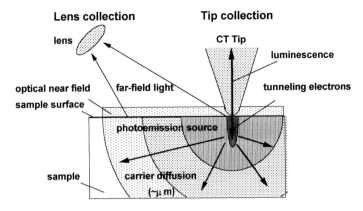

Fig. 6.1.  Improved spatial resolution for TL microscopy using the tip-collection method.

Figure 6.1 shows the behavior of the carriers in semiconductors during TL measurements. If the photoemission source is formed in a local region around an electron injection point just under the tip apex, the spatial resolution of the TL microscopy could reach the nanometer level. Unfortunately, the injected tunneling electrons or the secondary electrons and holes generated by the tunneling electrons usually diffuse away from the generation volume before their recombination. The photoemission source then expands so that its size equals the carrier diffusion length even if the generation volume is small. The lens collects luminescence from over the entire volume of the expanded photoemission source, so the spatial resolution is thus restricted by the size of the photoemission source.

The TL emitted from a very small local region is usually so weak, though, that high collection yields are required for precise TL spectroscopy and to obtain high-quality TL images.

Most of the luminescence emitted from the generation volume just under the tip apex is roughly orthogonal to the sample surface and directed toward the tip apex, particularly in high-refraction index materials such as GaAs. The lens angled off to the side and positioned far from the photoemission source has the small solid angle for luminescence collection, so the lens can collect only a small amount of the luminescence. Thus, collection yield is low and it is difficult to collect sufficient luminescence intensities for precise measurements.

These problems can be avoided by using "tip collection" where luminescence is collected at the tip apex, while tunneling electrons are simultaneously injected. [12] The tip apex is perpendicular to and very close to the sample surface (within several nanometers), so the most intense photoemission containing near optical fields can be picked up and a much higher luminescence collection yield is achieved. As the photoemission source narrows, the collection yields rise. Tip collection can pick up the luminescence from a small part of the photoemission source just under a tip apex sharpened to the nanometer level. Tip collection thus enables high spatial resolution despite the carrier diffusion. Thus, tip-collection TL microscopy can be used to characterize both the optical and electronic properties in nanometer-sized regions with high spatial resolution and a high collection yield.

## 6.4 TL Microscopy Using Tip Collection

An essential component in tip collection is the conductive transparent (CT) tip, which injects tunneling electrons and simultaneously collects luminescence at the tip apex. [13] For detailed characterization of semiconductors, luminescence measurements must be done at low temperatures to eliminate effects of thermal fluctuation such as peak-energy shifts and linewidth broadening. Semiconductor characterization also should be done in an ultrahigh vacuum (UHV) to avoid pinning effects due to surface contamination and oxidation. In this section, we describe the advantageous structures and characteristics of the CT tips and of the TL microscope with the CT tip when they are used at low temperatures in an ultrahigh vacuum.

### 6.4.1 Conductive transparent (CT) tip

Figure 6.2 illustrates the essential features of the CT tip which consists of a multimode optical fiber tapered to a point, a transparent electrode, and a thick coaxial metal plate. Precise spectroscopy requires high and flat transmittance

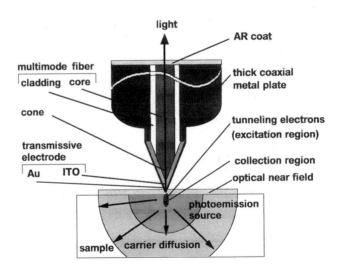

Fig. 6.2. Essential features of the CT tip.

over a wide spectral range, so we used a fiber with a core diameter of about 100 $\mu$m. The fiber was made of silica, which has a low outgassing rate in an ultrahigh vacuum and is stable over a wide temperature range from 430 K (for baking) to 10 K (for measurements). The back end of the fiber was ground optically flat and coated with antireflection layers.

The front end of the fiber was tapered to a point by stretching and chemical etching. [14] Figure 6.3 shows SEM micrographs of the taper at low and high magnifications. The taper has a flame-like shape (Fig. 6.3(a)), which is more effective than a sharp-cone shape for low-loss transmission and reduces the influence of optical interference between the sample surface and tip. At the apex, where the luminescence is picked up, the taper forms a cone (Fig. 6.3(b)). The cone angle is typically about 60° and the radius of the apex is less than 100 nm. As the apex becomes more acute, the spatial resolution for picking up luminescence increases. As the apex becomes more obtuse, the collection yield increases. The dimensions of the apex were thus optimized in consideration of the measurement conditions.

Since the silica fiber is not an electrical conductor, the taper was coated with a transparent electrode that supplied current to the apex. An indium-tin-oxide (ITO) film was used for the electrode because of its high conductivity and high optical transmittance. However, ITO films often consist of a loose pile-up of small grains several nanometers in diameter, and these degrade the STM spatial resolution. An ultrathin gold layer was therefore deposited over the ITO films to bind the grains and enable stable STM operation. The CT tip could then provide atomic-level spatial resolution during STM operation, the same level as provided by metal tips. The transmittance of the electrodes is over 80% for an ITO single layer and over 75% for the Au-on-ITO double-

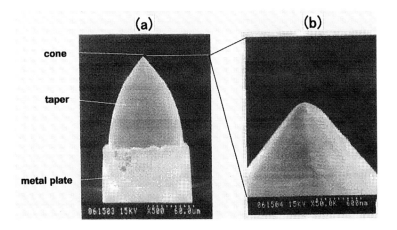

Fig. 6.3. SEM micrographs of the CT tip taper: (a) Overview, and (b) Apex of the tip.

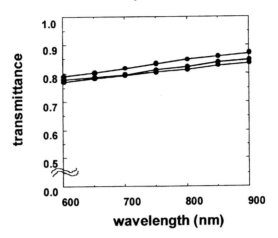

Fig. 6.4. Transmittance of a transparent electrode formed on the CT tip.

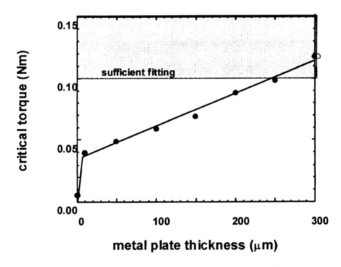

Fig. 6.5. Critical torque for screw fitting of the CT tip as a function of the coaxial metal plate thickness.

layer at wavelengths between 600 and 900 nm (Fig. 6.4). The sheet resistivity of the electrode is about 100 $\Omega/cm^2$, which is low enough for STM. If necessary, many kinds of wide-gap materials and ultra-thin metals can be applied to the transparent electrode in the same way as the ITO. This enables us to vary properties of the CT tip such as its work functions.

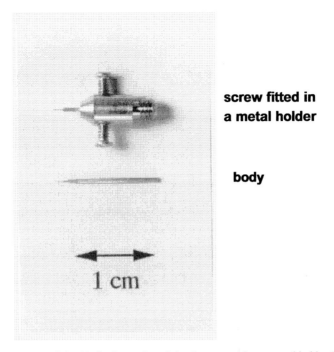

**screw fitted in
a metal holder**

**body**

1 cm

Fig. 6.6. Profiles of the CT tip (bottom), and the tip screwed into a metal holder (top).

For practical use, tip mounting must ensure stable operation, a reliable grip, and easy changing in a UHV chamber at both room temperature and low temperatures. Screw fitting meets these requirements. Since silica fibers are too weak to permit screw fitting, the shaft of the fiber was reinforced with thick coaxial metal plates. [15] Figure 6.5 shows the dependence of the critical torque on the metal plate thickness. [16] The thickness of 300 $\mu$m allows secure screw fitting and reliable electric contact to a metal holder. The thickness around the top of the tip that sticks out of the holder is less than 10 $\mu$m in order to prevent optical interference between the sample surface and the end of the metal plate.

Figure 6.6 shows the profile of the CT tip itself and of the tip installed in a metal holder with the screw fitting. The CT tip is roughly 12 mm long and its outer diameter, except for the top 2 mm of the tip, is 600 $\mu$m. These dimensions were modeled on the profile of the metal STM tips used for prealignment of the system. The CT tip and metal holder can be transferred as a unit in the UHV chamber and easily mounted on the front end of the piezo-tube scanner.

### 6.4.2 System set-up

Here, we describe the system set-up and the features of a TL microscope equipped with the CT tip and used for semiconductor characterization in a UHV chamber at low temperatures. This will demonstrate the effectiveness of the CT tip. [17]

### (a) Optical transmission from the CT tip to a monochromator

Figure 6.7 shows a schematic drawing of a low-temperature TL microscope using the CT tip in a cryostat and the fiber optics between the CT tip and the monochromator.

The CT tip and the metal holder are mounted as a unit on the top of a piezo-tube scanner. A sample is mounted on a piezo-driven sample stage with positioning accuracy of about 1 nm. The measurement temperatures can be set anywhere between room temperature and 10 K. With the conventional lens collection method, the thermal radiation through a wide window fused for

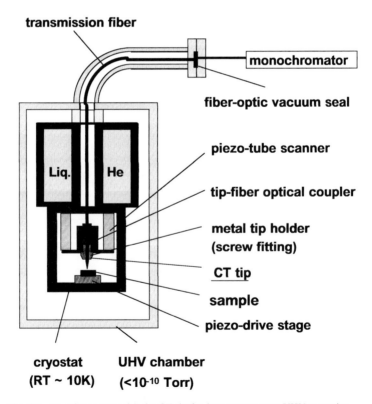

Fig. 6.7. TL microscope with the CT tip for low-temperature UHV operation.

luminescence collection in a cryostat is a significant consideration with respect to maintaining stable conditions at low temperatures. However, tip collection, however, enables luminescence to be collected through an optical fiber connected to the CT tip, so there is no need for a window on the cryostat. Thermal conduction through the optical fiber is negligible because the long thin fiber has a low thermal conductivity. Tip collection thus enables sufficient thermal isolation for stable measurement at low temperatures. The closed cryostat and the electron excitation method eliminate any background light, so a high signal-to-noise ratio is attained even when picking up weak TL. All these components are set in a UHV chamber evacuated to below $10^{-10}$ Torr to prevent surface contamination and oxidation of either the sample or the tip. Luminescence collected by the CT tip is transmitted to a monochromator by an optical transmission line passing out of the cryostat, through the UHV chamber, and into the monochromator. The transmission line mainly consists of an optical fiber in the UHV chamber and one in the outside air, with the two fibers connected by a fiber-optic vacuum seal. Effective optical coupling between the CT tip and the UHV optical fiber is automatically reproduced for each tip replacement. The fiber-optic vacuum seal consists of a small sapphire plate placed between the two fibers that are joined by FC-type connectors (Fig. 6.8). This vacuum seal provides sufficiently high transmittance of over 77% (for no antireflection coating) and sufficiently low gas leakage that is as low as that of metal conflat flanges. The monochromator and the vacuum chamber are each mounted on a vibration isolator and are optically connected with a long soft fiber so that any vibration noise is effectively isolated and the system layout can be flexibly arranged.

*(b) Optical detection and processing unit*

There is a serious problem, however, in the optical detection during TL measurements: the fluctuation in TL intensity due to unavoidable variation in the tunnel currents. Thus, to obtain precise TL measurements, we must

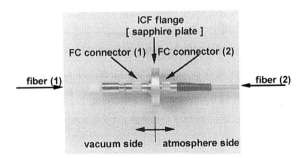

Fig. 6.8. Fiber-optic vacuum seal.

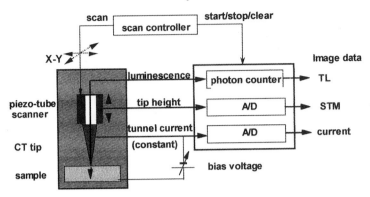

Fig. 6.9. Functional block diagram of the TL microscope. The TL intensity, tunnel current, and tip height (the topographic STM signal) are simultaneously measured at every pixel.

minimize the influence of the photoemission fluctuation. The main optical detection and processing functions needed for this purpose are whole spectral detection for spectroscopy and tip-scan-synchronized detection for microscopy. A photon-counting method is used for optical detection since TL intensities are usually very weak. Two types of cooled photon detectors are attached to the monochromator. A charge-coupled device (CCD) array is used for TL spectroscopy in combination with an optical multichannel analyzer. A photomultiplier tube (PMT) is used together with a tip-synchronized photon counter for TL microscopy.

For TL spectroscopy, the whole spectrum obtained with the CCD array is needed because emission spectra taken with a wavelength-scanning spectrometer are distorted by the TL intensity fluctuation.

For obtaining TL images, there are two ways to reduce the image distortion. Figure 6.9 is a block diagram of the TL imaging system. The TL intensity, tunnel current, and tip height (the STM topographic signal) are simultaneously measured at each pixel. The tip height is used for normal STM imaging of the sample surface. Although the system operates in a constant-current mode, tunnel currents fluctuate due to the necessary response times in the STM feedback operation. The TL intensities also fluctuate in proportion to the tunnel currents. The measured TL intensity is normalized with the tunnel current at every pixel to avoid this fluctuation. Output pulses from the PMT are counted with the digital photon counter at every pixel. This counter prevents crosstalk between adjoining pixels that would distort the TL images. For precise TL imaging, the luminescence signals at every pixel must be separated from those of adjoining pixels. The photon counter can be easily synchronized with the tip scan by using timing pulses applied at every pixel.

If an analog-output counter that operates independently of the tip scanning

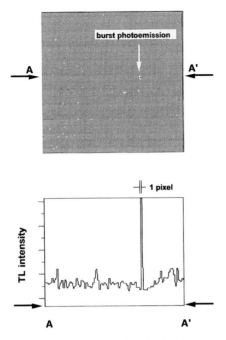

Fig. 6.10. TL image taken on GaI by the STM synchronized photon counter (top). The intensity profile along the horizontal scan line is marked A and A' and includes the burst photoemission pixel indicated by the white arrow.

is used—for example, a rate meter—distortions in the TL image will remain in many subsequent pixels even if the burst photoemission occurred at only one pixel. Figure 6.10 shows a TL image obtained with the STM-synchronized photon counter on a cleaved GaAs substrate. The bright pixel (indicated by the white arrow) is where burst photoemission occurred. The TL intensity profile is obtained along a horizontal scan line marked A and A' that includes the bright pixel. The counter effectively limits the effect of the burst photoemission to the source pixel.

## 6.5 Application: Characterization of Semiconductor Nanostructures

The performance of the tip-collection TL microscope was evaluated through low-temperature measuring of GaAs bulk and GaAs/AlAs multiple-quantum-wells (MQW) samples.

### 6.5.1 Emission spectral domain

First, we measured the dependence of the TL intensity on the tunneling-electron injection energy to investigate photon-emission processes caused by

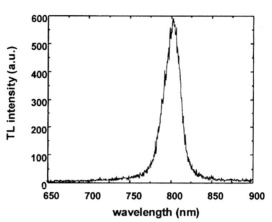

Fig. 6.11. TL emission spectrum from GaAs.

tunneling electrons in semiconductors. The sample was a p-type GaAs substrate with a Zn concentration of about $1 \times 10^{19}$ cm$^{-3}$. It was cleaved and measured in a UHV chamber at below $5 \times 10^{-10}$ Torr to avoid pinning effects due to surface contamination or oxidation. The measurement temperature was as low as 77 K to reduce thermal fluctuation. The doping concentration of the sample was so high that the Fermi level was at the top of the valence band. Therefore, the initial kinetic energy of the tunneling electrons $E_i$ corresponded to potential energy $eV_b$ at the tip bias $V_b$. Negative biases were applied to the tip relative to the sample in order to inject electrons from the tip into the sample. Figure 6.11 shows the TL spectrum obtained from GaAs at a bias voltage of –4 V and a tunnel current of 1 nA. The signal-to-noise ratio was sufficient for spectral analysis. Despite the tip bias, the TL emission spectrum had a single peak at a wavelength of 835 nm; this peak corresponds to the wavelength of band-acceptor radiative recombination in Zn-doped GaAs. This shows that tunneling electrons injected into GaAs relax nonradiatively and the GaAs eventually emits TL through band-to-acceptor radiative recombination.

### 6.5.2 Injection-energy domain

Figure 6.12 shows the dependence of TL intensity on bias voltage, that is, the photo-luminescence intensity—tip bias voltage (P-V) profile, measured in the bias range from 0.1 to 7 V while holding the tunnel current $I_t$ constant. The TL intensity increased stepwise, not monotonously, as the tip bias increased. [17] Below 1.5 V, no photoemission was observed despite stable STM operation. The TL emission started at 1.5 eV, which nearly coincides with the band gap of GaAs. The TL intensity increased quadratically in the bias range from the band gap to 2.3 eV, then was nearly constant between 2.3 and 5.0 eV

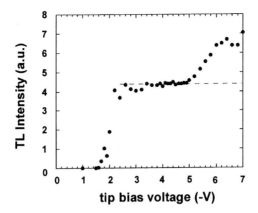

Fig. 6.12. Dependence of the TL intensity from GaAs on the tip bias voltage.

regardless of the tip bias. The TL intensity then quickly increased between 5 and 6 eV, but the rate of increased slowed in the energy range over 6 eV.

The P-V profile describes the relaxation and photoemission processes of the tunneling electrons in GaAs as shown in Fig. 6.13. The tunneling electrons cannot enter the conduction band when their injection energies are below the 1.5-eV band gap. They are lost by nonradiative recombination though continuous surface states. When the injection energy of the tunneling electrons is equal to the band gap, tunneling electrons can enter the conduction band and radiative recombination with acceptors begins. The intensity of the luminescence caused by the recombination of tunneling electrons with acceptors is determined by the number of electrons, not by the energy of the tunneling electrons. TL intensity is therefore constant, regardless of the tip bias above 2.3 eV. At much higher electron energies, excess electron-hole pairs are generated in GaAs though impact ionization. [18,19] The electron-hole pairs also cause radiative recombination, so the TL intensity rises. Above about 5 eV to 6 eV, the increased impact ionization results in observable increments in the TL intensity. Above about 6 eV, the increments in the TL intensity become much more gentle. This second saturation represents the saturation of single electron-hole-pair generation (single impact ionization) and the beginning of double electron-hole pair generation (double-impact ionization) in the primary electron relaxation. In this energy range, the impact ionization (I.I.) rate differs from the quantum yield, and the quantum yield agrees with the experimental values because more than one electron-hole pair can be generated during the relaxation. If one primary electron creates more than one electron-hole pair in the energy relaxation, the I.I. rates will differ from the quantum yields. The good agreement between the experimental and simulation results confirms that the P-V profile reflects the quantum yield.

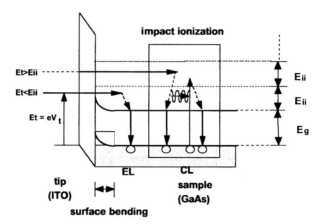

Fig. 6.13. Energy diagram of the tunneling-electron luminescence process in direct-gap semiconductors.

The P-V profile is thought to also be proportional to the injection energies of the tunneling electrons at much higher than 7.0 eV. The luminescence intensity is proportional to the primary electron energy in CL measurements where electrons with energies as high as the kilo-electron-volt order are used. Thus, the ability to scan the excitation energy continuously in a wide range below 10 eV enables effective characterization of semiconductors.

### 6.5.3 Spatial domain

Figure 6.14 shows an SEM image of a cross-section of a semiconductor MQW sample. This was obtained to evaluate the spatial resolution of the TL microscope. The MQW structure consisted of 50-nm-thick GaAs well layers and 50-nm-thick AlAs barrier layers.

This structure was examined because close-set layers in a cross-section of the MQW structure are difficult to individually distinguish when using conventional PL and CL methods. These structures are much smaller than the refraction limit of the excitation light and the barrier layer used for electron confinement is usually transparent to light that excites the well layers. Focused light, therefore, cannot separately excite the individual layers. Even in CL measurements, high-energy injection electrons scatter widely and excite both well and barrier layers.

The MQW structures consisted of six-period 50-nm-thick GaAs well layers and 50-nm-thick AlAs barrier layers, all of which were grown on a GaAs substrate by metalorganic chemical vapor deposition (MOCVD). All epitaxial layers were highly doped with Zn to a concentration of about $1 \times 10^{19}$ cm$^{-3}$. The samples were cleaved in a UHV chamber at below $5 \times 10^{-10}$ Torr to create clean

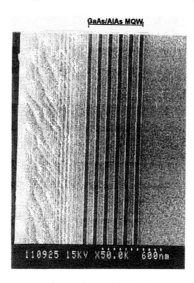

Fig. 6.14. Cross-sectional SEM image of the GaAs/AlAs MQW sample.

surfaces. At nanoampere-order tunnel currents, the TL intensity is thousands of counts per second, which is high enough for spectroscopy and microscopy. The TL from n-type semiconductors can be detected with a positive tip bias though hole injection.

Figure 6.15 shows an STM image and a TL microscopy image of an MQW cross-section simultaneously obtained at −3.0 V and 0.5 nA. [20] The TL image was created through direct radiative recombination of injected electrons with the majority of carriers since the injection energy was below the impact ionization energy of GaAs. In the TL image, luminous stripes and dark stripes appear clearly, and correspond respectively to the individual GaAs well layers and AlAs barrier layers. The GaAs and AlAs layers were both the same width as the layers revealed through SEM. The STM image, on the other hand, shows a rough surface morphology including a number of irregular steps generated during cleaving. The layered MQW structure was not clearly shown, and the steps in the STM image do not correspond to the layered structures of the GaAs/AlAs MQW structure observed in the TL image. These steps on the sample surface did not appear in the TL image, as the surface was transparent. This shows that the tunneling electrons emit luminescence from the subsurface region of the sample. Carriers at the surface rarely emit photons since surface states enhance the nonradiative recombination of carriers. Thus, TL microscopy enables us to measure subsurface nanostructures, while STM measurements only show the surface morphology. The high-resolution TL image thus suggests that TL microscopy provides a powerful means of characterizing high-density

**100 nm**

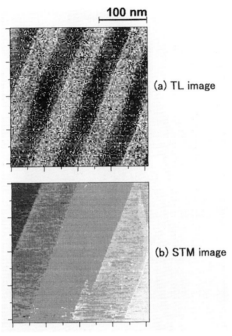

(a) TL image

(b) STM image

Fig. 6.15. Cross-sectional images of a GaAs/AlAs MQW sample simultaneously obtained by TL microscopy (a) and STM (b).

nanometer-scale structures. A one-dimensional TL profile transversal to the layers was obtained by averaging the TL intensities parallel to interfaces of the layers in the TL image. This TL profile is shown by the data points on a semi-log scale in Fig. 6.16 as a function of the distance between the electron-injection point and the center of the AlAs layer. Part of the TL profile measured from the AlAs, an indirect-gap material, is probably from the detection of emissions from adjoining GaAs layers since the electrons have likely diffused from the AlAs layer into adjoining GaAs layers. The TL intensity distribution in the AlAs layer was precisely fitted by a hyperbolic cosine function of $[\exp(-x/L) + \exp(x/L)]/2$ with a decay length L of 8 nm, where x is the distance to the electron-injection point from the center of the AlAs layer. In other words, the TL intensity measured in an AlAs layer decays exponentially from the edge of the AlAs layer with a decay length of 8 nm.

## 6.6 Conclusion

Tunneling-electron luminescence (TL) microscopy using the tip collection method enables real-space and multifunctional characterization of the optical and electronic properties of nanostructures with nanometer-level spatial

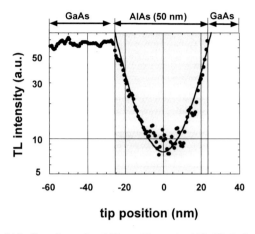

Fig. 6.16. One-dimensional TL profile on the AlAs/GaAs layers.

resolution. A conductive transparent (CT) probe is used to inject the tunnel current and to simultaneously collect the induced luminescence. The CT probe enables both stable tunnel current injections over a wide energy range and high-yield luminescence collection from nanometer-scale local regions. The screw fitting of the CT probes enables highly reliable and clean operation at a low temperature in a UHV chamber.

The TL microscope was used for low-temperature cross-sectional measurements of multiple-layer of GaAs/AlAs MQW structures in which each layer was 50 nm thick. It was able to clearly distinguish between the individual layers, measure the TL spectra from the individual layers, and measure the dependence of the TL intensity on the tunneling-electron energy that reflects the energy-band structure.

This microscope is thus an effective tool for the real-space and multifunctional characterization of optical and electronic properties in nanometer-scale regions. We plan to improve the performance of the microscope so that it can be used to evaluate many more materials and structures, including semiconductor devices, with nanometer-level spatial resolution.

[References]

1)  H. Ehrenreich and D. Turnbull, "*Solid State Physics*," Academic Press, Boston (1991).
2)  J. I. Pankov, "*Optical Processes in Semiconductors*," Dover, New York (1971), p. 249.
3)  B. G. Yacobi and D. B. Holt, "*Cathodoluminescence Microscopy of Inorganic Solids*," Plenum, New York (1990), p. 55.
4)  J. H. Coombs, J. K. Gimzewski, B. Reihl, J. K. Sass, and R. P. Schlittler, *J. Microscopy* **152**, 325 (1988).
5)  D. L. Abraham, A. Veider, Ch. Schonenberger, H. P. Meier, D. J. Arent, and S. F. Alvarado, *Appl. Phys. Lett.* **56**, 1564 (1990).

6)    S. Ushioda, Y. Uehara, and M. Kuwahara, *Appl. Surf. Sci.* **60/61**, 448 (1992).

7)    L. Samuelson, A. Gustafsson, J. Lindahl, L. Montelius, and M.-E. Pistol, *J. Vac. Sci. Technol.* **B12**, 2521 (1994).

8)    B. Akamatsu, P. Henoc, and A. C. Paradopoulo, *Scanning Electron Microscopy IV*, 1579 (1983).

9)    K. Wada, A. Kozen, H. Fushimi, and N. Inoue, *Jpn. J. Appl. Phys.* **27**, L1952 (1988).

10)   M. Ohtsu, *"Near-Field Nano/Atom Optics and Technology,"* Springer, Tokyo (1998).

11)   E. Burnstein and S. Lundqvist, *"Tunneling Phenomena in Solids,"* Plenum, New York (1969).

12)   T. Murashita and M. Tanimoto, *Jpn. J. Appl. Phys.* **34**, 4398 (1995).

13)   T. Murashita, *J. Vac. Sci. Technol.* **B15**, 32 (1997).

14)   T. Pangaribuan, K. Yamada, S. Jiang, H. Ohsawa, and M. Ohtsu, *Jpn. J. Appl. Phys.* **31**, L1302 (1992).

15)   M. Kuwabara, K. Kokura, and S. Ikegami, *Proceedings of the 40th International Wire and Cable Symposium*, IWCS, St. Louis (1991), pp. 167–171.

16)   T. Murashita, *J. Vac. Sci. Technol.* **B17**, 22 (1999).

17)   S. Sasaki and T. Murashita, *Jpn. J. Appl. Phys.* **38**, L4 (1999).

18)   H. K. Yung, K. Taniguchi, and C. Hamaguchi, *J. Appl. Phys.* **79**, 2473 (1995).

19)   M. V. Fichetti, N. Sano, S. E. Laux, and K. Natori, *Proceedings of SISPAD-96* (1996) p. 46.

*Chapter 7*

# Near-Field Optical Spectroscopy of Single Quantum Dots

*7.1 Introduction*

In the optical characterization of semiconductor materials and optoelectronic devices, near-field scanning optical microscopy (NSOM) allows studies of highly localized features and their distributions well beyond the diffraction limit of light [1–7]. The heart of NSOM is a near-field probe, illustrated in Fig. 7.1 which is a metal-coated optical fiber tapered to an opening (aperture) which is much smaller than the wavelength of light. When the probe end approaches a sample surface at a short distance, the object is illuminated only by near-field light generated by the aperture and the spatial resolution is not subject to diffraction. The reflected or reemitted light is locally collected by the same aperture (illumination-collection mode; I-C mode) or by a lens in the transmission configuration (illumination mode; I mode) as the probe scans the surface.

In the various applications of NSOM, near-field photoluminescence (PL) microscopy is the most promising technique, offering spectroscopic analysis able to achieve spatial resolution in the range of 10–100 nm. Advances in the fabrication of NSOM fiber probes have make remarkable contributions to the sensitive PL spectroscopy of nanometer-scale structures such as semiconductor quantum dots (QDs). The high resolution of the near-field probe enables us to observe individual QDs, whose ensemble shows inhomogeneous broadening due to their distribution in sizes and strains. Single-QD spectroscopy with NSOM can thus elucidate intrinsic nature of QDs including narrow optical transition arising from the atomic-like discrete density of sate. That is impossible to obtain using a conventional optical microscope. The single-QD technique provides a new physical insight into the spectral feature and the time trajectory of QDs: the determination of homogeneous linewidth in the spectral domain and the observation of blinking phenomena in the time domain as shown in Fig. 7.2.

The principle of single-QD spectroscopy involves significant challenges in the operation of NSOM. In the analysis of semiconductor materials, we

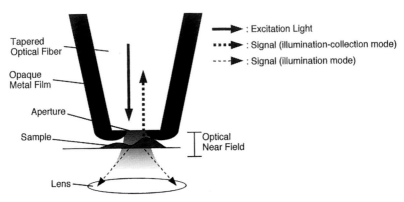

Fig. 7.1. Diagramatic illustration of near-field scanning optical microscope.

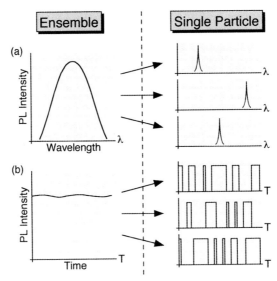

Fig. 7.2. Significance of single-particle spectroscopy. (a) determination of homogeneous linewidth and (b) observation of blinking phenomena.

frequently need to observe opaque samples or samples prepared on opaque substrates. Moreover, in near-field PL spectroscopy of semiconductors, reduction of the resolving power due to the carrier diffusion prior to recombination must be avoided. In these situations, the employment of I-C mode operation, in which the fiber probe is utilized not only for the excitation but also for the collection of the signal, is advantageous in obtaining the spatial resolution determined by the aperture diameter of the fiber probe [8].

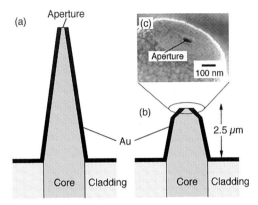

Fig. 7.3. (a) Single-tapered near-field fiber probe and (b) double-tapered probe. (c) Scanning electron micrograph of 80-nm aperture.

For the successful detection of weak signals, such as PL, Raman scattering, and the transmission change from extremely small QDs in I-C mode of NSOM, high efficiency in the collection of the signal through the fiber probe is indispensable. In addition to this requirement, it can be argued that high spatial resolution is also essentially important even in the case of observation of "artificial" samples, where QDs are prepared with extremely low density on the substrate. In the PL or Raman technique, limitation of the excited area minimizes the background level from the substrate or barrier layers. More seriously, in absorption spectroscopy, the ratio of absorption cross-section of QD to the illumination area should be greater than the signal to noise ratio of the measurement. The simultaneous achievement of high resolution and high sensitivity by optimizing the structure of the fiber probe is essential in satisfying such requirements [9].

In this paper, we firstly report on the successful fabrication of a highly sensitive near-field probe by combining a double-tapered structure and a well-defined aperture with a sharp edge. Through the PL imaging of single QDs, we evaluate the functional performance of many versions of the probe with different aperture diameters. Secondly, near-field PL spectroscopy of single InGaAs QDs at low and room temperature is described. The electronic structure information is obtained through the evolution of PL spectra with weak excitation intensity at 8 K. Owing to the high collection efficiency of the probe fabricated, we succeed in the room-temperature spectroscopy of single QDs, whose yields are much reduced with increase of temperature. Furthermore, by incorporating a femtosecond laser pulse, furthermore, time-resolved PL spectroscopy is performed to study trapping and relaxation processes of carriers in QDs. Finally, we describe modulated absorption spectroscopy of single QDs in combination with a standard pump-probe setup.

## 7.2 Fabrication of Near-Field Fiber Probe

Figure 7.3 shows schematic structures of the near-field fiber probe we fabricated based on a chemical etching technique using a hydrofluoric acid solution [10–12]. One of the attractive advantages of the chemically etched probe is its extremely short propagation length of light in a metal-cladding waveguide, which reduces the transmission efficiency of the probe. We propose to shorten the narrow metal-cladding region of strong optical losses by making a double-tapered structure with a large cone angle as shown in Fig. 7.3(b) [12]. Compared to a single-tapered probe with a steep cone angle (Fig. 7.3(a)), another advantageous feature of the double-tapered probe is that it has a sharp-edged aperture, which contributes to the effective generation and detection of the evanescent field in the vicinity of the nanostructures.

The chemical etching process employed in fabrication of the double-tapered probe is detailed in Ref. 12). The entire exterior surface of the etched probe was coated with an Au film using the sputtering coating method. A small aperture was created by pounding the Au-coated probe on a GaAs substrate (the sample itself) and squeezing the Au off to the side [9,13]. Figure 7.3(c) shows a scanning electron micrograph of the aperture 80 nm in diameter, taken after conducting PL imaging several times. Not only a smooth and flat end-face but also a round and well-defined aperture was obtained in this fabrication process.

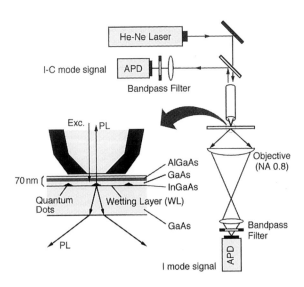

Fig. 7.4. Diagram of the sample structure, the optical near-field configuration, and a block diagram of the measurement.

## 7.3 Fundamental Performance of Near-Field Probe

In order to evaluate the functional performance of the probe developed, we performed PL imaging of single QDs at room temperature [9]. Figure 7.4 shows a diagram of the QD sample structure, the optical near-field configuration, and a block diagram of the measurement. $In_{0.5}Ga_{0.5}As$ self-assembled QDs were grown on a GaAs substrate by gas-source molecular beam epitaxy [14]. Defects-free and density-controlled QDs were obtained under the Stranski-Krastanow growth mode. The typical lateral size and height of the dots were 40 nm and 10 nm, respectively. There were 70-nm thick cap layers covering the QDs. These layers, composed of GaAs and AlGaAs, were introduced in order to prevent photoexcited carriers from leaking out to the surface. The PL wavelength of the QDs was 1 μm approximately. Similar samples were investigated in the various spectroscopic measurements described below.

For excitation of the carriers, a He-Ne laser ($\lambda = 633$ nm) was coupled into the cleaved fiber end. The QD sample on the scanning stage was illuminated with laser light through the aperture under shear-force feedback control [15,16]. Most of the photoexcited carriers were generated in the GaAs and

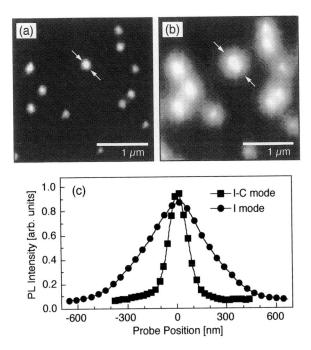

Fig. 7.5. Near-field PL images of single QDs, obtained using (a) I-C mode operation and (b) I mode operation, in the same scanning area. (c) Cross-sectional profiles of the PL signal intensity of the same dot indicated by arrows in (a) and (b).

AlGaAs barrier layers. After diffusing in the sub-$\mu$m region in the GaAs barrier layers and an InGaAs wetting layer, the carriers were captured by the confined states of the QDs. The resultant PL signal from a single QD was collected by the same aperture (I-C mode) and detected by means of a photon-counting avalanche photodiode (APD) through a band-pass filter (center wavelength $\lambda = 1$ $\mu$m, bandwidth $\Delta\lambda = 40$ nm). To evaluate the PL signal collection ability in I-C mode quantitatively, we simultaneously performed I mode operation in the transmission configuration. In this mode, the PL signal was collected by means of an objective lens with a numerical aperture (NA) of 0.8 in the backside of the sample. The same photodiode and filter as mentioned above were used for detection of the PL signal.

Figures 7.5(a) and (b) show the PL images of QDs obtained using I-C mode and the I mode, respectively, in the same scanning area. The aperture diameter was 100 nm. Each bright spot corresponds to the PL signal from a single QD. Cross-sectional profiles of the signal intensity for the same QD indicated by the arrows are plotted in Fig. 7.5(c). In the case of the I-C mode, the full width at half-maximum (FWHM) of the profile (PL spot diameter) was estimated as 130 nm, which is almost equal to the spatial resolution of this measurement. The PL spot diameter of 130 nm corresponds to $\lambda/8$ resolution, taking into account the PL wavelength of 1 $\mu$m. In the I mode, on the other hand, the PL spot diameter was about 450 nm, which is much wider than that in I-C mode. This large discrepancy is mainly due to the diffusion of the carriers in the barrier layers prior to the relaxation into the QD. In Fig. 7.5(c), the signal intensities are plotted on the same scale and can therefore be compared. It should be emphasized that the signal intensity in the I-C mode is as large as that in I mode, indicating that the collection ability of the probe with an aperture of 100 nm is comparable to that an objective lens with a NA of 0.8.

We prepared many versions of the probe with different aperture and carried out the same measurements as above, evaluating a throughput (a ratio of light power ejected from the aperture to light power coupled into the fiber), spatial resolution, and collection ability. The results are summarized in Fig. 7.6. The PL spot diameter and collection efficiency in the I-C mode imaging are plotted as a function of throughput and aperture diameter. The collection efficiency is defined as a ratio of the PL signal intensity in the I-C mode to that in the I mode. At the aperture diameter of around 100 nm, the collection efficiency was nearly equal to 1. With increasing aperture diameter, the collection sensitivity in I-C mode also increases monotonously and reaches a value five times larger than that in the I mode.

Regarding the spatial resolution, the PL spot diameter is not so strongly dependent on the aperture diameter compared to the throughput behavior and the collection efficiency behavior. The highest resolution in this measurement was limited to 130 nm since the QDs were covered with cap layers, which keep

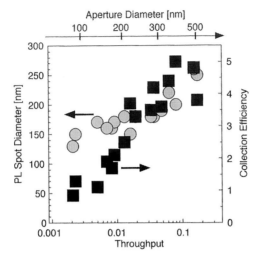

Fig. 7.6. Plots of PL spot diameter (circles) and collection efficiency (squares) as a function of throughput and aperture diameter.

the aperture-QD gap at 70 nm. Another notable result was that by using an aperture >200 nm, we obtained a PL spot much smaller than the aperture diameter. Even when we used a 500-nm aperture, for instance, a spot diameter as small as 250 nm is easily attained [4].

Providing further insight into the high collection efficiency of the probe, Fig. 7.7 shows the dependence of the signal intensity in the I and the I-C mode operation on the separation between sample surface and the aperture (150 nm in diameter) positioned just above a QD. When the separation is around 100 nm, the signal intensity in I-C mode is almost zero while that of I mode reaches 20% of its maximum. In 20 nm region from the surface, however, the collection efficiency in the I-C mode rapidly increases and finally becomes greater than that in the I mode. This result suggests that fabrication of the flat-end probe with well-defined aperture would makes a considerable contribution to the efficient interaction of the aperture and the evanescent field in the vicinity of the sample in 10 nm range.

## 7.4 Low-Temperature PL Spectroscopy of Single QDs

For the precise study of fine electronic structures of condensed matter, optical measurements at cryogenic temperature is indispensable for suppressing the energy broadening due to the interaction with phonons. In order to operate NSOM at a low temperature, we designed it to fit in a continuous He gas flow optical cryostat. Since we successfully utilize the I-C mode operation, the structure of the head is very simple, and does not employ any far-field

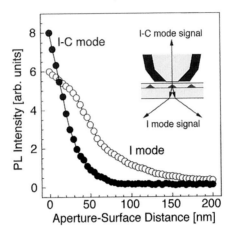

Fig. 7.7. Dependence of PL signal intensity on the separation between the aperture and the sample surface in I mode and I-C mode operations.

illumination or collection optics. Figure 7.8 shows near-field PL spectra with various excitation power densities [4]. The peak intensity of each spectrum is normalized by the excitation density. At the lowest excitation density of 1.3 W/cm$^2$, the excitation light of 2.4 nW through the aperture with a diameter of 500 nm generates $7 \times 10^9$ electron-hole (e–h) pairs per second within an excitation area of $\pi(250 \text{ nm})^2$. From the value of the QD density, the injection rate of e–h pairs into individual QDs is estimated to be $<2 \times 10^9$ e–h/s. In this estimation, we do not take into account the effect of carrier diffusion. Since the injection rate of e–h pairs is smaller than $1/\tau$, where $\tau$ is a ground state emission lifetime of about 300–500 ps, we conclude that the observed emission line originates from a recombination of the ground-state e–h pair.

The evolution of the single QD PL spectrum with excitation intensity gives us information on the intrinsic physical properties of a quantum confined system, information such as the quantization energy of electrons and the interaction energy of e–h pairs. For an excitation power lower than 5 W/cm$^2$, the PL spectra consist of a single line. As mentioned above, in such a power density, the single PL line originates from a recombination of the ground-state e–h pair. With increase in excitation power, other elementary features are observed in the PL spectra. Firstly, after the saturation of the ground-state emission, a new PL line appears 30 meV above the ground-state transition. This line corresponds to the emission from the first excited state, which arises from the state filling of the ground state; fast relaxation from the excited state to the ground state is hindered due to the occupation of ground state and emission from the excited state is observed. Furthermore, Fig. 7.8 shows another feature 2.2 meV below the ground-state (single e–h pair) emission.

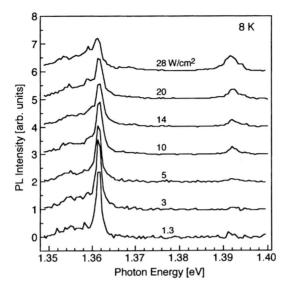

Fig. 7.8. Single-QD PL spectra with various excitation power densities at 8 K.

The emission intensity grows superlinearly with the excitation density. From the energetic position and the excitation power dependence, this emission line can be associated with the recombination of interacting two e–h pairs [6,17,18], the rate of which increases quadratically with the generation rate of the single e–h pair.

Another important use of single-QD spectroscopy is the observation of variations in optical properties between individual QDs. To study the distribution of emission intensities, we performed spectrally integrated PL imaging, where whole PL signals are detected without monochromating. Figures 7.9(a) and (b) are PL images made by exciting QDs at 1.42 and 1.96 eV, respectively. It can be seen that the distribution of PL spot intensity in Fig. 7.9(b) is much larger than that in Fig. 7.9(a). Such behavior is more clearly illustrated in the histograms of the PL intensities, shown in Figs. 7.9(c) and (d), which are obtained through the examination of several tens of QDs. When the sample is excited at 1.42 eV, below the absorption edge of WL, the carriers are generated in QDs directly. The dispersion of PL intensity in Fig. 7.9(c) mainly arises from the distribution of quantum efficiency. For a photon energy of 1.96 eV, on the other hand, most of the carriers are generated in the barrier layers. They relax into the WL, then migrate, and finally are captured by the confined states of QDs. The wider dispersion in Fig. 7.9(d) can be attributed to the difference of carrier flow rate into individual QDs in addition to that of quantum efficiency.

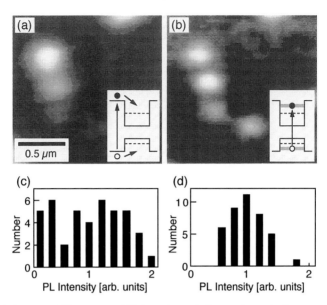

Fig. 7.9. PL images and histograms of PL intensity variations at the excitation energy of (a), (c) 1.96 eV and (b), (d) 1.42 eV, respectively.

## 7.5 Room-Temperature PL Spectroscopy of Single QDs

To date, several groups have reported various results for microscopic spectroscopy of single QDs, revealing atom-like spectra with discrete and extraordinary sharp lines at low temperature. From the aspect of device application, however, information on the optical properties of single QDs at room temperature is more urgently required. The measurement of a single QD PL spectrum at room temperature is technically difficult in comparison with that at low temperature, since the PL intensity from QDs decreases typically by 2–3 orders due to the escape of carriers to a nonradiative recombination path or the thermal excitation to the wetting layer. As well as high sensitivity for the detection of weak PL signals, high spatial resolution is essentially required to minimize the relatively strong background emission signal from the wetting layer and GaAs barrier layer. The most common micro-PL technique with spatial resolution of about 1 $\mu$m is not suitable for this purpose.

First of all, we mapped the PL spot distribution measuring the topographic features of the sample surface [19]. By comparing the PL and the topographic images, we could confirm that all QDs are efficiently emissive even at room temperature. Figure 7.10 shows single-QD PL spectra obtained by using a cooled CCD camera at various excitation densities. A single peak can be observed at 1080 nm under the lowest excitation condition. When the excitation

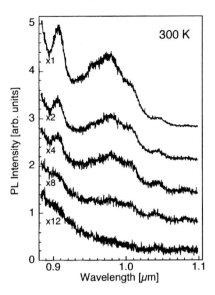

Fig. 7.10. Evolution of single-QD PL spectra with excitation power density at room temperature.

density is increased, higher energy peaks gradually appear and at the strongest excitation, as Fig. 7.10 shows, we can observe five peaks with an energy separation of about 35 meV [20]. To confirm the origin of the single emission in the weak excitation region, the integrated PL intensity and PL linewidth (FWHM) have been plotted as a function of the excitation density in Figs. 7.11(a) and (b), respectively [19]. Up to the excitation density of 0.8 kW/cm$^2$, the PL intensity linearly increases and the linewidth remains constant. Then, with more increase of excitation power, the PL intensity gradually saturates and the linewidth begins to broaden. Such PL behavior implies that the single emission line originates from a recombination of the ground-state e–h pair.

Figures 7.11(c)–(e) shows ground-state emission spectra at a low excitation condition (<0.8 kW/cm$^2$) for three different QDs [19]. The line shape of each spectrum can be reproduced by a Lorentzian function. By doing a precise examination of PL spectra of a large number of QDs at low excitation condition, the homogeneous linewidth of ground-state emission can be evaluated as 10–15 meV, which is smaller than that calculated for an ideal two-dimensional QW structure.

## 7.6  Time-Resolved PL Spectroscopy of Single QDs

Carrier relaxation via interaction with optical phonons in zero-dimensional QDs can be expected to be slowed down due to severe energy conservation problems originating from the atom-like discrete density of states. In addition,

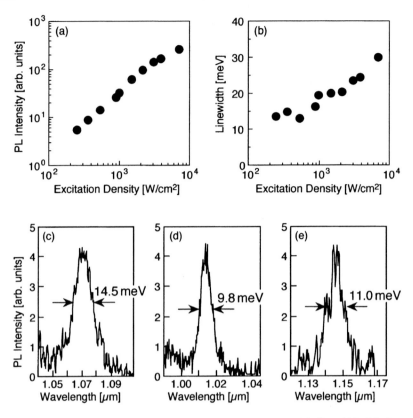

Fig. 7.11. Excitation power dependence of (a) integrated intensity and (b) linewidth of the lowest emission of a single QD. (c)–(e) PL spectra of the lowest emission lines of three different QDs in a weak excitation condition.

even at a low excitation density, carrier relaxation from excited states of QD to lower states is hindered due to the state filling effects (Pauli exclusion principle) arising from finite degeneracy of QD states. It is necessary to explain how the relaxation mechanism operates and evaluate the radiative decay rate to improve device performance and time-resolved PL spectroscopy of single QDs is the most effective tool for this [21,22].

As an excitation light source, a mode-locked fiber laser is used to generate carriers in the GaAs barrier layer [23]. The center wavelength, pulse width, and the repetition rate of the laser are 780 nm, 160 fs, and 48 MHz respectively. As in the other experiments, the near-field fiber probe is utilized both for the excitation of the sample and for the collection of the PL signal. The collected PL is dispersed by 10-cm monochromator with a spectrum resolution of about 15 meV and detected by the Si single photon APD (the same one as described

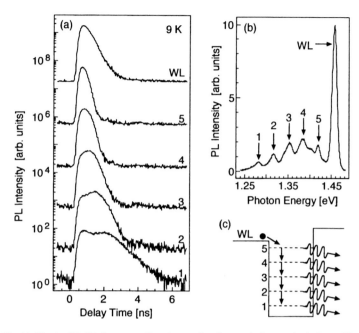

Fig. 7.12. (a) Single-QD PL decay profiles detected at five emission peaks indicated in (b) and (c). (b) Single-QD PL spectra obtained at high excitation density. (c) Diagramatic illustration of quantized energy levels of the QD.

above) using the time-correlated single photon counting method with temporal resolution of 270 ps. For a specification of the wavelength of each emission peaks, we used an optical multichannel analyzer with a spectral resolution of 1 meV. All the measurements were performed at 8 K.

The decay curves in Fig. 7.12(a) are the experimental result obtained by tuning the PL detection wavelength at five peaks, numbered in order from the lowest energy peak as indicated in Fig. 7.12(c). Comparison of the time response functions of different emission peaks reveals that the lower energy peaks exhibit slower decay and a nonexponential curve. Such behaviors can be explained using rate equations with rather simple assumptions: (1) The initial occupation of each levels is determined by a relaxation process faster than the time resolution of the measurement. (2) Cascade relaxation between neighboring levels is dominant. (3) Relaxation to a lower level is only possible if the lower level is empty (Pauli blocking) [23].

To determine important parameters such as radiative lifetime, interlevel relaxation rate, and so on, we systematically measured the PL decay profiles with various excitation densities. Numerical analysis employing the rate equations allowed us to conclude that the radiative lifetime and interlevel relaxation rate are 300-500 ps and 10–100 ps, respectively [23]. The initial state occupation that attributed to a fast relaxation process and a slower carrier feeding from a wetting layer were found to be important factors. These carrier relaxation properties vary dot-by-dot, which result would arise from the difference of level separation in each dot and spatial energy fluctuation in wetting layer.

### 7.7 Modulated Absorption Spectroscopy of Single QDs

QDs are quite attractive for optical devices used in information processing and transmission systems, since they have large optical nonlinearity, such as huge absorption change, due to the sharp transition originating from atomic-like discrete energy levels. For the precise observation of nonlinear phenomena in single QDs, the spatial resolution of conventional far-field optical microscopy is insufficient; the size of the focusing spot ($\lambda/2$) is much larger than the absorption or scattering cross section of the QD, and the nonlinear signal is buried in a large background.

To measure the nonlinear absorption change in the QDs, a continuous wave modulation spectroscopy (pump-probe technique) is employed [24,25]. As shown in Fig. 7.13, a probe light (Ti: Sapphire laser light, $\lambda = 900$–980 nm) was introduced into the back of the sample, and the transmitted light collected through the aperture of the fiber probe. The collected probe light was detected by an InGaAs photodiode. When the sample was illuminated by pump light (laser diode, $\lambda = 635$ nm) passing through the fiber probe, the ground states of the QDs were occupied with photoexcited carriers. The resultant change of the absorption coefficient of the QDs was measured by detecting the transmission

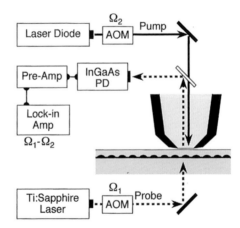

Fig. 7.13. Diagram of the experimental setup for the absorption modulation spectroscopy of single QD.

change in the probe light. To achieve high sensitivity in detection of the transmission change, both the pump light power and the probe light power were modulated, and the difference frequency component of the detected signal extracted by synchronous lock-in detection. The minimum detectable differential transmission ($T$) of this system is $\Delta T/T = 2 \times 10^{-5}$. All measurements were performed at 5–10 K.

Figure 7.14(a) shows the spatial distribution of the modulated absorption signal [25]. The wavelength of the probe light is fixed at 940 nm, and the power density of the pump light at the apex of the fiber probe is 300 W/cm². Figure 7.14(b) shows the distribution of the PL taken in the same scanning area. The center wavelength and bandwidth of the bandpass filter for the detection of PL signal are 940 nm and 10 nm, respectively. The power density of the excitation light is the same as the pump power density of the nonlinear absorption measurement. Since the nonlinear absorption change and PL are strong at the same positions indicated by circles, it could be confirmed that the nonlinear absorption change originates from single QDs. The full width at half maximum of the bright spots in Fig. 7.14(a) is 200 nm ($\lambda/5$).

To confirm that the nonlinear absorption change originates from the state filling of the ground state, the dependence of the nonlinear absorption change and PL intensity on the pump (excitation) power density was measured. The PL intensity increases linearly in the weak excitation region, and reaches saturation above about 300 W/cm². From this behavior, we could confirm that the PL is the ground state emission. The modulated absorption signal also increases linearly in the weak excitation region, and reaches saturation at the same pump power density as the PL intensity. We could therefore conclude that the

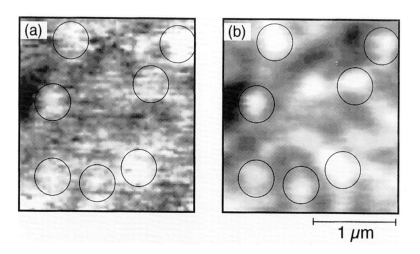

Fig. 7.14. Spatial distribution of (a) modulated absorption signal and (b) PL signal.

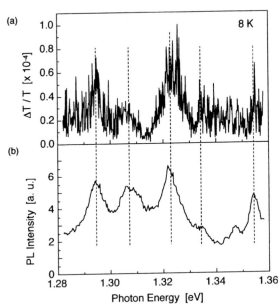

Fig. 7.15. (a) Modulated absorption spectrum and (b) PL spectrum obtained at the same probing position. In this measurement, there are more than five QDs in the observation area.

nonlinear absorption change originates from the state filling of the ground state.

Figure 7.15(a) shows the nonlinear absorption spectrum of single QDs obtained by changing the probe light wavelength continuously [25]. The pump power density is 300 W/cm². Fig. 7.15(b) shows the PL spectrum taken with the same excitation power. The peaks in the nonlinear absorption spectrum are also seen in the PL spectrum at the same wavelengths. These peaks correspond to the ground states of single quantum dots, as confirmed by observing the pump power dependence described above. The homogeneous broadening of the peak is about 5 meV. The maximum change in absorption cross section was also estimated from these results [25].

### 7.8 Summary

We developed a highly sensitive fiber probe with a double-tapered structure and a sharp-edged aperture at its apex, especially suitable for illumination-collection hybrid mode operation of near-field scanning optical microscope. It was found that a collection efficiency several times greater than that of an objective lens with a NA of 0.8 could be obtained, with resolution in the 100–200 nm range. The functional performance of the probe developed was demonstrated through the imaging and spectroscopy of single QDs at low and room temperature, including time-resolved photoluminescence spectroscopy and modulated absorption spectroscopy.

The single QD spectroscopy is powerful tool for the characterization of the quality of QD structures, as well as for the study of fundamental physics in a quantum confined system. The information obtained through the single QD observation is extremely useful in the search for better growth procedures and materials. From another aspect, the near-field approach also contributes to the efficient optical control of electron dynamics in QDs by means of short-range electromagnetic interaction. This technique will open up the realization of single QD devices with low energy consumption, such as optical memories, modulators, and switches, whose dimensions are much smaller than the wavelength of light.

## Acknowledgments

The author would like to thank K. Nishi and H. Saito for supplying the quantum dot sample. He is also grateful to his collaborators, K. Matsuda and T. Matsumoto.

[References]

1)  T. Saiki, in *Near-Field Nano/Atom Optics and Technology*, M. Ohtsu ed. (Springer-Verlag, Tokyo, 1998) Chap. 9.

2)  H. F. Hess, E. Betzig, T. D. Harris, L. N. Pfeiffer, and K. W. West, *Science* **264**, 1740 (1994).

3)  A. Richter, G. Behme, M. Süptitz, Ch. Lienau, T. Elsaesser, R. Nötzel, M. Ramsteiner, and K. H. Ploog, *Phys. Rev. Lett.* **79**, 2145 (1997)

4)  T. Saiki, K. Nishi, and M. Ohtsu, *Jpn. J. Appl. Phys.* **37**, 1638 (1998).

5)  Y. Toda, S. Shinomori, K. Suzuki, and Y. Arakawa, *Appl. Phys. Lett.* **73**, 517 (1998).

6)  A. Chavez-Pirson, J. Temmyo, H. Kamada, H. Gotoh, and H. Ando, *Appl. Phys. Lett.* **72**, 3494 (1998).

7)  H. D. Robinson, M. G. Müller, B. B. Goldberg, and J. L. Merz, *Appl. Phys. Lett.* **72**, 2081 (1998).

8)  T. Saiki, S. Mononobe, M. Ohtsu, N. Saito, and J. Kusano, *Appl. Phys. Lett.* **67**, 2191 (1995).

9)  T. Saiki and K. Matsuda, *Appl. Phys. Lett.* **74**, 2773 (1999).

10)  T. Pangaribuan, K. Yamada, S. Jiang, H. Ohsawa, and M. Ohtsu, *Jpn J. Appl. Phys.* **31**, L1302 (1992).

11)  S. Mononobe, M. Naya, T. Saiki, and M. Ohtsu, *Appl. Opt.* **36**, 1496 (1997); S. Mononobe, T. Saiki, T. Suzuki, S. Koshihara, and M. Ohtsu, *Opt. Commun.* **146**, 45 (1998).

12)  T. Saiki, S. Mononobe, M. Ohtsu, N. Saito, and J. Kusano, *Appl. Phys. Lett.* **68**, 2612 (1996).

13)  D. W. Pohl, W. Denk, and M. Lanz, *Appl. Phys. Lett.* **44**, 651 (1984).

14)  K. Nishi, R. Mirin, D. Leonard, G. Medeiros-Ribeiro, P. M. Petroff, and A. C. Gossard, *J. Appl. Phys.* **80**, 3466 (1996).

15)  R. Toledo-Crow, P. C. Yang, Y. Chen, and M. Vaez-Iravani, *Appl. Phys. Lett.* **60**, 2957 (1992).

16)  E. Betzig, P. L. Finn, and J. S. Weiner, *Appl. Phys. Lett.* **60**, 2484 (1992).

17)  L. Landin, M. S. Miller, M.-E. Pistol, C. E. Pryor, and L. Samuelson, *Science* **280**, 262 (1998).

18)  E. Dekel, D. Gershoni, E. Ehrenfreund, D. Spektor, J. M. Garcia, and P. M. Petroff, *Phys. Rev. Lett.* **80** 4991 (1998).

19)  K. Matsuda, T. Saiki, H. Saito, and K. Nishi, *Appl. Phys. Lett.* **76**, 73 (2000).

20)  A. Wójs, P. Hawrylak, S. Fafard, and L. Jack, *Phys. Rev. B* **54**, 5604 (1996).

21)  V. Zwiller, M.-E. Pistol, D. Hessman, R. Cederström, W. Seifert, and L. Samuelson, *Phys. Rev. B* **59**, 5021 (1999).

22)  U. Bockelmann, W. Heller, A. Filoramo, and Ph. Roussignol, *Phys. Rev. B* **55**, 4456 (1997).

23)  M. Ono, K. Matsuda, T. Saiki, K. Nishi, T. Mukaiyama, and M. Kuwata-Gonokami, *Jpn. J. Appl. Phys.* **38**, L1460 (1999).

24)  N. H. Bonadeo, A. S. Lenihan, G. Chen, J. R. Guest, D. G. Steel, D. Gammon, D. S. Katzer, and D. Park, *Appl. Phys. Lett.* **75**, 2933 (1999).

25)  T. Matsumoto, M. Ohtsu, K. Matsuda, T. Saiki, H. Saito, and K. Nishi, *Appl. Phys. Lett.* **75**, 3246 (1999).

# Chapter 8

# Chemical Vapor Deposition of Nanometric Materials by Optical Near-Fields: Toward Nano-Photonic Integration

## 8.1 Introduction

Optical near-fields have been applied to various fields, including spatially high resolution microscopy, spectroscopy, ultra high density optical memory, and atom manipulation [1–3]. In the case of microscopy, high-resolution beyond the diffraction limit has been enabling observation of a 4 nm-width image of a single strand DNA [4]. Spatial Fourier analysis has estimated that the resolution has reached as high as 0.8 nm [5]. The basic process governing near-field optics is the short-range electromagnetic interaction between the probe tip and the sample in the optical frequency region. If this interaction is sufficiently strong, the sample structure and conformation can be optically modified, and thus this modification can lead to new applications, such as the fabrication of nanometric materials and the manipulation of atoms.

Recently, there has been an interest in application to nano-structure fabrication because of the possibility of realizing nano-photonic integration [6]. For the realization of a device which uses the optical near-field as a carrier for signal transmission, various materials with nanometric size must be integrated laterally on a substrate. For this integration, we need an advanced nano-structure fabrication technique, able to realize spatially high resolution, precise control of size and position, and be applicable for various materials.

Conventional techniques popularly used for nano-structure fabrication are X-ray or electron beam lithography and the self-organized growth technique [7]. The former makes it easy to produce the desired pattern, but is disadvantaged by the complexity associated with a multi-step process that involves pattern definition and transfer. The latter easily produces high-quality nano-scale dots, but control over position and size is difficult.

To date, lithographic technology has been utilized for the fabrication of submicron structures using optical near-field [8]. However, because of the complexity of the process and the damage done to substrate and film by etchants, there are a lot of problems that still need to be solved. As a technology with the potential to solve those problems, the photo-enhanced chemical vapor deposition (PE-CVD) method has been attracting attention. Based on photochemical reaction, PE-CVD offers not only the possibility for the lateral integration of different structures (different sequence of layers, materials, thickness and dopants) in a truly single growth run [9,10] but also the option of chemical selective growth by varying the wavelength of the light source used.

Since the optical near-field energy is concentrated within nanometric dimensions smaller than the wavelength of light [1–3], we are able to deposit various materials of nanometric dimensions by utilizing the photo-decomposition of chemical gases. The combination of optical near-field technology with the PE-CVD process thus appears to be the most suitable technology for the integration of nanometer scale elements, because it not only allows us to fabricate nanostructures but also gives dual advantage of *in-situ* measurement of the optical properties of the fabricated nanostructures.

With these considerations as the background, this chapter reviews the recent progress of authors' work towards fabricating nanometric materials to enable the realization of a nanometric photonic integrated circuit [11].

## 8.2 Principles

PE-CVD combined with an optical near-field, i.e., a near-field optical-CVD (NFO-CVD), was carried out by utilizing the optical near-field generated from the subwavelength aperture at the tip of a fiber probe introduced into a vacuum chamber. The scanning of the probe was performed by a typical near-field optical microscope system. As NFO-CVD is based on a photodissociation reaction, it is necessary for a reactant molecule to absorb photons with a higher energy than its dissociation energy. Hence, as the light source, light with a higher energy than the dissociation energy of a reactant molecule should be used.

Two deposition mechanisms using the optical near-fields are represented in Fig. 8.1. One is the prenucleation method for fabricating prenuclei by decomposing only adsorbed molecules on the substrate with an optical near-field (Fig. 8.1(a)) [12]. The process consists of two steps. In the first step, metalorganic gas is put into the vacuum chamber for a few minutes and then evacuated, a process which will leave a few adsorbed monolayers on the substrate surface. By decomposing the adsorbed molecules with the optical near-field, nuclei for growth are formed. This happens patterning of the prenuclei is executed by scanning the probe. In the second step, conventional propagating light is directed onto the prenucleated area in the presence of a parent gas and then the decomposed atoms are selectively deposited on the pre-existing nuclei. This method has the advantage of being free from the deposition at the probe tip, but otherwise has the drawback that the lateral integration of various materials is not easy due to using the propagating light in the second step.

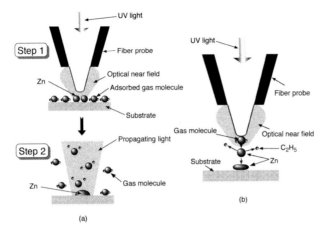

Fig. 8.1. Principle of NFO-CVD. (a) Prenucleation method. (b) Gas phase direct photodissociation method.

The other mechanism is to directly deposit atoms by gas phase photodissociation of metalorganic gas with the optical near-field (Fig. 8.1(b)) [13]. The advantage lies in the possibility of selective deposition of various materials by changing parent gases, which is useful for lateral integration.

## 8.3 Depositing Zinc

Let us discuss the deposition of Zn as an example of NFO-CVD by using a diethlzinc (DEZ) as a parent gas. This molecule absorbs light with a photon energy higher than 4.76 eV (wavelength below $\lambda = 260$ nm), and the photodissociation reaction occurs as follows:

$$Zn(C_2H_5)_2 + 2.256 \text{ eV} \rightarrow ZnC_2H_5 + C_2H_5,$$

$$ZnC_2H_5 + 0.9545 \text{ eV} \rightarrow Zn + C_2H_5.$$

Thus the second harmonic light (SH light) of an Ar$^+$ laser ($\lambda = 244$ nm) and an ArF excimer laser ($\lambda = 193$ nm) were used as the light source for photodissociation of DEZ gas. The SH light was generated in a BBO crystal placed in the built-up cavity. For the purpose of generating the UV optical near-field with sufficient power density to decompose DEZ gas, an UV fiber with transmission loss as low as 1.1 dB/m at 244 nm was used to fabricate a probe [14]. An UV probe was coated with 200 nm thick Al film after being tapered by chemical etching. The shape of this probe is shown in Fig. 8.2. The throughput of the probe was $1 \times 10^{-4}$ and the power density at the probe tip with

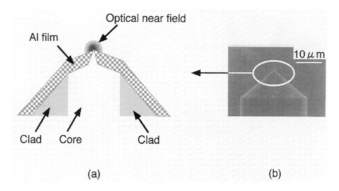

Fig. 8.2. Cross sectional profile (a) and SEM image (b) of the fiber probe for UV light.

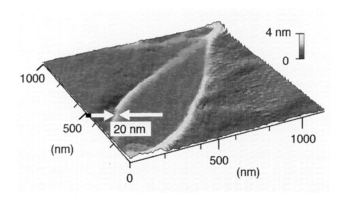

Fig. 8.3. Shear force image of a loop-shaped Zn deposited on a glass substrate by the prenucleation method.

a sub-100 nm aperture for 1 mW of incident UV light was as high as 1 kW/cm$^2$. The separation between the probe and the substrate was maintained within several nanometers by a shear-force technique using a tuning fork based probe [15].

Figure 8.3 shows the shear force image of the loop-shaped Zn pattern on a glass substrate produced by the prenucleation method [12]. The vacuum chamber was evacuated to below 10$^{-5}$ Torr prior to the prenuclei fabrication stage, then filled with about 10 Torr of DEZ gas, maintaining the pressure for 20 min. Next, the chamber was re-evacuated to the pressure of 10$^{-5}$ Torr, leaving a few adsorbed monolayer on the substrate surface. Prenucleation was

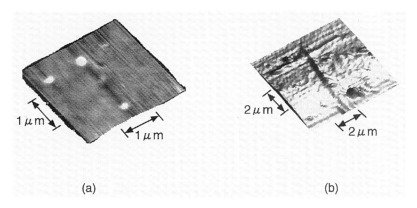

(a)                                                    (b)

Fig. 8.4. Shear force image of deposited Zn dots (a) and a T-shaped pattern (b) deposited on a glass substrate by gas phase direct photodissociation method.

performed by delivering the SH light on the substrate covered with adsorbed molecules using a probe. Nuclei of Zn were formed by the decomposition of DEZ gas adsorbed on the substrate with the optical near-field at the probe tip. In the growth stage after nuclei fabrication, DEZ gas was refilled in the chamber with a few Torr and the unfocused ArF excimer laser with the maximum energy of 10 mJ was directly irradiated on the prenucleated substrate. Then growth proceeded only on the pre-existing nuclei.

As Fig. 8.3 shows, the minimum width of the pattern is as small as 20 nm. The width achieved here is two orders smaller than the minimum width reported so far by conventional PE-CVD using a far-field light [16]. Since the measured width includes the resolution of a vacuum shear-force microscope(VSFM) [17] (depending on the shape of the used probe), the intrinsic width can be smaller than the value estimated from Fig. 8.3.

We will now discuss the results from the direct gas phase photodissociation method. Figures 8.4(a) and (b) show the shear force image of deposited Zn dots and a T-shaped pattern. The gas pressure and input power of SH light were 1 mTorr and 10 mW, respectively. In the fabrication of the dots, the optical near-field on the probe tip was illuminated over five spots at an interval of 800 nm on the substrate for a few seconds. As Fig. 8.4(a) shows, the position of the dots are spaced by 800 nm in excellent agreement with the spacings of the illuminated points, which establishes a high controllability of the positioning when fabricating nanostructures by this technique. The T-shaped pattern was prepared by scanning the substrate at a speed of 10–50 nm/s. A glance at Fig. 8.4(b) is sufficient to show that NFO-CVD makes it possible to fabricate subwavelength-scale structures with control of their size and position. One disadvantage of this method is that the probe tip is also gradually covered with the depositing materials while it is fabricating a nano-structure on the substrate.

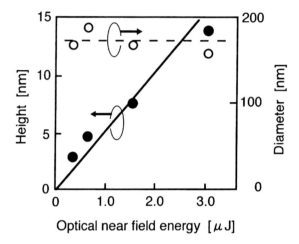

Fig. 8.5. Variation in the height (closed circles) and diameter (open circles) of the deposited Zn dots.

In our experience, however, this only becomes a problem after a few hours of operation, while only a few seconds are necessary to fabricate a nano-structure. It is therefore not a serious problem. NFO-CVD technique also allows us to fabricate nanostructure of oxides, insulators, and semiconductors as well as metals containing Zn, Al, Cr, and W.

In order to examine the effect of optical near-field energy on the growth, an experiment was performed by varying the illumination time at a constant gas pressure of 1 mTorr and an input SH light power of 15 mW. The SH light power at the probe tip was estimated to be 300 nW. The value of the optical near-field energy was evaluated by measured SH light power × illumination time. The substrate was kept at room temperature.

Figure 8.5 represents the variation in the height and diameter of the shear force image of the dots as a function of optical near-field energy. This figure shows that the Zn dots grew as the optical near-field energy increased, i.e., the height increased linearly at the rate of about 5 nm/$\mu$J in proportion to the increase of energy, while the diameter was almost constant with respect to the optical near-field energy. This means that the dot size depends on the spatial distribution of the optical near-field in a direction lateral to the substrate surface, while depending on the optical near-field energy in the normal direction. Thus the aspect ratio of the dot, i.e., the ratio of its height to diameter increased proportionally with the increase of optical near-field energy at a ratio of about 0.03/$\mu$J.

One of the most attractive features of this technique is its high spatial resolution. The lateral size of the fabricated pattern depends on the spatial

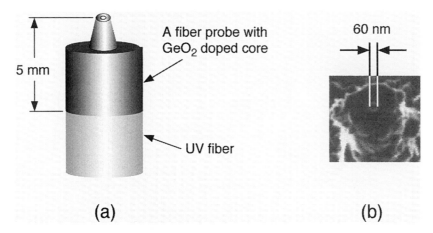

Fig. 8.6. Cross sectional profile (a) and SEM image (b) of the spliced fiber probe for UV light.

distribution of the optical near-field energy and its reproducibility also depends on the reproducibility of fabricating probes. Currently used probes are made of an optical fiber with a $GeO_2$ doped core, and have shown high resolution and high reproducibility. But the transmission loss for the light of 244 nm (96 dB/m) is so high that the fiber cannot be used in the UV region. To increase the reproducibility and spatial resolution of the UV fiber probe, we spliced the fiber probe with a $GeO_2$ doped core to the UV fiber, as is diagramatically explained by Fig. 8.6(a). The fiber probe was fabricated by the two-step etching process [18]. During the first step, the cladding diameter was reduced by immersing it in a buffered HF solution with a volume ratio of [40wt.% $NH_4F$ aqueous solution] : [50%wt.% HF acid] : [deionized water] = 1.7:1:1. In the next step, the fiber was selectively etched in the same kind of solution with a volume ratio of 10:1:1, while the temperature was kept at 25°C. After the etching process was completed, the probe was coated with 200 nm thick Al by vacuum evaporation. In order to form an aperture, the Al film was removed from the probe tip by focused ion beam. Figure 8.6(b) shows an SEM image of a fabricated probe with an aperture diameter of 60 nm. This probe was fabricated with a highly reproducible aperture size of less than 100 nm. Figures 8.7(a) and (b) show the shear force image of dots deposited by using a probe with an aperture diameter of 60 nm and a cross sectional profile along the dashed line, respectively. Two dots with a diameter of 60 nm and 70 nm (full width at the half maximum of the cross sectional profile) were fabricated at a very close distance of 100 nm. The diameter of the dots was comparable with the aperture diameter of the used probe, which suggests that smaller dots can be fabricated by using a fiber probe with a smaller aperture. Since the measured diameter of the dot image includes the resolution of VSHM (depending on the

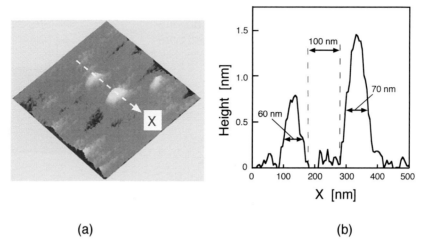

(a)                                                    (b)

Fig. 8.7. (a) Shear force image of Zn dots fabricated by using a spliced fiber probe with an aperture diameter of 60 nm. (b) Cross sectional profiles of the dots along the dashed line in (a).

shape of the used probe), the intrinsic diameter may be smaller than the value estimated from Fig. 8.7.

## 8.4 Depositing Zinc Oxide

One of the advantages of this NFO-CVD is the fact that there is no limitation in regard to substrate and deposited materials. Another technique in which CVD is combined with scanning tunneling microscopy (STM) has been reported to be able to produce nanostructures with dimensions close to the atomic level [19]. However, the main drawback comes from the impossibility of using non-conductive substrates and growing patterns of non-conductive materials. However, this limitation is eliminated by using an optical near-field. As an example, let us demonstrate the nanofabrication by NFO-CVD of ZnO on an $\alpha$-Al$_2$O$_3$ substrate as insulator [20].

As a preliminary experiment, ZnO films were deposited by the PE-CVD method by using a propagating far-field light. Here, we used the reaction between oxygen and DEZ conveyed by a carrier gas (Ar) into the chamber during the irradiation by the propagating SH light ($\lambda = 244$ nm) of an Ar$^+$ laser. (0001)$\alpha$-Al$_2$O$_3$ was used as a substrate for the epitaxial growth of ZnO. The reaction chamber was initially evacuated to a pressure in the low $10^{-5}$ to $10^{-7}$ Torr range, then filled with the reactant gases at the ratio of DEZ:O of 1:10 at a working pressure of 10 mTorr. The chamber pressure was maintained at 10 mTorr during the growth. In order to find the optimal growth conditions, the crystallinity, stoichiometry, optical transmission, and photoluminescence were

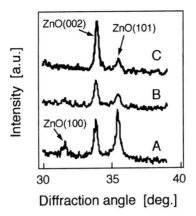

Fig. 8.8. X-ray diffraction patterns of ZnO films deposited at the substrate temperatures of 150°C (A), 200°C (B), and 300°C (C).

evaluated. The crystallinity and the stoichiometry were evaluated by X-ray diffraction (XRD) measurement using Cu K$\alpha$ radiation and X-ray photoelectron spectroscopy (XPS) using Al K$\alpha$ X-ray. The optical transmission measurements were carried out by a double-beam monochromator in an atmospheric ambient at wavelengths from 300 to 1000 nm to measure the optical energy band gap of films. The photoluminescence was measured at room temperature using a cw He-Cd laser as an excitation light source.

The PE-CVD of ZnO was carried out for 10 min within a range of the substrate temperature from room temperature to 300°C. The energy density of the laser source and the spot size were 10 mW and 600 $\mu$m, respectively. The films were grown on only the irradiated area and the photo-irradiation effect for the deposition rate was clearly observed for all substrates.

Figure 8.8 shows X-ray diffraction patterns of ZnO films deposited at various substrate temperatures. The rise in the surface temperature of the substrate from UV light was negligible because the light power was as low as 0.2 mW/cm$^2$, which means that the deposition resulted from a photochemical reaction between DEZ and oxygen. The film deposited at room temperature exhibits no XRD reflection lines, which implies that it is essentially amorphous. Crystalline films were deposited at substrate temperatures over 100°C and the films with c-axis oriented crystalline, exhibiting the (002) XRD lines, were grown at substrate temperatures above 150°C. With the increase of substrate temperature above 150°C, the intensity of the X-ray peak from (002) planes became stronger, while peaks such as the (101) peak became weaker. The c-axis lattice constant was estimated to be 0.5207 nm from the peak position of the (002) line, which is very close to the c-axis value of 0.5206 nm reported for bulk ZnO [21]. This indicates that these films have a high-quality crystalline

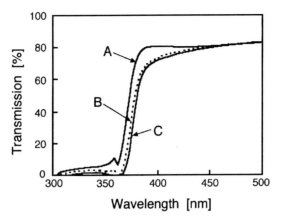

Fig. 8.9. Optical transmission spectra of ZnO films deposited at the substrate temperature of 150°C (A), 200°C (B), and 300°C (C). They were measured at room temperature.

structure. The stoichiometry was confirmed from the XPS spectrum for the planar films. For films deposited at a substrate temperature above 150°C, the atomic ratios of Zn:O were also 1.00:1.00 within an accuracy of a few percent.

The optical properties were also investigated. Figure 8.9 shows the optical transmission spectra of films deposited at substrate temperatures from 150°C to 300°C. Transmission fell off steeply at around 380 nm, a characteristic of high-quality ZnO film. From the plot of the wavelength vs. the absorption coefficient, optical band gap energies ranging from 3.26 to 3.31 eV were estimated, which is identical to the value recorded for high-quality ZnO films [22]. The photoluminescence spectra were measured using the 325 nm line of a cw He-Cd laser. Figure 8.10 shows the emission spectra measured at room temperature from films deposited at the substrate temperature from 150°C to 300°C. The emission peak position is coincident with the expected energy of the free exciton, and even at room temperature, a strong free exciton emission at 380 nm can be clearly observed [23,24]. This confirms that if we use PE-CVD method at a low temperature, a ZnO film emitting UV light at room temperature can be fabricated. This was the first observation of room temperature UV emission from ZnO films deposited by PE-CVD.

Under the growth conditions studied by the preliminary experiments mentioned above, the ZnO nanostructure fabrication was carried out by NFO-CVD. The fabrication was performed by introducing the SH light of an Ar⁺ laser onto the $\alpha$-Al$_2$O$_3$ substrate surface through the fiber probe which is shown in Fig. 8.6(a). The maximum power of the SH light was 10 mW, which means that the light power density at the probe tip was of the order of several kW/cm². During the deposition, substrate-probe tip separation was maintained at 10 nm by the shear force technique.

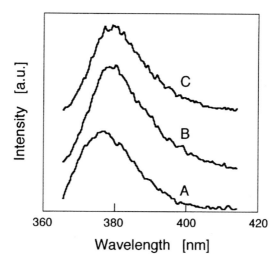

Fig. 8.10. Photoluminescence spectra of ZnO films deposited at the substrate temperature of 150°C (A), 200°C (B), and 300°C (C). They were measured at room temperature.

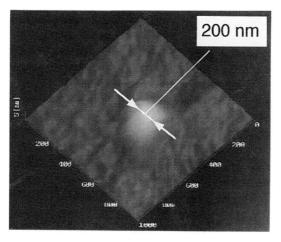

Fig. 8.11. Shear force image of a ZnO dot deposited on the (0001) $\alpha$-Al$_2$O$_3$ substrate by NFO-CVD.

Figure 8.11 shows the VSFM image of fabricated nanometric scale ZnO. The dot was of a 200 nm diameter and a 5 nm height. The diameter is smaller than the wavelength of the irradiating light source. However, because this value includes broadening due to the resolution of VSFM, the real diameter should be much smaller than that observed. We have actually succeeded in making Zn dots and wires below 100 nm with the same probe by optimizing growth conditions, as Fig. 8.8 shows.

## 8.5 Toward Nano-Photonic Integration

Future optical transmission systems require an advanced photonic integrated circuit (IC) for increasing speed and capacity. To meet this requirement, its size should become much smaller than that of a conventional diffraction-limited photonic IC, which is composed of diode lasers, optical switches, optical waveguides, and so on. The concept of such a nano-photonic IC is shown in Fig. 8.12, where metallic wires, light emitters, optical switches, input/output terminals, and photo-detectors are all controlled by nano-scale single dots and lines [6]. These devices use the optical near-field as a carrier for signal transmission. For this integration, we need an advanced nano-structure fabrication technique, able to realize spatially high resolution and precise control of size and position. Furthermore, various materials of a nanometric size must be integrated laterally on a substrate, i.e., not a point-contact type but a planar-type near-field optical system should be realized.

As has been demonstrated in the previous sections, the NFO-CVD constitutes a very promising tool for *in-situ* patterning of nanostructures for this integration. Nanoscale Zn dots and lines with the size of sub-100 nm have been successfully realized in a single growth step. We have also shown that

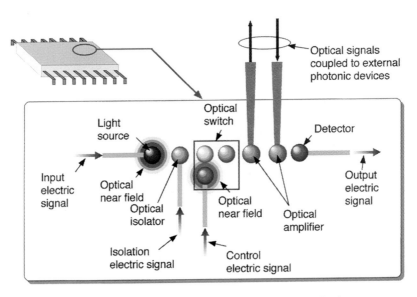

Fig. 8.12. Concept of a planar nano-photonic integrated circuit.

ZnO nanostructures emitting UV light at the room temperature can be fabricated. Furthermore, this technique exhibits extraordinarily high controllability and reproducibility in fabricating nanostructures at desired position. This has never been demonstrated by any other conventional self-organized growth technique for semiconductor nanostructures. What is excellent is that as it is based on a photodissociation reaction, selective growth of various materials, i.e., metals, insulators, and semiconductors, can be accomplished by the choice of light source, which allows us to realize a nano-photonic IC composed of nanostructures.

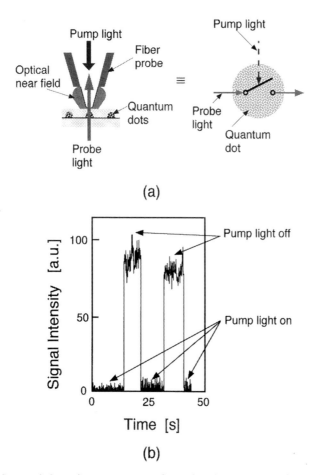

Fig. 8.13. Saturated absorption spectroscopy of a semiconductor quantum dot. (a) Schematic explanation of the principle. (b) An experimental result of optical switching properties of a single InGaAs quantum dot.

To operate a nano-photonic IC, several nonlinear and resonant optical phenomena such as the dynamic Stark effect, optical bistability utilizing nonlinearity and plasmon-like field enhancement in a quantum dot [25], and so on, should be utilized. In order to demonstrate, e.g., a nonlinear optical switching capability of a single quantum dot, we measured the nonlinear absorption change in a single InGaAs quantum dot [26]. For this experiment, we used self-assembled quantum dots grown on a (100) GaAs substrate by gas-source molecular beam epitaxy. Atomic force microscope studies confirmed that the average dot diameter was 30 nm and the height was 15 nm. The dot density was about $2 \times 10^{10}$ dots/cm$^2$. These dots were covered with cap layers with a total thickness of 180 nm. As diagramatically explained in Fig. 8.13(a), a probe light ($\lambda = 900$–980 nm) was introduced into the back of the sample, and the transmitted light was collected and detected by a high throughput fiber probe with a double tapered structure [27] placed in the vicinity of the sample surface using the shear force technique. When the sample was illuminated by a pump light ($\lambda = 635$ nm) passing through the fiber probe, carriers were generated in the barrier layer, and flew into the quantum dots. The ground states of the quantum dots were occupied by the carriers. The resultant reduction of the absorption of the quantum dots was measured by detecting the transmission change in the probe light. Experiments were carried out in a liquid helium cryostat, where a double modulation/demodulation technique was used for detecting very weak signals.

Figure 8.13(b) shows an experimental result demonstrating a deep modulation of the transmitted probe light power due to irradiation of the pump light. This result confirms that a single quantum dot works like an optical switch, and moreover, the switching operation can be detected by a conventional optical signal detection technique, which is advantageous for application to nano-photonic IC. Faster switching operation can be expected by utilizing nonlinear optical phenomena involving virtual transitions. Problems having to be solved for realizing a practical nano-photonic IC are coupling with input/output ports, signal isolation, estimating the limits of bandwidth, gain/efficiency, switching time and power, and so on.

By combining the technique reviewed here with atom manipulation by the optical near-field [28], further progress in depositing novel materials and operating more advanced photonic IC can be expected. For this progress, theoretical works are required to establish the design criteria for optical near-field systems, for which a quantum theoretical treatment of optical near-fields has been developed to establish an intuitive model of Yukawa potential [29].

## 8.6 Summary

This chapter reviewed the recent progress of application of interactions between optical near-fields and nanoscale materials. Photochemical vapor deposition of Zn dots and lines with a size of sub-100 nm was realized by using an UV optical near-field. Deposition of nanoscale ZnO was also shown. Optical near-field technology offers the opportunity of modifying surfaces and developing new nanostructures that may exhibit a quantum effect due to their extremely small size. Utilizing the very advanced potential of this technology, the concept of nano-photonic IC was proposed. The optical switching operation of a single InGaAs quantum dot was shown to be able to be used for nano-photonic devices.

## Acknowledgments

The authors would like to thank Mr. Y. Yamamoto (Tokyo Institute of Technology) and Dr. V. Polonski (Japan Science and Technology Corporation) for their devoted collaboration. They acknowledge Prof. H. Hori (Yamanashi University), Dr. K. Kobayashi (Japan Science and Technology Corporation), and Dr. M. Kourogi (Tokyo Institute of Technology) for their valuable comments and suggestions.

[References]

1)  M. Ohtsu, *Near-Field Nano/Atom Optics and Technology*, Springer-Verlag, Berlin, Tokyo, New York (1998)

2)  M. Ohtsu, *J. Lightwave Technol.*, **13**, 1200 (1995)

3)  M. Ohtsu and H. Hori, *Near-Field Nano-Optics*, Kluwer Academic/Plenum Publishers, New York (1999)

4)  Uma Maheswari Rajagopalan, S. Mononobe, K. Yoshida, M. Yoshimoto and M. Ohtsu, *Jpn. J. Appl. Phys.*, **38**, 6713 (1999)

5)  R. Uma Maheswari, H. Kadono and M. Ohtsu, *Opt. Commun.*, **131**, 133 (1996)

6)  M. Ohtsu, *Technical Digest of the 18th Congress of the International Commission for Optics*, SPIE, **3749**, 478 (1999)

7)  D. Leonald, M. Krishnamurthy, C. M. Reaves, S. P. Denbaas and P. M. Petroff, *Appl. Phys. Lett.*, **63**, 3203 (1993)

8)  I. I. Smolyaninov, D. Mazzoni and C. C. Davis, *Appl. Phys. Lett.*, **67**, 3859 (1995)

9)  E. Maayan, O. Kreinin, G. Bahir, J. Salzman, A. Eyal and R. Beserman, *J. Crystal Growth*, **135**, 23 (1994)

10) D. Bauerle, *Appl. Surface Sci.*, **106**, 1 (1996)

11) M. Ohtsu, K. Kobayashi, H. Ito and G. H. Lee, *Proc. IEEE* (2000) (to be published)

12) V. V. Polonski, Y. Yamamoto, M. Kourogi, H. Fukuda and M. Ohtsu, *J. Microscopy*, **194**, 545 (1999)

13) Y. Yamamoto, M. Kourogi, M. Ohtsu, V. Polonski and G. H. Lee, *Appl. Phys. Lett.*, **76**, 2173 (2000)

14) S. Mononobe, T. Saiki, T. Suzuki, S. Koshihara and M. Ohtsu, *Opt. Commun.*, **146**, 45 (1998)

15) G. T. Ruiter, *Appl. Phys. Lett.*, **71**, 28 (1997)

16) D. Ehrlich, R. M. Osgood, Jr. and T. F. Deutch, *J. Vac. Sci. Technol.*, **21**, 23 (1982)

17) V. V. Polonski, Y. Yamamoto, J. D. White, M. Kourogi and M. Ohtsu, *Jpn. J. Appl. Phys.*, **38**, L826 (1999)

18) T. Pangaribuan, K. Yamada, S. Jiang, H. Ohsawa and M. Ohtsu, *Jpn. J. Appl. Phys.*, **31**, L1302 (1992)

19) R. Wiesendanger, *Appl. Surf. Sci.*, **54**, 271 (1992)

20) G. H. Lee, Y. Yamamoto, M. Kourogi and M. Ohtsu, *SPIE 44th Annual Meeting*, 3791-18, Denver, CO, July 18–23 (1999)

21) O. Madelung, *Landolt-Bernstein New Series*, Vol. 17b, Springer-Verlag, Berlin (1982)

22) T. Y. Ma, G. C. Park and K. W. Kim, *Jpn. J. Appl. Phys.*, **35**, 6208 (1996)

23) Z. T. Tang, G. K. L. Wong and P. Yu, *Appl. Phys. Lett.*, **72**, 3270 (1998)

24) D. B. Bagnall, Y. F. Chen, Z. Zhu and T. Yao, *Appl. Phys. Lett.*, **70**, 2230 (1997)

25) S. Schmitt-Rink, D. A. B. Miller and D. S. Chemla, *Phys. Rev. B*, **35**, 8113 (1987)

26) T. Matsumoto, M. Ohtsu, K. Matsuda, T. Saiki, H. Saito and K. Nishi, *Appl. Phys. Lett.*, **75**, 3246 (1999)

27) T. Saiki, S. Mononobe, M. Ohtsu, N. Saito and J. Kusano, *Appl. Phys. Lett.*, **68**, 2612 (1996)

28) H. Ito, K. Sakaki, M. Ohtsu and W. Jhe, *Appl. Phys. Lett.*, **70**, 2496 (1997)

29) K. Kobayashi and M. Ohtsu, *J. Microscopy*, **194**, 249 (1999)

*Chapter 9*


# Noncontact Atomic Force Microscopy


## 9.1 Introduction—Historical Background

The atomic force microscope (AFM), invented by Binnig *et al.* in 1986 [1], has been used for developing a novel technique for obtaining high-resolution surface images of both conductors and insulators. For several layered and nonlayered materials [2–5], atomic resolution has been achieved in the contact mode. However, the question has been raised as to whether the AFM operating in the contact mode is really a microscope like the scanning tunneling microscope with a "true" atomic resolution. That is, most of the reported data obtained with the AFM has shown either perfectly ordered periodic atomic structures or defects only on a large lateral scale [4], but no atomic-scale defects routinely observed by a scanning tunneling microscope (STM). Moreover, the usual contact mode imposes a repulsive force of 0.5–5 nN [4,6] or higher [5,6], which is greater than the ~0.1 nN acceptable for a single atom on the tip apex of an AFM cantilever and for a single atom on the sample surface. Furthermore, the AFM in the contact mode mainly measures so-called atomic-scale friction with a lattice periodicity [6] rather than the topography. The contact-mode AFM thus seems to have a large contact area between the tip and the sample [6], although in a few experiments measured under restricted conditions such as in solutions [7,9] or at low temperature [8], monoatomic step lines [7–9] and atomic-scale point defects [9] could be observed even in the contact mode. Research into the true atomic resolution of the AFM has therefore focused on resolving two problems. These are (1) to observe atomic-scale defects and (2) to observe a Si(111)7×7 reconstructed surface which is the standard sample of the ultra-high vacuum STM (UHV-STM).

In order to measure the monoatomic tip-sample force interaction using the AFM, the possibility of true atomic resolution in the noncontact mode has been studied in a solution. [9] Although the noncontact-mode AFM has the advantage of avoiding damage to the sharp tip and sample as a result of mechanical contact, atomic lateral resolution has not been achieved until recently in an ultra-high vacuum (UHV), because of the technical difficulty of measuring the weak distance-dependence of the attractive force. In 1995, nearly ten years after the invention of the AFM [1], a few groups [10–12] achieved true atomic resolution with a noncontact AFM in UHV using a frequency modulation (FM) detection method. [10] They succeeded in obtaining atomically resolved images of a Si(111)7×7 reconstructed surface [10,11] and an InP(110) cleaved surface [12] with atomic-scale point defects for the first time. Thus, these results resolved the above two problems of how to demonstrate the true atomic resolution of the AFM. However, the noncontact AFM in the early stage was unstable and showed sudden contrast changes during lateral scanning [10,11,13] due to the positional changes of atoms at the tip apex. This phenomenon was commonly observed with the STM in the early stage and is evidence of true atomic resolution. By improvements to the noncontact AFM system such as the FM detection circuit, and by the tip-cleaning of the AFM cantilever, the noncontact AFM became stable and reproducible in a step by step way, enabling us to observe even defective motion on an InP(110) [14] surface, and to observe insulators [15] and even a metal [16] with true atomic resolution.

In the following sections, the principles behind this, and, recent achievements and their significance are described. The aim of this is to show how to make observations, what can be imaged and what kind of information can be obtained using the noncontact AFM with the FM detection method.

## 9.2 Guidelines for Spatial Resolution [17]

Noncontact AFM with the FM detection method measured under UHV is rapidly developing as a scientific tool for true atomic resolution. However, many important problems such as the conditions needed to achieve true atomic resolution and the imaging mechanism remain. In this section, we will describe the conditions needed to achieve true atomic resolution with the noncontact AFM by deriving equations of vertical and lateral resolutions, and the decay length of the noncontact AFM.

### 9.2.1 Equation of vertical resolution in noncontact AFM

Here, the information $f(z)$ measured by the noncontact AFM, that is, the so called frequency shift, was assumed to be proportional to $\exp(-z/L)$, where $L$ is the decay length of the frequency shift. This means that by increasing the tip-to-sample distance from $z$ to $z + \delta z$, the frequency shift will decrease from $f(z)$ to $f(z + \delta z) = f(z)\exp(-\delta z/L)$. The signal-to-noise ratio of $f(z)$ was then defined as $k = S/N$. This means that the smallest measurable change of $f(z)$ is given by $\delta f(z) = f(z)/k$, which is equivalent to the noise level. If the smallest distance change controlled by a feedback loop is $\delta z$, the vertical resolution can be defined as $\delta z$. This also means that $f(z) - f(z + \delta z)$ is equivalent to the noise level $\delta f(z)$. We can thus obtain the relationship of $k = f(z)/\delta f(z) = f(z)/[f(z) -$

$f(z + \delta z)] = 1/[1 - \exp(-\delta z/L)]$. From this relationship, we derived the equation of vertical resolution

$$\delta z = L\ln[k/(k - 1)] \qquad (9.1)$$

as a function of the decay length $L$ and the signal-to-noise ratio $k$. By assuming $k \gg 1$, that is, $\delta z/L \ll 1$, Eq. (9.1) can be approximated as

$$\delta z \fallingdotseq L/k. \qquad (9.2)$$

Equations (9.1) and (9.2) suggest that we can obtain better vertical resolution by decreasing the decay length $L$, i.e., by making the tip-to-sample distance dependence of the frequency shift stronger, and by increasing the signal-to-noise ratio $k$.

### 9.2.2 Equations of lateral resolution

#### 9.2.2.1 Large tips ($R \gg z$)

By assuming that the radius $R$ of the tip curvature is much larger than the tip-to-sample distance $z$, i.e., $R \gg z$, the lateral resolution $\delta x$ is determined by the vertical resolution $\delta z$ as shown in Fig. 9.1, because the ambiguity of the tip-to-sample distance is $\delta z$. To be exact, the lateral resolution of the microscope is defined as the minimum distance required for two bumps to be separately

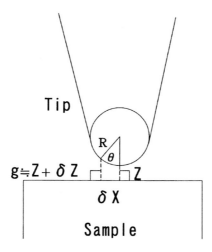

Fig. 9.1. Schematic simple one-dimensional model of the tip-to-sample configuration for $R \gg Z$. Here, $R$ and $Z$ are the tip radius and the tip-to-sample distance, respectively. Using this model, the relationship between vertical and lateral resolutions was derived for $R \gg Z$.

recognized, so that the present definition is an approximated simple one. From Fig. 9.1, we can obtain the relationship of $\delta z = R(1 - \cos\theta)$ and $\delta x = R\sin\theta$. From these relationship, we can derive the equation of lateral resolution

$$\delta x \fallingdotseq [\delta z(2R - \delta z)]^{1/2}. \tag{9.3}$$

From Eq. (9.3), we can obtain better lateral resolution by decreasing the radius $R$ of the tip curvature, and by improving the vertical resolution, that is, by decreasing the decay length $L$, and by increasing the signal-to-noise ratio $k$. However, from the assumption of $R >> z \gtreqqless z_c \sim 0.3$ nm [$z_c$; contact point which will be discussed in section 9.2.3.2], the radius $R$ of tip curvature cannot become smaller than ~3 nm. Therefore, the assumption of $R >> z$ corresponds to a rather worse lateral resolution.

By assuming $R >> \delta z$, we can rewrite Eq. (9.3) as

$$\delta x \fallingdotseq [2R\delta z]^{1/2}. \tag{9.4}$$

*9.2.2.2 Single atom tips (R << z)*

By assuming that the radius $R$ of the tip curvature is much smaller than the tip-to-sample distance $z$, i.e., $R << z$, the lateral resolution $\delta x$ can be determined by the vertical resolution $\delta z$ as shown in Fig. 9.2, because the ambiguity of the tip-to-sample distance is $\delta z$. From Fig. 9.2, we can obtain the relationship ($z +$

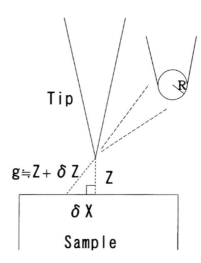

Fig. 9.2. Schematic simple one-dimensional model of the tip-to-sample configuration for $R <<$ Z. Using this model, the relationship between vertical and lateral resolutions was derived for $R << Z$.

$\delta z)^2 = z^2 + \delta x^2$. From this relationship, we can derive the equation of the lateral resolution

$$\delta x \fallingdotseq [\delta z(2z + \delta z)]^{1/2} \qquad (9.5)$$

as a function of the tip-to-sample distance $z$ and the vertical resolution. By assuming $z \gg \delta z$, we can rewrite Eq. (9.5) as

$$\delta x \fallingdotseq [2z\delta z]^{1/2}. \qquad (9.6)$$

Equations (9.5) and (9.6) suggest that we can obtain better lateral resolution by decreasing the tip-to-sample distance $z$ and by improving the vertical resolution $\delta z$. By interpolating from Eq. (9.4) to Eq. (9.6), we can derive the equation

$$\delta x \fallingdotseq [2(R + z)\delta z]^{1/2}. \qquad (9.7)$$

However, the "effective" radius $R$ will depend on both the vertical resolution $\delta z$ and the tip-to-sample distance $z$, so that Eq. (9.7) might be more complicated than it looks.

If we assume a tip-to-sample distance of $z = 1$ nm and a vertical resolution $\delta z = 0.01$ nm, we can obtain an approximate lateral resolution of $\delta x \fallingdotseq 0.14$ nm from Eq. (9.5). This result means that if we can achieve the vertical resolution of $\delta z = 0.01$ nm by combining a small decay length $L$ and a large signal-to-noise ratio k, we can achieve true atomic resolution with the lateral resolution of $\delta x \fallingdotseq 0.14$ nm at the tip-to-sample distance $z = 1$ nm. Here, $\delta z = 0.01$ nm and $\delta x \fallingdotseq 0.14$ nm seem to indicate that the topmost atom of the tip detects atomic force exerted by its nearest neighbor atom on the sample surface. In this case, the "effective" radius $R$ of the tip becomes the radius of the topmost atom, i.e., the single atom probe because of the monoatomic tip, and the condition of $R \ll z$ will be nearly satisfied.

From the above discussion, we can obtain guidelines for the achievement of true atomic resolution with the noncontact AFM as follows. At first, we should decrease the decay length $L$ and increase the signal-to-noise ratio $k$ to improve the vertical resolution $\delta z \fallingdotseq L/k$ down to nearly 0.01 nm. We can achieve then true atomic resolution with the lateral resolution of $\delta x \fallingdotseq 0.14$ nm at $z = 1$ nm. These guidelines for the achievement of true atomic resolution seem to be very useful because they can be applicable to all kinds of scanning probe microscopes (SPM). It should be also noted that the experimental decay length $L$ becomes smaller by decreasing the effective radius $R$ of the tip and also by removing the contamination on the tip apex. By decreasing the oscillation amplitude used by the frequency modulation (FM) detection method in noncontact atomic force microscopy, the experimental decay length $L$ also becomes smaller.

### 9.2.3 The equation of decay length

#### 9.2.3.1 The equation of decay length in the frequency shift

Now we will deduce the equation of decay length in the frequency shift. Firstly, we will assume the tip-to-sample distance dependence of the frequency shift to be $f(z) = A/z^n$ ($n$: integer). Next, we will estimate the tip-to-sample distance dependence of the frequency shift as being $f(z_0) = B\exp(-z_0/L)$ around the distance $z_0$, where we will measure the frequency shift. By increasing the tip-to-sample distance from $z_0$ to $z_0 + \delta z$, $f(z_0) = A/z_0^n$ decreases to $f(z_0 + \delta z) = A/(z_0 + \delta z)^n$, while $f(z_0) = B\exp(-z_0/L)$ decreases to $f(z_0 + \delta z) = B\exp(-[z_0 + \delta z]/L)$. Under the assumption of $\delta z \ll z_0$, we can obtain $f(z_0 + \delta z)/f(z_0) = z_0^n/(z_0 + \delta z)^n \fallingdotseq 1 - n\delta z/z_0$ and $f(z_0 + \delta z)/f(z_0) = \exp(-\delta z/L) \fallingdotseq 1 - \delta z/L$, respectively. By combining both equations, we can obtain the decay length

$$L \fallingdotseq z_0/n. \tag{9.8}$$

Thus, for a smaller distance $z_0$ and for a force interaction with a larger integer $n$, we can obtain a smaller decay length.

In this section, the effect due to the large oscillation amplitude used by the frequency modulation (FM) detection method in noncontact atomic force microscopy is neglected for the simplicity and a small oscillation amplitude is assumed.

#### 9.2.3.2 The distance between atoms at the contact point

If we would like to derive a possible decay length using Eq. (9.8), we have to determine the tip-to-sample distance where true atomic resolution can be achieved. Under experimental conditions, true atomic resolution can be achieved only between 0 nm~0.4 nm just before the contact point, as will be shown in section 9.5.3. Then we should, roughly estimate the distance between atoms at the contact point in the frequency shift measurement where the distance dependence of the frequency shift begins to weaken, that is, the oscillation amplitude begins to decrease, as will be discussed in section 9.5.2. Using the Lennard-Jones potential $U = 4\varepsilon[(\sigma/z)^{12} - (\sigma/z)^6]$, we can derive the equilibrium distance of $z_a = \sigma(2)^{1/6} \fallingdotseq 1.12\sigma$ between atoms where $f = -\partial U/\partial z = 0$ is satisfied. This distance $z_a$ may be roughly approximated by the spacing between the nearest neighbor atoms, that is, ~0.235 nm for Si crystal, and 0.244~0.254 nm for GaAs and InP crystals. However, the distance dependence of the force gradient will begin to weaken below the deflection point of the force gradient. This deflection point is defined as $\partial^2 f/\partial z^2 = -\partial^4 U/\partial z^4 = 0$. Thus, if the measured frequency shift is assumed to be proportional to the force gradient, the contact point $z_c$ will be given by $z_c = \sigma(65/6)^{1/6} \fallingdotseq 1.49\sigma$, that is, about 1.3 times the equilibrium spacing between atoms. Thus, the estimated distances between atoms at the contact point for Si(111)7×7 and for GaAs(110) and InP(110) are roughly ~0.3 nm.

### 9.2.3.3 Possible decay length in frequency shift

If we assume a single atom probe, the van der Waals potential $U = B/z^m$ ($m$: integer) will be given by $m = 6$. Its force $f = -\partial U/\partial z$ and force gradient $f' = \partial f/\partial z = -\partial^2 U/\partial z^2$ will be given by $m + 1 = 7$ and $n = m + 2 = 8$, respectively. The distance $z_0$ in Eq. (9.8) will be given by summing up the distance at the contact point $z_c$ and the measured distance 0~0.4 nm just before the contact point. Thus, by approximating the distance $z_0$ in Eq. (9.8) as $z_0 \fallingdotseq 0.3$~0.7 nm, we obtained the possible decay length of the force gradient due to the van der Waals potential of L $\fallingdotseq 0.04$~0.09 nm. However, the electrostatic (Coulomb) potential and its force gradient will be given by $m = 1$ and $n = m + 2 = 3$, respectively. Thus, at the distance $z_0 \fallingdotseq 0.3$~0.7 nm, the decay length of the force gradient due to the electrostatic potential becomes L $\fallingdotseq 0.1$~0.23 nm. Therefore, we can expect to achieve a true atomic resolution by the noncontact AFM with a signal-to-noise ratio of 0.75~1.75 times for the van der Waals force gradient and 2.0~4.7 times for the electrostatic force gradient compared with that of the scanning tunneling microscope, whose decay length is about 0.05 nm.

The smallest decay length of the force gradient obtained experimentally with n-GaAs(110) was about 0.3 nm for the van der Waals potential and about 0.5 nm for the electrostatic potential respectively, as will be shown in section 9.8, which roughly agrees with the values derived above.

However, in case of the large oscillation amplitude $A_0 \gg z_0$, used by the frequency modulation (FM) detection method in the noncontact atomic force microscopy, the detected frequency shift is not proportional to the force gradient. According to the result given by Giessibl, [18] the decay length in Eq. (9.8) will be given by $n \fallingdotseq 6.5$ for the van der Waals potential and by $n \fallingdotseq 1.5$ for the electrostatic (Coulomb) potential in Eq. (9.8), respectively, [18] which increases the possible decay length of the frequency shift up to L $\fallingdotseq 0.05$~0.11 nm for the frequency shift due to the van der Waals potential and up to L $\fallingdotseq 0.2$~0.47 nm for the frequency shift due to the electrostatic potential, respectively. Thus, this effect decreases the difference between the possible decay length and the one experimentally obtained.

### 9.3 Problems in AFM Measurement under the Contact Mode

Most of the high-resolution data obtained with the contact-mode AFM showed only periodic atomic structures, and no atomic-scale defects. Moreover, the usual contact mode imposes a repulsive force greater than the ~0.1 nN acceptable for a single atom on a tip apex and on the sample surface. Thus, the contact-mode AFM seems to have a large contact area between the tip and the sample as Fig. 9.3(a) shows. Therefore, in order to achieve true atomic resolution, the possibility of the monoatomic tip measurement using the AFM was studied in the noncontact mode, as Fig. 9.3(b) shows.

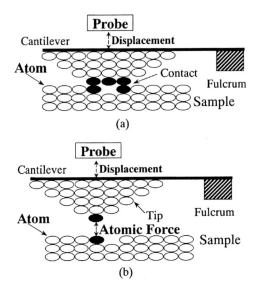

(a)

(b)

Fig. 9.3. Schematic model of (a) contact AFM and (b) noncontact AFM imaging. The force probing region shown by closed circles in the contact AFM is larger than that in the noncontact AFM. Only the noncontact AFM has the possibility of obtaining true atomic resolution because of the possibility of doing monoatomic tip measurement.

### 9.4 Noncontact Atomic Force Microscopy (Experimental Method)

In our experiment, the vertical displacement of the cantilever was detected by an optical fiber interferometer [6] as Fig. 9.4 shows, although in most of the other groups, the deflection of the cantilever was detected with an optical lever method. [6] As Fig. 9.4 shows, the cantilever was vibrated by the piezoelectric tube scanner and the mechanical resonant frequency $v$ of the cantilever was detected by the frequency modulation (FM) detection method. [19] When the force from the sample surface can be neglected, the mechanical resonant frequency $v$ of the cantilever becomes

$$v_0 = (k/m_0)^{1/2}/2\pi \tag{9.9}$$

by assuming the cantilever oscillation of deflection to be a one-dimensional harmonic oscillator where $k$ and $m_0$ are the spring constant of deflection and the effective mass of the cantilever, respectively. Using the vertical displacement $\delta Z$ and the spring constant $k_a$ between the cantilever and the sample, the increase in the attractive force $-\delta f$ between the cantilever and the sample can be written as $-\delta f = k_a \delta Z$. Therefore, the spring constant $k_a$ between the cantilever and the sample is given by $k_a = -\delta f/\delta Z \fallingdotseq -\partial f/\partial Z$. By assuming that

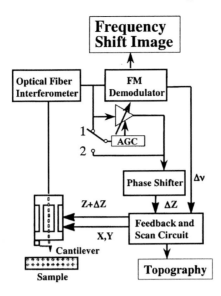

Fig. 9.4.  Schematic model of the noncontact AFM using the FM detection method.

the effective spring constant of the cantilever under the attractive force $-\delta f$ can be written as $k_{eff} \doteqdot k + k_a$, the mechanical resonant frequency $v$ of the cantilever under attractive force can be rewritten as

$$v = ([k - \partial f/\partial Z]/m_0)^{1/2}/2\pi. \tag{9.10}$$

By assuming the weak interaction $k >> |\partial f/\partial Z|$, the frequency shift $\Delta v = v - v_0$ becomes

$$\Delta v \doteqdot -(\partial f/\partial Z)v_0/2k \tag{9.11}$$

as Fig. 9.5 shows. Thus, under the assumptions of a weak interaction and a small oscillation amplitude, the frequency shift measured by the FM detection method is proportional to the force gradient, i.e., the $Z$ derivative of force between the tip of the cantilever and the sample. Therefore, the FM detection allows us to detect a weak force gradient with high sensitivity because of a high Q value of the cantilever in UHV.

The two operation modes (constant-vibration mode and constant-excitation mode) can be changed by switching the input signal into the automatic gain control (AGC) circuit. [20] In the constant-vibration mode (switch 1 is "ON"), the input signal to the AGC circuit is the vertical displacement signal of the

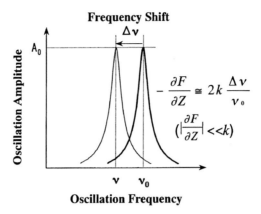

**Fig. 9.5.** Oscillation amplitude as a function of the oscillation frequency, and the frequency shift $\Delta v$ due to the tip-to-sample surface interaction.

cantilever, as Fig. 9.4 shows. The AGC circuit controls the gain of the variable-gain amplifier to maintain the oscillation amplitude of the cantilever constant. In the constant-excitation mode (switch 2 is "ON"), the input signal to the AGC circuit is the output signal of the variable-gain amplifier. The AGC circuit maintains the excitation voltage supplied to the piezoelectric tube scanner constant.

When the force gradient acting on the tip exceeds the spring constant of the cantilever, the cantilever with a weak spring constant jumps into contact with the sample surface. However, to obtain the smallest decay length and the smallest vertical resolution, we have to bring the tip toward the sample surface to decrease the tip-to-sample distance $z$ under the noncontact condition as much as possible. Moreover, in order to obtain the reproducible distance-dependence of the frequency shift which enables us to investigate the force interaction, it is necessary to avoid destruction of the tip apex and the sample surface as a result of a crash caused by the jump into contact. We therefore used a stiffer cantilever than that (typically less than 1 N/m) used in the contact mode. As a force sensor, we used a conductive Si cantilever with a sharpened tip. Its spring constant and mechanical resonant frequency were about 27–41 N/m and 151–172 kHz, respectively. The nominal radius of the tip apex was 5–10 nm. The Q factor of the cantilever was estimated to be about 38,000 in UHV. The initial surface of the tip was considered to be contaminated and had to be covered by a thin oxide layer. To remove the contamination and the oxide layer on the tip apex, we used Ar ion sputtering.

Noncontact AFM measurements were performed at room temperature under a pressure of less than $4 \times 10^{-10}$ Torr.

## 9.5 Experimental Results on Compound Semiconductors

Samples of compound semiconductors were cut from a (001) wafer and those were then cleaved parallel to the {110} faces in UHV at pressures lower than $(1–4) \times 10^{-10}$ Torr. Cleaved surfaces on compound semiconductors were used to avoid the chemical bonding between the Si tip and the surface, because it has no surface state in the band gap and is chemically inert compared with the Si(111)7×7 surface.

### 9.5.1 Defects on a semi-insulating InP(110) cleaved surface [21]

The samples were cut from an Fe-doped semi-insulating InP(001) wafer. It was then cleaved parallel to the {110} faces in UHV and its surface was then exposed to oxygen gas. No Ar ion sputtering was used and no bias voltage was applied between the tip and the sample. Figure 9.6 shows noncontact AFM topographies of the InP(110) cleaved surface which were sequentially imaged at 80 seconds' interval. Here, the feedback circuit in Fig. 9.4 was used to maintain the frequency shift constant, and bright and dark sites in the topographies show the high and low areas. Scan areas were 9.2 nm × 9.2 nm. The atomic-scale point defects as well as the rectangular lattice are clearly resolved. The InP(110) surface is characterized by quasi-one-dimensional zigzag chains. The chains consist of alternating In and P atoms, as shown in Fig. 9.6. The electrons in In dangling bonds transfer partially to P atoms, and the P atoms move out from the surface, resulting in buckling. The distances between the protrusions along [001] and [1 $\bar{1}$ 0] directions are in agreement with the length of 0.587 nm and 0.415 nm for the unit cell. At present, we are unable to clearly resolve the quasi-one-dimensional zigzag chains consisting of alternating In and P atoms.

Interestingly, motion of the atomic-scale point defects was observed during the measurement. In the sequential images of Fig. 9.6, the separation between the point defects (labeled A and B) clearly changed from image to image. By comparing the positions of these two point defects with those of other point defects, it was found that only the point defect A was moving around. In Fig. 9.6(a), two point defects (labeled A and B) can be seen with a separation of three atomic rows and one unit cell along the [001] and the [1 $\bar{1}$ 0] directions, respectively, but in Fig. 9.6(b), two point defects (labeled A and B) can be seen with a separation of three atomic rows and one unit cell toward the opposite side along the [001] and the [1 $\bar{1}$ 0] directions, respectively. This indicates that from Figs. 9.6(a) to (b), the point defect A has moved two unit cells along the [1 $\bar{1}$ 0] direction. Furthermore, in Fig. 9.6(c), two point defects (labeled A and B) can be seen with a separation of two atomic rows and two unit cells along the [001] and the [1 $\bar{1}$ 0] directions, respectively. This means that from Figs. 9.6(b) to (c), the point defect A moved one row and two unit cells toward the opposite side along the [001] and the [1 $\bar{1}$ 0] directions,

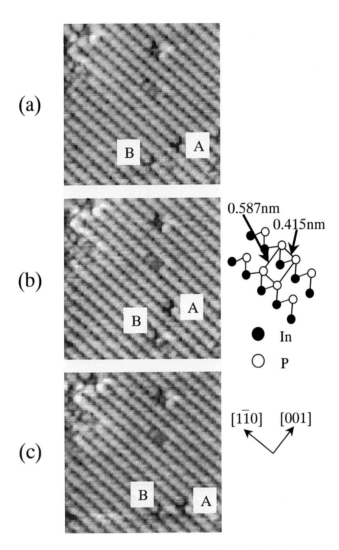

Fig. 9.6. Series of the noncontact AFM topography on the semi-insulating Fe-doped InP(110) surface exposed to 300 L of O₂ gas. Scan areas were 9.2 nm × 9.2 nm. The topographies were obtained in intervals of about 80 s. Several point defects as well as the rectangular lattice are clearly observed. A schematic model of quasi-one-dimensional zigzag chains on InP(110), consisting of alternating In and P atoms, is attached.

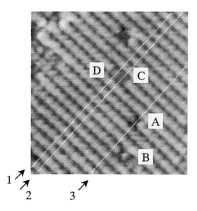

Fig. 9.7. The same noncontact AFM topography as in Fig. 9.6(a). Defects A, C, D show three kinds of defects with different depths.

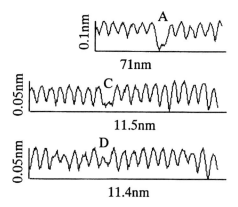

Fig. 9.8. Three line profiles through three kinds of defects, A, C and D, respectively, which were measured along the white lines shown in Fig. 9.7. These line profiles show the depth of three defects measured with the noncontact AFM.

respectively. Thus, there are two types of point defects; one is movable and the other is immovable. Furthermore, the movable point defect A changes its position not only within zigzag chains but also between zigzag chains. In Fig. 9.6, there are other kinds of defects with a different dark contrast. This means that these defects show different depths for the noncontact AFM topography measurement. To measure the depth of defects in Fig. 9.6(a), we investigated line profiles along the white lines in Fig. 9.7. As Fig. 9.8 shows, these line profiles made it clear that the deep defect A, the little shallow defect C and the

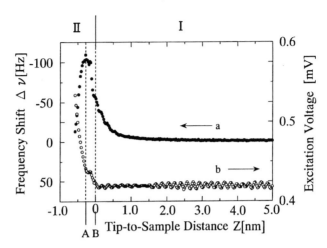

Fig. 9.9. (a) Closed circles show an approaching trace of the frequency shift and (b) open circles show the simultaneously measured excitation voltage supplied to the piezoelectric tube scanner as a function of the tip-to-sample distance Z in the constant-vibration mode. The frequency shift curve reverses at point A. Contact occurs at point B where the excitation voltage curve starts to increase.

very shallow defect D show 0.05–0.06 nm, 0.02–0.03 nm and ~0.01 nm depths, respectively. There are thus three types of point defects with different depths.

### 9.5.2 An experimental definition of the contact point [20]

The sample used was cut from a Zn-doped p-type GaAs(001) wafer with a carrier concentration of $1.4 \times 10^{19}$/cm$^3$. Figure 9.9 shows simultaneously measured approaching traces of (a) the frequency shift and (b) the excitation voltage supplied to the piezoelectric tube scanner in the constant-vibration mode. Hereafter, we will call these curves the frequency shift curve and the excitation voltage curve, respectively. In this mode, the switch in Fig. 9.4 was connected with terminal 1. Figure 9.10 shows simultaneously measured approaching traces of (a) the frequency shift and (b) the oscillation amplitude of the cantilever in the constant-excitation mode. Hereafter, we will call the latter curve the oscillation amplitude curve. In this mode, the switch in Fig. 9.4 was connected with terminal 2. The initial oscillation amplitude was set at ~18 nm in both operation modes. The sample bias was set to be $V_{DC} = + 1$ V to compensate for the contact potential difference between the tip and the sample surface. To remove the contamination and the oxide layer on the tip apex, we used Ar ion sputtering.

In the constant-vibration mode in Fig. 9.9, the excitation voltage curve was almost constant down to point B (region I) and then quickly increased

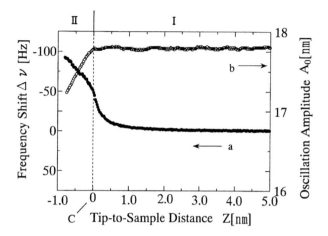

Fig. 9.10. (a) Closed circles show an approaching trace of the frequency shift and (b) open circles show the simultaneously measured oscillation amplitude as a function of the tip-to-sample distance Z in the constant-excitation mode. Contact occurs at point C where the oscillation amplitude curve starts to decrease and the gradient of the frequency shift curve changes abruptly.

(region II) with decreasing tip-to-sample surface distance. An increase in the excitation voltage indicates an increase in the energy loss of the oscillating cantilever as a result of the cyclic repulsive force due to the cyclic contact between the tip and the sample surface. Therefore, point B at which the excitation voltage curve starts to increase is the contact point between the tip and the surface, and hence regions I and II are the noncontact and the cyclic-contact regions, respectively. On the other hand, the frequency shift curve, firstly decreased gradually and then quickly in region I and became a minimum at point A. It then increased rapidly from point A in region II with decreasing tip-to-sample surface distance. It should be noted that in the cyclic-contact region II, the distance dependence of the frequency shift curve reverses at point A ($Z = -0.3$ nm).

In the constant-excitation mode shown in Fig. 9.10, the oscillation amplitude curve of the cantilever was almost constant up to point C (region I) and then decreased almost linearly (region II) with decreasing tip-to-sample surface distance. A decrease in the oscillation amplitude means an increase in the energy loss of the oscillating cantilever because of the cyclic repulsive force due to the cyclic contact between the tip and the sample surface. Therefore, point C at which the oscillation amplitude curve starts to decrease is the contact point between the tip and the surface, and hence regions I and II are the noncontact and the cyclic-contact regions, respectively. On the other hand, the frequency shift curve gradually and then quickly decreased down to point C (region I), and decreased rather slowly from point C (region II). It should be noted that in the cyclic-contact region II, the distance dependence of the frequency shift curve does not reverse even when the tip approaches the surface until it is 1 nm ($Z = -1$ nm) from the contact point C.

In the cyclic-contact region, the rate of decrease in the distance between the tip apex and the surface effectively weakens in the constant-excitation mode because of the decrease in the oscillation amplitude, while it does not weaken in the constant-vibration mode. Therefore, if the tip accidentally touches the surface during the noncontact AFM imaging, the repulsive force between the tip and the surface will weaken in the constant-excitation mode, whereas it will not weaken in the constant-vibration mode. This means that the constant-excitation AFM mode is much more gentle than the constant-vibration AFM mode at sudden repulsive contact, so that it is quite effective in avoiding degradation of the initial sharp tip and destruction of the sample surface.

The comment should be made that the smallest decay lengths in both the constant-vibration mode and the constant-excitation mode were $L \doteqdot 0.35$ nm and 0.33 nm, respectively, obtained around the points just before the contact point.

### 9.5.3  Distance dependence of noncontact AFM image on a p-GaAs(110) surface

Using a Zn-doped p-type GaAs(110) cleaved surface with a carrier concentration of $1.4 \times 10^{19}/cm^3$, we investigated the distance dependence of the noncontact AFM image. To remove the contamination and the oxide layer on

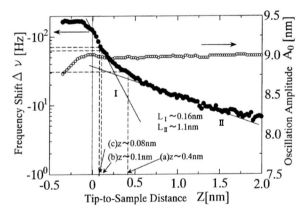

Fig. 9.11. Closed circles show the frequency shift curve and open circles show the simultaneously measured oscillation amplitude curve in the constant-excitation mode. Contact occurs at the point $Z = 0$ nm where the oscillation amplitude curve starts to decrease and the gradient of the frequency shift curve changes abruptly. The decay lengths in the regions I and II were 0.16 nm and 1.1 nm, respectively.

the tip apex, we used Ar ion sputtering. The scan areas were 10 nm × 10 nm. Firstly, we measured the frequency shift and the oscillation amplitude under the constant-excitation mode as a function of tip-to-sample surface distance, as Fig. 9.11 shows. Next, we determined the contact point at where the oscillation amplitude curve begins to decrease by decreasing the tip-to-sample distance, i.e., $Z = 0$ nm in Fig. 9.11. We then obtained noncontact AFM images at three different distances: (a) $Z \sim 0.4$ nm, (b) $Z \sim 0.1$ nm and (c) $Z \sim 0.08$ nm, just before the contact point, as Fig. 9.11 shows. It should be noted that the frequency shift in the noncontact region can be divided into two parts with the decay lengths of $L \sim 1.1$ nm in region II and $L \sim 0.16$ nm in region I. Region II with a large decay length seems to be the macroscopic region where the tip and the sample can be approximated as a continuous body and the volume integration effect decreases the distance dependence. Region I with a small decay length seems to be the microscopic region where the tip and the sample should be treated as a discontinuous body made from atoms. Here, the monoatomic interaction between the topmost atom of the tip and the nearest neighbor atom of the sample surface dominates the distance dependence of force. With this in mind, we chose the above three measured points, that is, (a) $Z \sim 0.4$ nm, located close to region II, while (b) $Z \sim 0.1$ nm and (c) $Z \sim 0.08$ nm, located in region I.

Figures 9.12(a), (b) and (c) show measured noncontact AFM images and line profiles along the white lines in the noncontact AFM images obtained at (a) $Z \sim 0.4$ nm, (b) $Z \sim 0.1$ nm and (c) $Z \sim 0.08$ nm just before the contact point, respectively. The variable frequency shift mode which measures the frequency

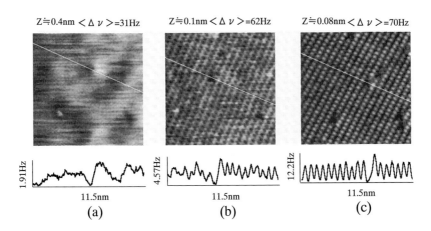

Fig. 9.12. Frequency shift images and line profiles along the white lines shown in the noncontact AFM images measured at (a) $Z \fallingdotseq 0.4$ nm, (b) $Z \fallingdotseq 0.1$ nm and (c) $Z \fallingdotseq 0.08$ nm far from the contact point $Z = 0$ nm as shown in Fig. 9.11. The scan area was 10 nm × 10 nm.

shift image (see Fig. 9.4) was used. That is, during the scanning, the distance between the tip and the surface was controlled to keep the mean frequency shift $<\Delta v>$ at a constant level under the weak feedback condition, and AFM images were obtained from the change in the frequency shift $\Delta v$. From Fig. 9.12(a) measured at $Z \sim 0.4$ nm, we found that the frequency shift image close to region II shows only a large-scale contrast, perhaps, due to defects, but no atomic-scale contrast. On the other hand, from Fig. 9.12(b) measured at $Z \sim 0.1$ nm, we found that the frequency shift image in region I shows atomic-scale point defects as well as periodic lattice structures, although the image seems a little vague. Furthermore, from Fig. 9.12(c) measured at $Z \sim 0.08$ nm, we found that the frequency shift image in region I suddenly becomes very clear. Thus, the distance dependence of the noncontact AFM image is very strong where the very small distance change of 0.02 nm changes the image contrast drastically. This strong distance dependence of the noncontact AFM image agrees qualitatively with the expectation predicted from the distance dependence of the lateral resolution given by Eq. (9.6) and of the decay length given by Eq. (9.8). From these results, we determined that true atomic resolution can be experimentally achieved only between 0 nm–0.4 nm just before the contact point.

## 9.6 Experimental Results on Si Semiconductors

Noncontact AFM with the FM detection method measured under UHV is rapidly developing as a scientific tool for a true atomic resolution. However, there still remain many important problems such as the conditions needed to achieve true atomic resolution and the imaging mechanism. In this section, we will describe the imaging mechanism related to the existence of the dangling bond, i.e., covalent bonding interaction, on a Si(111)7×7 surface, and the distance dependence of the imaging mechanism on Si(111) $\sqrt{3} \times \sqrt{3}$ -Ag.

Samples of Si semiconductor were cut from a (111) wafer. Then in UHV, a Si (111) sample was thoroughly outgassed at 520°C for a few days and then flashed at 1250°C to remove the native oxide and to clean the surface using direct resistive heating. The $7 \times 7$ reconstructed surface was then prepared by cooling slowly from 900°C down to 500°C. Finally, the sample was transferred from the sample preparation chamber to the AFM chamber.

### 9.6.1 The covalent bonding effect on the noncontact AFM of a Si(111)7×7 surface [22]

#### 9.6.1.1 The discontinuity phenomenon of the frequency shift curve on Si(111)7×7

Firstly, we measured the distance dependence of the frequency shift and oscillation amplitude, respectively, on a Si(111)7×7 surface in the constant-excitation mode. Here, Ar ion sputtering to clean the tip apex was not used. As shown in Figs. 9.13(a) and 9.14(a), two types of the distance-dependence of the

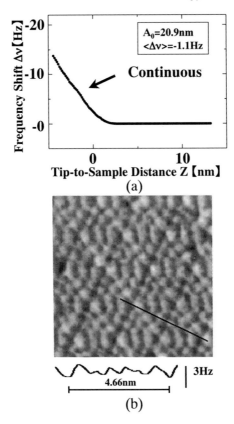

Fig. 9.13. (a) The frequency shift curve without discontinuity on a Si(111)7×7 surface. Ar ion sputtering to clean the tip was not used. $Z = 0$ nm corresponds to the contact point. The constant-excitation mode was used. (b) Corresponding frequency shift image for Si(111)7×7 measured at $<\Delta v> = -1.1$ Hz. The line profile was measured along the solid line in the AFM image. The scan area was 9.9 nm × 9.9 nm.

frequency shift curve were observed (types I and II). In these figures, $Z = 0$ nm corresponds to the contact points where the oscillation amplitude curves begin to decrease as shown in Figs. 9.10 and 9.11 when the tip-to-sample distance is decreased. In type I of Fig. 9.13(a), the frequency shift curve is nearly constant above $Z \gtreqqless 2$ nm, but it begins to only decrease continuously below $Z < 2$ nm, as Fig. 9.13(a) shows. In type II of Fig. 9.14(a), the frequency shift curve is nearly constant, then it begins to decrease slowly when the tip-to-sample distance is decreased, in a similar way to type I. However, it decreases discontinuously at the distance $Z \fallingdotseq 0$ nm and again decreases slowly, as Fig. 9.14(a) shows. We could not control the frequency shift in this jump region at the distance $Z = 0 \pm 0.0025$ nm, and hence discontinuity of the frequency shift

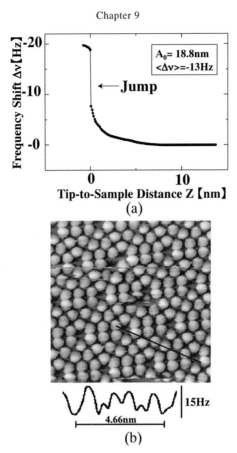

Fig. 9.14. (a) The frequency shift curve with discontinuity on a Si(111)7×7 surface. Ar ion sputtering to clean the tip was not used. Z = 0 nm corresponds to the contact point. The constant-excitation mode was used. (b) Corresponding frequency shift image for Si(111)7×7 measured at $\langle \Delta v \rangle = -13$ Hz. The line profile was measured along the solid line in the AFM image. The scan area was 9.9 nm × 9.9 nm.

occurred within ±0.0025 nm. Here, the uncontrollable tip-to-sample surface distance of ±0.0025 nm was estimated from the Z sensitivity of the piezoelectric scanner and the noise level of the scanning voltage. This discontinuity in the frequency shift indicates rapid change and/or rapid increase of the attractive interaction. But when we took a static measurement of the force itself as a function of the tip-to-sample distance, we were able to confirm that this discontinuity in the frequency shift curve did not originate from the spatial jump of the AFM tip to the Si(111)7×7 surface. Moreover, such a discontinuity in the frequency shift curve has never been observed for the relatively inert InP(110) and GaAs(110) cleaved surfaces without dangling bonds near the

Fermi level, as Figs. 9.9–9.11 show. Therefore, these results strongly suggest that this discontinuity in the frequency shift curve is induced by the presence of the reactive dangling bond on the Si(111)7×7 surface.

In all experiments, at the beginning of the AFM measurement, only the frequency shift curve without discontinuity (type I) was observed. After several accidental contacts between the tip and the Si surface during the measurement, the frequency shift curve with discontinuity suddenly began to be evident. After that, both types of frequency shift curves (types I and II) were observed. This result suggests that interaction between the tip and the reactive Si surface was changed by the contact between the tip and the Si surface during the measurement. We then used Ar ion sputtering to clean the tip before the measurement. In this case, we observed a frequency shift curve with discontinuity (types II) from the beginning of the AFM measurement. We thus confirmed that the oxidized Si tip creates a frequency shift curve without discontinuity, but the clean Si tip with dangling bond creates a frequency shift curve with discontinuity. This result strongly suggests that this discontinuity in the frequency shift curve is induced by the covalent bonding effect between the reactive dangling bonds on the Si(111)7×7 surface and on the clean Si tip apex. This result may mean that by changing the topmost atom of the tip apex from (for example) an oxygen atom without a dangling bond to a silicon atom with a dangling bond, we can change and control the force interaction mechanism between the tip and the sample surface, that is, exercise a kind of atomic force control.

We next investigated the measurement condition for atomic resolution imaging by adjusting the frequency shift. This controls the nearest distance between the tip and the surface. In the case of the frequency shift curve without discontinuity (type I), no atomic images were observed at a distance of $Z > 1$ nm in Fig. 9.13(a), but adatoms and corner holes could be detected at the distance of $-5$ nm $< Z < 1$ nm. In the case of the frequency shift curve with discontinuity (types II), no atomic images were observed at a distance of $Z > 1$ nm (see Fig. 9.14(a)). At a distance of 0 nm $< Z < 1$ nm, atomic corrugation can be faintly seen and corner holes with a separation of 2.69 nm could be detected slightly, but adatoms could not be resolved. However, near the distance of $Z = 0$ nm in Fig. 9.14(a), adatoms and corner holes could be clearly detected. At a distance of $Z < 0$ nm, no atomic images could be observed because both the frequency shift and the oscillation amplitude became unstable during scanning even under the constant-excitation mode, which seems to be due to intensive chemical interaction, i.e., the covalent bonding effect, between the tip and the sample surface.

Figures 9.13(b) and 9.14(b) show the atomically resolved images of the Si(111)7×7 surface obtained for the frequency shift curves without discontinuity (type I) and with discontinuity (types II), respectively. The scan areas were 9.9 nm × 9.9 nm. The variable frequency shift mode which measures the frequency shift image (see Figs. 9.4 and 9.12) was used. Here, the mean frequency shifts were set to $\langle \Delta v \rangle = -1.1$ Hz and $-13$ Hz, respectively. We can see not only adatoms and corner holes but also missing adatoms. As shown in the line profiles in Figs. 9.13(b) and 9.14(b), along the long diagonal of the $7 \times 7$ unit cell, indicated by the black solid lines in the frequency shift images of Figs. 9.13(b) and 9.14(b), respectively, the maximum change of the frequency shifts was estimated to be about 3 Hz in Fig. 9.13(b) and about 15 Hz in Fig. 9.14(b), respectively. We can see that the image contrast in Fig. 9.14(b) is clearly stronger than that in Fig. 9.13(b). Here, the bright and dark areas of the AFM image in Fig. 9.14(b) almost correspond to the top and the bottom of the jump region in the frequency shift curve of Fig. 9.14(a). This means that the interaction with discontinuity of the frequency shift curve provides a very significant contribution to the image contract of the noncontact AFM image on Si(111)7×7.

### 9.6.1.2 Site-dependent frequency shift curve

To explain the covalent bonding effect between the reactive dangling bonds on the Si(111)7×7 surface and on the clean Si tip apex in more detail, we simultaneously measured the noncontact AFM image and the site-dependent frequency shift curve on an atomic scale. In this experiment, we used Ar ion sputtering to clean the tip before the measurement. The variable frequency shift mode was used and the mean frequency shift was set to $\langle \Delta v \rangle = -5$ Hz. The scan area was 9.8 nm $\times$ 8.9 nm. We found that the frequency shift curve shows discontinuity only over adatoms but does not show this over gaps between adatoms, as Fig. 9.15(a) shows. This strong site dependence of the frequency shift curve agrees with the characteristics of the dangling bond which exists on adatoms at the topmost layer, and of the covalent bonding formation which is strongly directional. This result is direct evidence that the discontinuity in the frequency shift curve is induced by the covalent bonding effect between the reactive dangling bonds of adatoms on the Si(111)7×7 surface and on the clean Si tip apex. The simultaneously observed noncontact AFM image showed a clear Si(111)7×7 surface, as Fig. 9.15(b) shows. The line profile in Fig. 9.15(b) along the long diagonal of the $7 \times 7$ unit cell, which was indicated by the black solid line in Figs. 9.15(b), showed about 8 Hz change in the frequency shift between these over adatoms such as $S_a$ and over gaps between adatoms such as $S_h$, indicated by the white arrows in the noncontact AFM image. This 8 Hz change in the frequency shift image agrees quite well with the change of the frequency at $Z_0$, where the discontinuity occurred, as Fig. 9.15(a) shows. This method for measuring the site-dependent frequency shift curve from site to site on an atomic scale enables us to map the spatial distribution of the atomic force interaction, that is, a kind of the atomic force spectroscopy. We named this method a vertical mapping of the atomic force, because the frequency shift curve measures the tip-to-sample (vertical) distance dependence of the frequency shift.

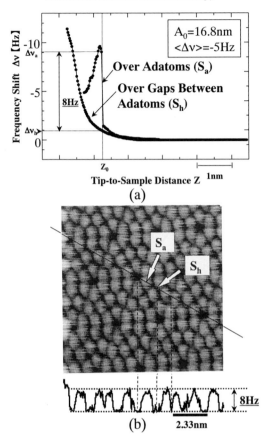

Fig. 9.15. (a) The site dependent frequency shift curves measured over adatoms ($S_a$) and over gaps between adatoms ($S_h$), respectively. (b) The simultaneously obtained frequency shift image and line profile along the solid line shown in the noncontact AFM image. The scan area was 9.8 nm × 8.9 nm.

It should be noted that the corrugation amplitude of adatoms ~0.14 nm obtained from the noncontact AFM topography of Fig. 9.16 is higher than that of 0.08 nm–0.1 nm obtained with the STM, although the depth of corner holes obtained with the noncontact AFM is almost the same as that observed with the STM. Moreover, in noncontact AFM images, the corrugation amplitude of adatoms was frequently larger than the depth of the corner holes. The origin of such a large corrugation of adatoms may be due to a peculiar contrast mechanism where the effect of the discontinuity in the frequency shift curve dominates even the topography obtained with the noncontact AFM.

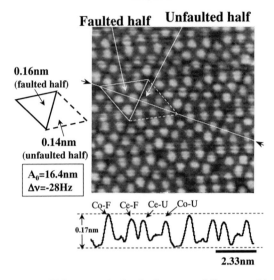

Fig. 9.16. Noncontact AFM topography for the frequency shift curve with discontinuity shows a contrast between the faulted and the unfaulted halves, while a line profile along the white line written in the AFM image shows height differences among inequivalent adatoms. The scan area was 8.9 nm × 8.9 nm.

### 9.6.1.3 The contrast between inequivalent halves and among inequivalent adatoms

Interestingly, in the case of the frequency shift curves with discontinuity (types II), we also observed a contrast between inequivalent halves and also among inequivalent adatoms of the 7 × 7 unit cell; namely, as the noncontact AFM topography of Fig. 9.16 shows, the faulted halves surrounded by a white solid triangle show spots brighter than those of the unfaulted halves surrounded by a white broken triangle. Here, the positions of the faulted and unfaulted halves were decided from the step direction. [23] A similar contrast is generally found in a filled-state scanning tunneling microscope (STM) image. [24] As the line profile along the long diagonal of the 7 × 7 unit cell in Fig. 9.16 shows, the heights of the corner adatoms are slightly higher than those of the adjacent center adatoms of the unit cell. The measured height differences of corrugation are in the following decreasing order:

$$Co\text{-}F > Ce\text{-}F > Co\text{-}U > Ce\text{-}U,$$

where Co-F and Ce-F indicate the corner and center adatoms in faulted halves, and Co-U and Ce-U indicate the corner and center adatoms in unfaulted halves, respectively. By averaging over several units, the corrugation height differences are estimated to be 0.025, 0.015 and 0.005 nm for Co-F, Ce-F and Co-U, respectively, referring to Ce-U. This tendency for the heights of the corner adatoms to be higher than those of the center adatoms is consistent with the experimental results of using a silicon tip by Nakagiri *et al.* [25], although they could not assign faulted and unfaulted halves of the unit cell in the measured AFM images. However, this tendency is completely contrary to the experimental results obtained using a tungsten tip by Erlandsson *et al.* [26] This difference may be caused by the difference in the materials of the tip, which seems to affect the interaction between the tip and the reactive sample surface. Another possibility is that (as pointed out by Erlandsson *et al.*), the tip is in weak contact with the surface during a small fraction of the oscillating cycle.

We concluded that the contrast among inequivalent adatoms is not caused by tip artifacts for the following reasons: (1) Each adatom, corner hole and defect was clearly observed and (2) the same contrast in several images of the different tips was observed. In view of the discontinuity of the frequency shift and its site dependence on an atomic scale, the chemical reactivity of adatoms seems to be most probable as the contrast mechanism. Further, the discontinuity in the frequency shift curve seems to be induced by the crossover between the physical and chemical bonding interactions. Here, the physical bonding interaction indicates interactions such as the van der Waals and/or electrostatic interactions, while chemical interaction indicates the covalent bonding interaction. On the other hand, a weak contrast image in case of the frequency shift curve without discontinuity seems to be induced by the interactions such as the van der Waals and/or electrostatic interactions.

### 9.6.2 The distance dependence of noncontact AFM image on Si(111) $\sqrt{3} \times \sqrt{3}$ - Ag surface [27]

In this section, we investigate force interaction between a Si tip and Si(111) $\sqrt{3} \times \sqrt{3}$ -Ag surface. Firstly, a Si(111)7×7 reconstructed surface was prepared, and then Si(111) $\sqrt{3} \times \sqrt{3}$ -Ag surface was formed in the same sample treatment chamber by the usual processes of Ag deposition and annealing. [28] After sufficient cooling, the sample was transferred from the sample treatment chamber to the AFM chamber.

For a Si(111) $\sqrt{3} \times \sqrt{3}$ -Ag structure, the honeycomb-chained trimer (HCT) model has been accepted as being an appropriate model. As Fig. 9.17 shows, this structure contains Si trimers at the first layer 0.075 nm below the Ag trimer at the topmost layer. In this model, Ag atoms and Si atoms at the first layer form covalent bonds, that is, the Ag atom terminates the Si dangling bond. The interatomic distance between the nearest neighbor Ag atoms and the nearest neighbor Si atoms at the first layer forming trimers are 0.343 nm and 0.231 nm [29], respectively, as shown in Fig. 9.18(a). Moreover, the distance between the centers of Ag trimers, which build hexagonal rings to form a honeycomb

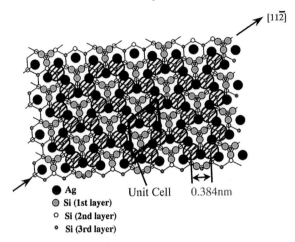

$[11\bar{2}]$

● Ag
◎ Si (1st layer)
○ Si (2nd layer)
◦ Si (3rd layer)

Unit Cell    0.384nm

Fig. 9.17. Top view of the honeycomb-chained trimer (HCT) model. The rhombus indicates a $\sqrt{3} \times \sqrt{3}$ unit cell where each corner is located at the center of the Ag trimer shown by hatched open circles. This figure shows that the center of the Ag trimer forms a hexagon pattern which constructs the honeycomb structure, and that there are two Ag trimer centers per $\sqrt{3} \times \sqrt{3}$ unit cell.

arrangement, is 0.384 nm as the hatched open circles in Fig. 9.17 shows. There are three Ag atoms and three Si atoms at the first layer per $\sqrt{3} \times \sqrt{3}$ unit cell as Fig. 9.18(a) shows, but only two centers of Ag trimers per $\sqrt{3} \times \sqrt{3}$ unit cell as Fig. 9.17 shows. From the first principle calculation for the HCT model, [30] it was proved that unoccupied surface states were mainly distributed in the center of the Ag trimer, where the bright spots were observed at a positive sample bias with the STM.

As the noncontact AFM topography of Fig. 9.19(a) shows, we observed two kinds of adjacent $\sqrt{3} \times \sqrt{3}$-Ag domains higher or lower than the Si(111)7×7 terrace. This result means that Si adatoms in lower $\sqrt{3} \times \sqrt{3}$-Ag domains transfer to higher $\sqrt{3} \times \sqrt{3}$-Ag domains to form lower and higher $\sqrt{3} \times \sqrt{3}$-Ag domains, respectively, as Fig. 9.19(b) shows. [31] In contrast to the frequency shift curve with discontinuity on the Si(111)7×7 terrace, a frequency shift curve without discontinuity was observed on the Si(111) $\sqrt{3} \times \sqrt{3}$-Ag domains. It suggests that the force interaction mechanism on the Si(111) $\sqrt{3} \times \sqrt{3}$-Ag surface differs from that on the Si(111)7×7 surface.

We performed atomic-scale imaging on the Si(111) $\sqrt{3} \times \sqrt{3}$-Ag surface by gradually changing the tip-to-sample distance. Figures 9.20(a), (b) and (c) show noncontact AFM topographies measured nearly at the tip-to-sample distance of 0.25 nm ($\Delta v = -32.4$ Hz), 0.05 nm ($\Delta v = -40.2$ Hz) and 0.03 nm ($\Delta v = -51.8$ Hz) just before the contact point $Z = 0$ nm, respectively. From these

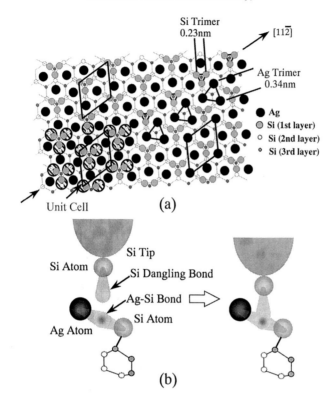

Fig. 9.18. (a) Top view of the honeycomb-chained trimer (HCT) model. Three rhombuses indicate a $\sqrt{3} \times \sqrt{3}$ unit cell where each corner located on a Ag atom, a Si atom and the site of covalent bond between the Si and Ag atoms indicated by hatched open circles is shown, respectively. This figure shows that the Ag atom, the Si atom and the site of covalent bond between Si and Ag atoms form triangle patterns, and that three of them exist per $\sqrt{3} \times \sqrt{3}$ unit cell, respectively. Triangles of Si atom and the site of the covalent bond between the Si and the Ag atoms line up in the [11 $\bar{2}$] direction, but Ag triangles do not. (b) Schematic model of orbital hybridization interaction between a Si tip and Ag-Si covalent bond, that is, covalent bond hybridization.

figures, it is apparent that the contrast between noncontact AFM images changes remarkably when the tip-to-sample distance is changed by only about 0.22 nm. At the small distance of $Z \fallingdotseq 0.03$ nm, the contrast of noncontact AFM images became very clear.

In Fig. 9.20(a) measured at $Z \fallingdotseq 0.25$ nm, the rhombus indicated by the solid line shows a $\sqrt{3} \times \sqrt{3}$ unit cell consisting of two triangles. The four corners of the rhombus are located on the bright spots. Moreover, one half seems to be brighter than the other so that an additional bright spot will be located at the center of the brighter half. Thus, in the $\sqrt{3} \times \sqrt{3}$ unit cell, there

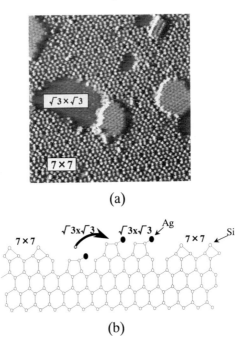

(a)

(b)

Fig. 9.19. (a) Noncontact AFM topography image which shows the coexistence of Si(111)7×7, Si(111) $\sqrt{3} \times \sqrt{3}$ -Ag and Si(111)3×1-Ag. The scan area was 30 nm × 30 nm. (b) Schematic side view which shows the formation process of Si(111) $\sqrt{3} \times \sqrt{3}$ -Ag.

are two bright spots which consist of hexagonal rings and the honeycomb structure, as indicated by open circles in Fig. 9.20(a). As Fig. 9.20(b), measured at $Z \fallingdotseq 0.05$ nm, shows, dark lines in the image create a trefoil, shown by three solid lines. On the other hand, in Fig. 9.20(c), measured at $Z \fallingdotseq 0.03$ nm, the periodic structure consisting of the triangle made from three bright spots can be clearly observed. The minor distortion of the hexagon in Fig. 9.20(a) and of the triangle in Fig. 9.20(c) is due to thermal drift during the AFM measurement at room temperature.

The distance between the bright spots in Fig. 9.20(a) is 0.37 ± 0.04 nm, which approximately agrees with the distance between the centers of the Ag trimer of 0.384 nm and with the interatomic distance between the nearest neighbor Ag atoms of 0.343 nm. From the hexagon pattern of bright spots, we attributed the bright spots in Fig. 9.20(a) to the centers of the Ag trimer, because only the center of the Ag trimer constructs hexagons, as Fig. 9.17 shows, but the nearest neighbor Ag atoms construct triangles, as Fig. 9.18(a) shows. At the centers of the Ag trimers, the main tip-to-sample interaction contributing to imaging is perhaps the van der Waals force or the electrostatic

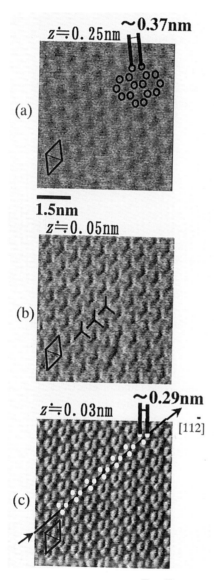

Fig. 9.20. Noncontact AFM topographies of Si(111) $\sqrt{3} \times \sqrt{3}$ -Ag measured at (a) $Z \fallingdotseq 0.25$ nm, (b) $Z \fallingdotseq 0.05$ nm and (c) $Z \fallingdotseq 0.03$ nm. The scan areas were 6 nm × 6 nm.

force. As for the former, there is a possibility that the van der Waals force acts on the AFM tip most strongly at the center of the Ag trimer, composed of three Ag atoms at the topmost layer. As for the latter, there is a possibility that the electrostatic force acts on the electron localized at the dangling bond out of the tip apex Si atoms, because a positive charge seems to be induced at the center of the Ag trimer.

On the other hand, the distance between the bright spots in Fig. 9.20(c) is $0.29 \pm 0.03$ nm, which differs from any of the distances of the trimers in the HCT model such as the Ag trimer with a 0.343 nm interatomic distance and the Si trimer with a 0.231 nm interatomic distance. As Fig. 9.20(c) shows, all triangles are composed of three bright spots lines up in the $[11\bar{2}]$ direction decided from the step direction. The most appropriate site corresponding to the three bright spots seems to be the site of the orbit of the covalent bond between the Si atoms and the Ag atoms, because only this site satisfies the experimental size of triangles and the direction as shown by the hatched open circles in Fig. 9.18(a). A probable explanation is that, when a Si tip with a dangling bond approaches closely enough to the covalent bond between the Si atoms and the Ag atoms, the orbit of the dangling bond out of the tip apex Si atom hybridizes with the orbit of the covalent bond between the Si atoms and the Ag atoms, and then strong tip-to-sample interaction contributing to imaging occurs as Fig. 9.18(b) shows. Finally, the contrast to Fig. 9.20(b) can be explained by regarding it as a synthesis of the hexagon pattern in Fig. 9.20(a) and the triangle pattern in Fig. 9.20(c), because this contrast appears in the intermediate distance where the bright spots change from the hexagon pattern to the triangle pattern.

Thus, in the Si(111) $\sqrt{3} \times \sqrt{3}$ -Ag surface, the force interaction mechanism changes from the van der Waals and/or electrostatic force interaction to an orbital hybridization interaction, that is, covalent bond hybridization, when the tip-to-sample distance decreases. This method of measuring the tip-to-sample distance dependence of the noncontact AFM image on an atomic scale enables us to map the spatial distribution of the atomic force interaction, that is, a kind of atomic force spectroscopy. We named this method a lateral mapping of the atomic force, because the noncontact AFM image measures the lateral change of the frequency shift.

## 9.7 Noncontact AFM Imaging on an Ag(111) Surface [16]

The growth mode for the thin Ag film on a Si(111) surface at room temperature is characterized by island formation, and its structure must be either pure islands of Ag on a Si surface or on a thin two-dimensional silicide layer, as Fig. 9.21(b) shows. [32] Firstly, a Si(111)7×7 reconstructed surface was prepared. Ag was then deposited on the sample at room temperature at $4.8 \times 10^{-11}$ Torr. Crystallinity was confirmed by the observation of clear LEED patterns generated by the Ag(111) surfaces. Silver films on the Si(111)7×7

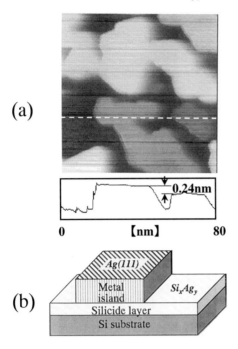

(a)

0.24nm

0            【nm】            80

(b)

Ag(111)

Metal island    $Si_xAg_y$

Silicide layer

Si substrate

Fig. 9.21. (a) Noncontact AFM topography of the Ag (111) film evaporated on the Si(111)7×7 surface. Scan area was 80 nm × 80 nm. The frequency shift was set to be $\Delta v = -10$ Hz. The oscillation amplitude was about $A_0 = 11.4$ nm. (b) Nanostructure model of Ag(111) islands evaporated on the Si(111)7×7 surface.

showed only sharp hexagonal spots.

The noncontact AFM images were taken in the constant frequency shift mode under the constant excitation condition of cantilever oscillation. The Si cantilever was cleaned by Ar ion sputtering.

Figure 9.21(a) shows a noncontact AFM topography of the Ag film evaporated on the Si(111)7×7 surface. The scan area was 80 nm × 80 nm. The frequency shift was set to be $\Delta v = -10$ Hz. Atomically flat islands with sizes between 10 nm and 30 nm were observed. The heights of the islands were estimated to be about 1.2 to 2 nm. The difference in the height of the islands was an integer multiple of the monoatomic step height of 0.236 nm on the Ag(111) surface, as a line profile along the broken white line in the AFM image of Fig. 9.21(a) shows. This result supports the idea that the atomically flat islands in Fig. 9.21(a) are a pure metallic Ag crystal with an Ag(111) surface.

The contact potential difference between the tip and the sample weakens the distance-dependence of the frequency shift due to weak distance-dependence of the electrostatic force interaction and hence degrades the spatial resolution

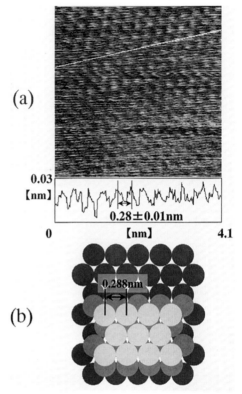

Fig. 9.22. (a) Noncontact AFM topography on the atomically flat surface of Ag(111). The scan area was 4 nm × 4 nm. A trigonal pattern with a lattice spacing of 0.28 ± 0.01 nm can be clearly observed. (b) Top view of the Ag(111) structure. The lattice spacing is 0.288 nm.

of the noncontact AFM image. Compensation of the contact potential difference between the tip and the sample using bias voltage is therefore essential to obtain an atomic resolution image. For high-resolution imaging on the pure metallic island of Ag(111), a bias voltage of $V_{DC} = -1.2$V was applied to the sample to compensate for the contact potential difference.

Figure 9.22(a) shows the noncontact AFM topography on the atomically flat surface of Ag(111). The scan area was 4 nm × 4 nm. The trigonal pattern can be clearly seen. No image processing was performed for Fig. 9.22(a). The distance between the protrusions in the pattern was estimated to be 0.28 ± 0.01 nm, which is in good agreement with the lattice spacing of the Ag(111) surface shown in Fig. 9.22(b). Thus, atomic resolution imaging on a metallic surface was achieved by using the noncontact AFM in UHV. From the line profile along the white line in the noncontact AFM image, the measured corrugation height was estimated to be 0.01–0.02 nm. The measured corrugation height is almost equal to that measured by the STM, [32] although this depends on imaging parameters such as the oscillation amplitude and the frequency shift.

## 9.8 Simultaneous Imaging of Topography and Electrostatic Force on $n^+$-GaAs(110) [33]

The noncontact AFM (NC-AFM) method was recently applied to achieve high resolution imaging of the electrostatic force using the double frequency modulation method with mechanical resonance (167–169 k Hz) and alternating bias voltage (300 Hz). [34] In that experiment, deterioration of the spatial resolution of the noncontact AFM topography was prevented by isolating the electrostatic force interaction from the van der Waals interaction. The result was that simultaneous imaging of the noncontact AFM topography and the frequency shift due to electrostatic force with electrostatic force microscope (EFM) function achieved an electrostatic force sensitivity of ~0.1 pN and a spatial resolution of ~1.5 nm on silicon oxide. In this section, we will introduce a novel method of applying the noncontact AFM system to image the frequency shift due to electrostatic force on an atomic scale. Using the newly developed combined NC-AFM/EFM system, we achieved simultaneous imaging of the noncontact AFM topography and the EFM (frequency shift image due to electrostatic force) on a Si doped $n^+$-GaAs(110) cleaved surface with true atomic resolution.

A highly doped $n^+$-GaAs(110) cleaved surface was used as a sample. The doped density of Si impurities measured by the SIMS was $N_{Si} = 2.8 \times 10^{18}/cm^3$, while the carrier density determined from the Hall coefficient was $n_H = (1.6–2) \times 10^{18}/cm^3$.

A bias voltage $V_{DC}$ was applied to a sample while a conductive Si cantilever was earthed. To compensate for the contact potential difference $V_C = + 0.375$ V between the tip and the sample, when the noncontact AFM topography was imaged, a bias voltage of $V_{DC} = -V_C = -0.375$ V was applied. By compensating for the electrostatic force due to the contact potential difference, noncontact AFM topography in the constant frequency shift mode due to the van der Waals force could thus be imaged. On the other hand, when the frequency shift due to the electrostatic force was imaged, a relatively large bias voltage of $V_+ = 0.625$ V or $V_- = -1.234$ V was applied to enhance the electrostatic force between localized charges on the sample surface and induced charges on the tip due to the effective bias voltage $\Delta V_\pm = V_\pm + V_C$, i.e., + 1.0 V and –0.859 V, respectively. $-V_C$ and $V_\pm$ bias voltages were alternately applied by a square-wave voltage with a frequency of 500 Hz. The feedback loop to control the Z-piezo was closed at $V_{DC} = -V_C$ (to measure the noncontact AFM topography by maintaining the frequency shift due to the van der Waals force constant), and opened at $V_{DC} = V_\pm$ (to measure the frequency shift due to the electrostatic force). The frequency shift due to the electrostatic force was then imaged using a variation in the frequency shift from $-V_C$ to $V_\pm$ with the frequency of 500 Hz.

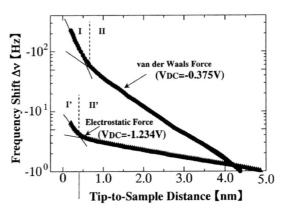

Fig. 9.23. Frequency shift curves for the van der Waals force and the electrostatic force, respectively. The former was obtained by compensating the contact potential difference with bias voltage $V_{DC} = -0.375$V, while the latter was obtained by enhancing the electrostatic force with bias voltage $V_{DC} = -1.234$V.

### 9.8.1 Frequency shift curves due to van der Waals and electrostatic forces

Figure 9.23 shows frequency shift curves due to the van der Waals force and the electrostatic force, respectively. The former curve was obtained by eliminating the electrostatic force due to the contact potential difference $V_C = +0.375$ V by the application of bias voltage $V_{DC} = -V_C = -0.375$ V, while the latter curve was obtained by enhancing the electrostatic force by the application of bias voltage $V_{DC} = V_- = -1.234$ V. Our result showed that each frequency shift curve can be divided into two regions, that is, the regions I and II for the van der Waals force, and the regions I′ and II′ for the electrostatic force, respectively. Furthermore, each region can be approximated as $\exp(-Z/L)$ with the decay lengths of $L_I = 0.3$ nm, $L_{II} = 0.9$ nm, $L_I′ = 0.5$ nm and $L_{II}′ = 3.2$ nm, respectively. It should be noted that the atomically resolved noncontact AFM and EFM images can be obtained only in the regions I and I′, respectively, so that regions I(and I′) and II (and II′) seem to correspond to microscopic and macroscopic regions, respectively.

### 9.8.2 Atomically resolved imaging of charged point defects

Figure 9.24(a) with a scan area of 15 nm × 15 nm shows the noncontact AFM topography of a Si doped $n^+$-GaAs(110) cleaved surface measured under a bias voltage of $V_{DC} = -V_C = -0.375$ V. As this figure shows, we made clear observations of many atomic-scale point defects as well as a periodic lattice structure. As the noncontact AFM topography shows, there are two kinds of defects, small and shallow defects, and large and deep defects. The former defect is rich compared with the latter defect. The line profile of Fig. 9.24(c)

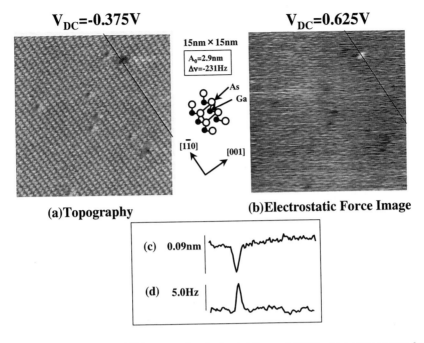

Fig. 9.24. (a) Noncontact AFM topography obtained at $V_{DC} = -0.375V$, which compensates for the contact potential difference. (b) Simultaneously obtained frequency shift image due to the electrostatic force (EFM) at $V_{DC} = 0.625V$. The scan area was 15 nm × 15 nm. (c) Line profile of corrugation along the solid line shown in the noncontact AFM topography and (d) line profile of frequency change along the solid line in the electrostatic force image, respectively.

shows the line profile of a large and deep defect along the solid line in Fig. 9.24(a). Figure 9.24(b) with a scan area of 15 nm × 15 nm shows a simultaneously obtained EFM image, while the line profile of Fig. 9.24(d) shows the line profile of a large and deep defect along the solid line in Fig. 9.24(b).

Figure 9.25 was imaged 8 minutes later than Fig. 9.24. Noncontact AFM topographies measured under $V_{DC} = -V_C$ clearly show that small and shallow defects change relative position within 8 min. as Figs. 9.24(a) and 9.25(a) show. This result means that small and shallow point defects move around on the GaAs(110) surface even at room temperature. In the EFM image of Fig. 9.24(b) measured under $V_+ = 0.625$ V, the large and deep point defect was imaged as a bright protrusion, indicating a peak in the frequency shift due to attractive electrostatic force, as the line profile of Fig. 9.24(d). On the other hand, in the EFM image of Fig. 9.25(b), measured under $V_- = -1.234V$, the large and deep defect changed to a dark depression indicating a depression in the frequency shift due to attractive electrostatic force, as the line profile of Fig. 9.25(d). Thus, only the large and deep defect showed strong dependence

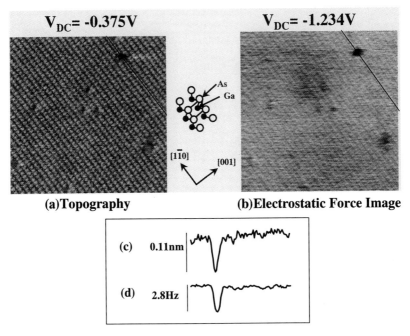

**Fig. 9.25.** (a) Noncontact AFM topography obtained at $V_{DC} = -0.375V$, which compensates for the contact potential difference. (b) Simultaneously obtained frequency shift image due to the electrostatic force (EFM) at $V_{DC} = -1.234V$. The scan area was 15 nm × 15 nm. (c) Line profile of corrugation along the solid line shown in the noncontact AFM topography and (d) line profile of frequency change along the solid line in the electrostatic force image, respectively.

on bias voltage in the EFM image. When $V_+ = 0.625V$ was applied to the sample, negative charges were induced at the tip of the cantilever, as Fig. 9.26(a) shows. Consequently, if the large and deep defect has positive charges, we will observe the charged point defect as a peak in the frequency shift as a result of electrostatic force in the EFM image because of increase in attractive electrostatic force. On the other hand, when $V_- = -1.234$ V was applied to the sample, positive charges were induced at the tip of the cantilever, as Fig. 9.26(b) shows. Consequently, if the defect has positive charges, we will observe the charged point defect as a depression in the frequency shift due to electrostatic force in the EFM image because of decrease in attractive electrostatic force. We thus succeeded in imaging a charged point defect with positive charges on an atomic scale. On the other hand, with the scanning tunneling microscope (STM), spatial variation in local density of states at the Fermi level (LDOS) will be detected, so that only spatial variation in electron density of states at the Fermi level screening the fixed positive charge will be imaged. A fixed charge such as a charged point defect cannot therefore be

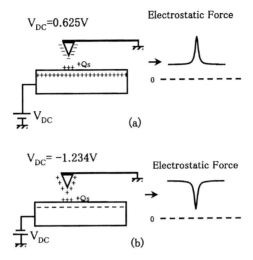

Fig. 9.26. Model to explain the bias voltage dependence of contrast of the EFM images for (a) $V_{DC} = 0.625$V and (b) $V_{DC} = -1.234$V in Figs. 9.24. and 9.25, respectively.

directly observed in the STM image. In contrast to the large and deep defect, small and shallow defects as well as the periodic lattice structure do not show clear bias voltage dependence. Small and shallow point defects would seem to be neutral or have rather small charges. The small and shallow defects as well as the periodic lattice structure shown in the EFM images may thus be ghosts. From the noise level of the frequency shift due to the electrostatic force, the minimum detectable electrostatic force was estimated to be ~0.04 pN, while from the line profiles, the spatial resolution of the electrostatic force was estimated to be ~0.4 nm.

### 9.8.3 Electrostatic force imaging for screening by electron-cloud

Figures 9.27 and 9.28 show wider images of simultaneously obtained noncontact AFM topography and EFM with a scan area of 40 nm × 40 nm. In the noncontact AFM topography, a periodic lattice structure cannot be clearly observed and the lattice raw along [1 $\overline{1}$ 0] can be barely observed. There are many point defects, though we can not discriminate large and deep defects from small and shallow ones in the noncontact AFM images. However, in the EFM images, defects denoted by C, D and E showed strong dependence on the bias voltage. Thus only these seem to be large and deep defects with positive charges. As C, D and E defects with positive charges showed nearly the same change in the frequency shift due to electrostatic force, as shown in the peaks or depressions in the line profiles in the EFM images, these defects seem to have nearly the same charges. From the carrier density $n_H = (1.6-2) \times 10^{18}$/cm$^3$

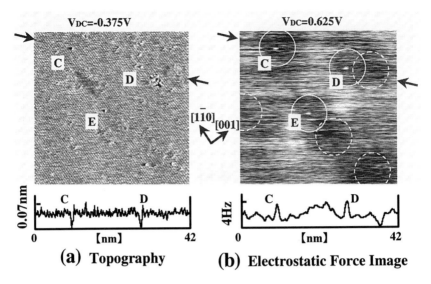

**(a)** Topography             **(b)** Electrostatic Force Image

Fig. 9.27. (a) Noncontact AFM topography obtained at $V_{DC} = -0.375$V, which compensates for the contact potential difference, and the line profile of corrugation along the line indicated by arrows. (b) Simultaneously obtained frequency shift image due to the electrostatic force (EFM) at $V_{DC} = 0.625$V, and the line profile of the frequency change along the line indicated by arrows. The scan area was 40 nm × 40 nm. Solid and broken open circles in the EFM image show the regions of the Thomas-Fermi screening with screening length ~4.5 nm around positive charges at the first layer and perhaps at the lower layers, respectively.

which is equivalent to the number of the ionized Si$^+$ donor, we calculated the expected number of the ionized Si$^+$ donor at the surface to be ~1.3 in a 40 nm × 40 nm area by assuming the thickness of ~0.4 nm of GaAs monolayer. Number of charged point defects of 3 observed in the 40 nm × 40 nm area seems to be a little rich. However, there are a smaller number of charged point defects in the other area. The average number of observed charged point defects may therefore roughly agree with the calculated number of ionized Si$^+$ donors at the surface. However, if the origin of the large and deep defect is a simple Si$^+$ ionized donor, it raises questions as to why the ionized Si$^+$ donor could not be observed as a peak in the noncontact AFM image and why this kind of defect is large and deep in the noncontact AFM image. The origin of the large and deep defect should thus be investigated in more detail. From the measured frequency shift 2–3 Hz for positively charged point defects, the electrostatic force between positively charged point defects and induced negative charges at the tip of the cantilever was estimated to be ~0.5 pN. However, by assuming 1 nm separation, the electrostatic force between single charges was calculated to be ~230 pN. Such smallness in the measured electrostatic force compared

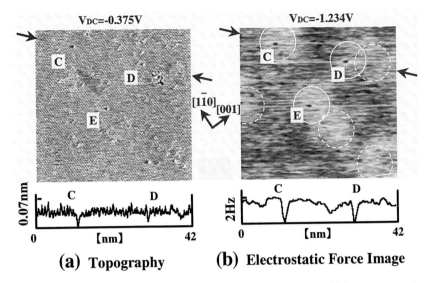

**(a)** Topography    **(b)** Electrostatic Force Image

Fig. 9.28. (a) Noncontact AFM topography obtained at $V_{DC} = -0.375V$, which compensates for the contact potential difference, and the line profile of corrugation along the line indicated by arrows. (b) Simultaneously obtained frequency shift image due to the electrostatic force (EFM) at $V_{DC} = -1.234V$, and the line profile of the frequency change along the line indicated by arrows. The scan area was 40 nm × 40 nm. Solid and broken open circles in the EFM image show the regions of the Thomas-Fermi screening with screening length ~4.5 nm around positive charges at the first layer and perhaps at the lower layers, respectively.

with the calculated one was tentatively explained as an averaging effect due to the large oscillation amplitude of the cantilever for the FM detection. On the other hand, the number of small and shallow defects which will be neutral or may have rather small charges seems to be larger than that of the Si impurity as well as that of the ionized $Si^+$ donor. The origin of the small and shallow defect should therefore also be investigated in more detail.

In the EFM images of Figs. 9.27(b) and 9.28(b), we found that the bright protrusions of the charged point defects C, D, E measured at $V_+ = 0.625$ V are surrounded by dark areas, while the dark depressions of the charged point defects C, D, E measured at $V = -1.234$ V are surrounded by bright areas. In other words, as the line profiles of Figs. 9.27(b) and 9.28(b) show, peaks in the frequency shift due to electrostatic force originating from charged point defects are located at the center of hollows, while depressions in the frequency shift due to the electrostatic force originating from the charged point defects are located at the center of hills. This indicates that areas surrounding the charged point defects show the opposite change of the frequency shift due to electrostatic force and that positively charged point defects may be surrounded by negative charges. We therefore calculated the screening length due to free electrons as follows: Firstly from the carrier density $n_H = (1.6-2) \times 10^{18}/cm^3$, the Fermi temperature was estimated to be ~1000 K so that the electrons degenerate at room temperature. We then calculated the Thomas-Fermi screening length as ~4.5 nm. In solid open circles with radius of ~4.5 nm, shown in Figs. 9.27(b) and 9.28(b), the calculated Thomas-Fermi screening length seems to agree with the observed surrounding areas with negative charges. Areas with negative charges surrounding the positively charged point defects may therefore be a direct image of the spatial variation in the total charge due to the Thomas-Fermi screening effect. As the broken open circles with a radius of ~4.5 nm in the EFM images of Figs. 9.27(b) and 9.28(b) show, we also found similar areas though there are no positively charged point defects at the center. In this case, positively charged point defects may be located below the surface, so that only the Thomas-Fermi screening effect may be imaged at the surface.

## 9.9 Summary

With an atomic force microscope (AFM) operating under the noncontact mode in an ultrahigh vacuum (UHV), atomically-resolved imaging of semiconductor and metal surfaces based on the frequency modulation (FM) detection method was successfully achieved. Atomic-scale defects as well as periodic lattices were clearly observed on cleaved compound semiconductor surfaces such as InP(110) and GaAs (110), on reconstructed silicon semiconductor surfaces such as Si(111)7×7 and Si(111)$\sqrt{3} \times \sqrt{3}$-Ag, and on metal surfaces such as Ag(111). Moreover, motion of point defects and different kinds of point defects with different depths were found in the compound semiconductor surface images. Furthermore, site-dependent atomic force microscopy on an atomic scale showed that the atomically-resolved image of Si(111)7×7 measured at the jumping-point of the frequency shift curve shows the spatial distribution of the dangling bond of adatom on an atomic-scale. For Si(111)$\sqrt{3} \times \sqrt{3}$-Ag surfaces, the force interaction mechanism changes by changing the tip-to-sample distance from the van der Waals and/ or electrostatic force interaction to the orbital hybridization interaction, that is, covalent bond hybridization. On a Ag(111) surface, atomic resolution imaging on the metallic surface was achieved by using the noncontact AFM in UHV. From the line profile, the measured corrugation height was estimated to be 0.01–0.02 nm. Finally, by using the electrostatic force microscope (EFM) combined with the noncontact UHV-AFM based on the FM detection method, topography and electrostatic force images of Si-doped n-GaAs(110) cleaved surface were simultaneously obtained on an atomic scale. One result of this was that the electrostatic force image on an atomic scale showed that there are two kinds of point defects with and without charge, and that the charge is positive.

The achievement of true atomic resolution with the noncontact AFM will show us how to achieve true atomic resolution with various kinds of SPM. Moreover, these experimental results show what can be imaged with the noncontact AFM. Furthermore, present results enable us to map the spatial distribution of the atomic force interaction on an atomic scale, i.e., atomically resolved atomic force spectroscopy, showing the site-dependent frequency shift curve from site to site or tip-to-sample distance dependence of the noncontact AFM. The present results made that there is the clear possibility that we can change and control the force interaction mechanism and the atomic force itself between the tip and the surface by changing the topmost atom of the tip apex.

## Acknowledgments

We would like to thank our colleagues Dr. M. Ohta, Dr. T. Uchihashi, Dr. M. Abe, Dr. T. Okada, Dr. F. Osaka, Dr. S. Ohkouchi, Dr. M. Suzuki, Mr. K. Hontani, Mr. H. Ueyama, Mr. T. Tsukamoto, Mr. T. Minobe, Mr. S. Orisaka, Mr. S. Mishima, Mr. H. Nagaoka, Mr. Y. Yanase and Mr. T. Shigematsu. This work was supported by a Grant-in-Aid for Scientific Research from the Ministry of Education, Science, Sports and Culture of Japan.

[References]

1)  G. Binnig, C. F. Quate and Ch. Gerber, *Phys. Rev. Lett.*, **56**, 930 (1986).
2)  G. Binnig, Ch. Gerber, E. Stoll, T. R. Albrecht and C. F. Quate, *Europhysics Lett.*, **3**, 1281 (1987).
3)  G. Meyer and N. M. Amer, *Appl. Phys. Lett.*, **56**, 2100 (1990).
4)  Y. Sugawara, M. Ohta, K. Hontani, S. Morita, F. Osaka, S. Ohkouchi, M. Suzuki, H. Nagaoka, S. Mishima and T. Okada, *Jpn. J. Appl. Phys.*, **33**, 3739 (1994).
5)  Y. Sugawara, M. Ohta, H. Ueyama and S. Morita, *Jpn. J. Appl. Phys.*, **34**, L462 (1995).
6)  S. Morita, S. Fujisawa and Y. Sugawara, *Surface Science Reports*, **23**, 1 (1996).
7)  S. Manne, P. K. Hansma, J. Massie, V. B. Elings and A. A. Gewirth, *Science*, **251**, 183 (1991).
8)  G. Binnig, *Ultramicroscopy*, **42–44**, 281 (1992).
9)  F. Ohnesorge and G. Binnig, *Science*, **260**, 1451 (1993).
10) F. J. Giessibl, *Science*, **267**, 68 (1995).
11) S. Kitamura and M. Iwatsuki, *Jpn. J. Appl. Phys.*, **34**, L145 (1995).
12) H. Ueyama, M. Ohta, Y. Sugawara and S. Morita, *Jpn. J. Appl. Phys.*, **34**, L1088 (1995).
13) M. Ohta, H. Ueyama, Y. Sugawara and S. Morita, *Jpn. J. Appl. Phys.*, **34**, L1692 (1995).
14) Y. Sugawara, M. Ohta, H. Ueyama and S. Morita, *Science*, **270**, 1646 (1995).
15) M. Bammerlin, R. Luthi, E. Meyer, A. Baratoff, J. Lu, M. Guggisberg, Ch. Gerber, L. Howald and H.-J.Guntherodt, *Probe Microscopy*, **1**, 3 (1997).
16) S. Orisaka, T. Minobe, T. Uchihashi, Y. Sugawara and S. Morita, *Appl. Surf. Sci.*, **140**, 243 (1999).
17) S. Morita and Y. Sugawara, *Appl. Surf. Sci.*, **140**, 406 (1999).
18) F. J. Giessibl, *Phys. Rev. B*, **56**, 16010 (1997).
19) T. R. Albrecht, P. Grutter, D. Horne and D. Rugar, *J. Appl. Phys.*, **69**, 668 (1991).
20) H. Ueyama, Y. Sugawara and S. Morita, *Appl. Phys. A*, **66**, S295 (1998).

21)  Y. Sugawara, H. Ueyama, T. Uchihashi, M. Ohta, S. Morita, M. Suzuki and S. Mishima, *Appl. Surf. Sci.*, **113/114**, 364 (1997).

22)  T. Uchihashi, Y. Sugawara, T. Tsukamoto, M. Ohta, S. Morita and M. Suzuki, *Phys. Rev. B*, **56**, 9834 (1997).

23)  T. Hasegawa, M. Kohno, S. Hosaka and S. Hosoki, *J. Vac. Sci. Technol. B*, **12**, 2078 (1994).

24)  R. J. Hamers, R. M. Tromp and J. E. Demuth, *Surf. Sci.*, **181**, 346 (1987).

25)  N. Nakagiri, M. Suzuki, K. Okiguchi and H. Sugimura, *Surf. Sci. Lett.*, **373**, L329 (1997).

26)  R. Erlandsson, L. Olsson and P. Martensson, *Phys. Rev. B*, **54**, R8309 (1996).

27)  T. Minobe, T. Uchihashi, T. Tsukamoto, S. Orisaka, Y. Sugawara and S. Morita, *Appl. Surf. Sci.*, **140**, 298 (1999).

28)  R. J. Wilson and S. Chiang, *Phys. Rev. Lett.*, **58**, 369 (1987).

29)  T. Takahashi and S. Nakatani, *Surf. Sci.*, **282**, 17 (1993).

30)  S. Watanabe, M. Aono and M. Tsukada, *Phys. Rev. B*, **44**, 8330 (1991).

31)  A. Shibata, Y. Kimura and K. Takayanagi, *Surf. Sci.*, **275**, L697 (1992).

32)  G. Meyer and K. H. Rieder, *Surf. Sci.*, **331–333**, 600 (1995).

33)  Y. Sugawara, T. Uchihashi, M. Abe and S. Morita, *Appl. Surf. Sci.*, **140**, 371 (1999).

34)  T. Uchihashi, M. Ohta, Y. Sugawara, Y. Yanase, T. Shigematsu, M. Suzuki and S. Morita, *J. Vac. Sci. Technol. B*, **15**, 1543 (1997).

# Chapter 10

# Correlation between Interface States and Structures Deduced from Atomic-Scale Surface Roughness in Ultrathin SiO$_2$/Si System

## 10.1 Introduction

Since MOSFETs with a gate oxide film thickness of 1.5 nm were shown to operate at room temperature [1], extensive studies have been performed on the reliability of ultrathin gate oxides. [2] Because the thickness of the structural transition layer is in the order of 1 nm [3,4] and the amount of electronic defect states in the structural transition layer must be larger than that in bulk SiO$_2$, the reliability of gate oxides must be mainly determined by the chemical structures of compositional and structural transition layers. In spite of the extensive studies on the interface states and structures, the structural origin of interface states has not yet been explained. [5] This is because of the difficulty of two-dimensional atomic-scale observation of interface structures and the two-dimensional atomic-scale detection of interface states. In the present study the dependence of the atomic-scale surface roughness of ultrathin oxide on oxide film thickness was measured. This enabled the effect of the interface structure on the atomic-scale surface roughness to be clarified on an atomic-scale. The effect of the interface structure on the interface electronic states and the valence band structure were also studied.

## 10.2 Experimental Details

Because the formation of native oxide can be suppressed by terminating the Si surface with hydrogen atoms, [6,7] atomically flat hydrogen-terminated Si surfaces (abbreviated as H-Si surfaces hereafter) were used as initial surfaces before the oxidation. In the experiment, an atomically flat H-Si(111)-1×1 surface was obtained by treatment [8] in 40% NH$_4$F solution and an atomically flat H-Si(100)-2×1 surface was obtained by the surface reconstruction [9] of Si(100) surface at high temperature (>1100°C) in a H$_2$ atmosphere at 1 bar.

The oxide films studied in the present paper were prepared as follows. In order to preserve the flatness of the initial Si surface during the oxidation at high temperature, the nearly 0.5 nm thick preoxides were formed at 300°C [10] in 1–4 Torr dry oxygen by oxidizing atomically flat hydrogen-terminated Si(111)-1×1 and Si(100)-2×1 surfaces without breaking the Si-H bonds. Through these preoxides, oxidation at 600–900°C in 1 Torr dry oxygen was performed. A rather high pressure of oxygen was used in order to minimize the effect of impurities in the oxidizing atmosphere on the oxidation process. The amount of water vapor in the oxygen gas used was below 37 ppb. In order to heat Si wafers in oxygen gas under high pressure, Si wafers were only heated optically.

Interface and valence band structures of ultrathin silicon oxides were studied by measuring photoelectron spectra excited by monochromatic AlK $\alpha$ radiation with an acceptance angle of 3.3 degrees, using ESCA-300 manufactured by Scienta Instrument AB, [11] while surface structures of ultrathin silicon oxides were studied by observing noncontact-mode atomic force microscope (NC-AFM) images with a force constant of 39 N/m and resonant frequency of about 300 kHz, and a single-crystalline silicon probe, using an instrument manufactured by OMICRON Vakuum Physik GmbH. [12]

The interface state densities were measured using the method developed by Lau and Wu [13] Firstly, organic molecules of 2-propanol were absorbed on oxide film. Secondly, oxide was charged by electron beam irradiation with electron kinetic energy of 2 eV to produce a voltage drop across the oxide film. Thirdly, the charging-induced changes in electrical potentials at the surface and interface which are necessary for the determination of interface state densities were obtained from the measurements of chemical shifts in the C1s level of organic molecules and Si2p core level of Si substrate. The organic molecules were decomposed during the oxidation. Other experimental details and analytical procedure of Si2p photoelectron spectra have been described elsewhere. [14]

## 10.3 SiO₂/Si Interface Structures

### 10.3.1 Structural origin

Figure 10.1 shows changes in Si2p3/2 spectra with the oxide film thickness (abbreviated as thickness hereafter) as a parameter. [15] In this figure the spectral intensities of the silicon substrate were adjusted to be equal to each other in order to show the oxidation-induced changes in the interface structure consisting of intermediate oxidation states (abbreviated as intermediate states hereafter), $Si^{1+}$, $Si^{2+}$ and $Si^{3+}$. Here, $Si^{1+}$, $Si^{2+}$ and $Si^{3+}$ denote a Si atom bonded to one oxygen atom and three Si atoms, a Si atom bonded to two oxygen atoms and two Si atoms and a Si atom bonded to three oxygen atoms and one Si atoms, respectively. [16] As Fig. 10.1 shows, the amounts of $Si^{1+}$ and $Si^{3+}$ for two

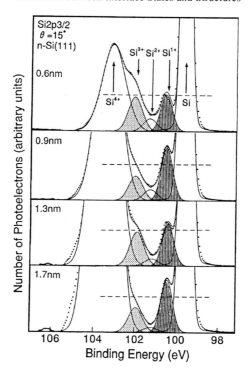

Fig. 10.1. Oxidation-induced changes in Si2p spectra obtained for a photoelectron take-off angle of 15 degrees with oxide film thickness as a parameter. The dashed line on each spectrum shows an average of the amounts of $Si^{1+}$ and $Si^{3+}$.

thicknesses of 0.6 and 1.3 nm, and those for two thicknesses of 0.9 and 1.7 nm are almost the same. This implies that the interface structure changes periodically with the progress of oxidation. Figure 10.2(a) shows the spectral intensity of $Si^{4+}$ and that of intermediate states normalized by the Si2p spectral intensity of silicon substrate as a function of thickness. Here, $Si^{4+}$ denotes a Si atom bonded with four oxygen atoms. Figure 10.2(a) shows that the normalized spectral intensity of intermediate states saturates at a thickness of nearly 0.5 nm, while the normalized spectral intensity of $Si^{4+}$ does not saturate at this thickness. Furthermore, the saturated normalized spectral intensity of intermediate states is in good agreement with that calculated for an abrupt compositional transition at the interface [15], which is shown by the dashed line in Fig. 10.2(a). Here, an abrupt compositional transition is defined as the existence of intermediate states only at the interface. Therefore, once the interface is formed, the amount of intermediate states is not affected by further oxidation. The abrupt compositional transition was also realized on a Si(100) surface by oxidation at

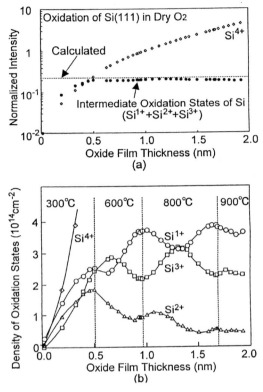

Fig. 10.2. (a) Normalized spectral intensity of Si⁴⁺ and summation of intensities for all intermediate oxidation states are shown as a function of oxide film thickness. (b) Dependence of areal densities of $Si^{1+}$, $Si^{2+}$, $Si^{3+}$ and $Si^{4+}$ on thickness.

600-800°C through nearly 0.5 nm thick preoxide formed by the oxidation of H-Si(100)-2×1 at 300°C. [17] Figure 10.2(b) shows the areal densities of $Si^{1+}$, $Si^{2+}$, $Si^{3+}$ and $Si^{4+}$ as a function of thickness. According to this figure, although the total amount of intermediate states does not change with the progress of oxidation, the areal density of $Si^{1+}$ and that of $Si^{3+}$ repeatedly increases and decreases with a period in thickness of nearly 0.7 nm for a thickness less than 1.7 nm. Furthermore, with progress of oxidation, the areal density of $Si^{1+}$ changes in opposite phase with the areal density of $Si^{3+}$. These findings clearly demonstrate that the interface structure changes periodically with progress of oxidation for thickness less than 1.7 nm.

If the oxidation reaction occurs at atomically flat interface, the $SiO_2$/Si(111) interface structures consisting of $Si^{1+}$ and $Si^{3+}$ should appear alternately as a result of periodic change in the bonding nature of Si crystal at the interface.

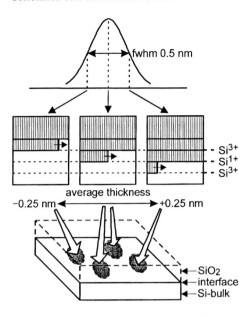

Fig. 10.3. Distribution of oxide film thickness expressed by a Gaussian function.

The period in a thickness of 0.7 nm in Fig. 10.2(b) is in excellent agreement with the thickness of two molecular layers of silicon dioxide calculated assuming an oxidation-induced volume expansion factor of bulk Si. [15] Therefore, at 800°C, oxidation reactions occur monolayer by monolayer at the interface until a 1.7-nm-thick oxide film is formed. For atomically flat interfaces, the minimum amounts of $Si^{1+}$ and $Si^{3+}$ must be zero and the amount of $Si^{2+}$ must be close to zero at every stage of oxidation. However, this is not the case. In order to explain the coexistence of $Si^{1+}$ and $Si^{3+}$ in Figs. 10.1 and 10.2(b), it is necessary to consider the existence of monoatomic steps at the interface at every stage of oxidation. The decrease in areal density of $Si^{2+}$ with the increase in thickness in Fig. 10.2(b) implies that the monoatomic step density decreases with the progress of oxidation. In order to explain Fig. 10.2(b), it is necessary to consider non-uniform oxide film as shown in Fig. 10.3. If we describe this distribution of thickness by a Gaussian function, Fig. 10.4 is obtained for various values of full-width at half-maximum (FWHM). The best fit of thickness dependence of calculated density of oxidation states to experimental data in Fig. 10.2(b) is obtained for FWHM of 0.5 nm. [18] Almost the same amount of changes in $Si^{1+}$ and $Si^{3+}$ are obtained at oxide film thicknesses of 1.0 and 1.7 nm. This implies that the non-uniformity is not enhanced by the subsequent oxidation.

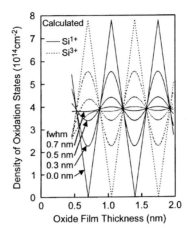

Fig. 10.4. Density of oxidation states calculated for a Gaussian distribution of oxide film thickness with FWHM as a parameter.

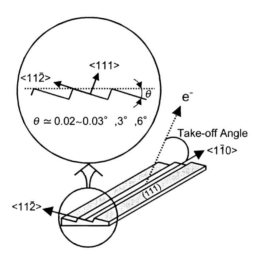

Fig. 10.5. Schematic diagram of vicinal surfaces.

### 10.3.2 The lateral size of an atomically flat interface

The lateral size of an atomically flat oxidized region on Si(111) surface was determined as follows from the effect of the terrace width on the layer-by-layer oxidation. [19] For this study, the vicinal (111) 0.02–0.03 degree, 3 degree and 6 degree surfaces with a terrace width of 360–600 nm, 6 nm and 3

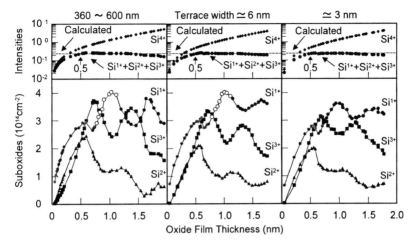

Fig. 10.6. The upper part shows normalized spectral intensity of Si⁴⁺, and the sum of intensities of all intermediate oxidation states are shown as functions of the oxide film thickness, while the lower part shows the dependence of densities of Si¹⁺, Si²⁺ and Si³⁺ on thickness for three kinds of vicinal surfaces.

nm, which are illustrated in Fig. 10.5, respectively, were used. The spectral intensity of Si⁴⁺ and that of intermediate oxidation states normalized by the Si2p3/2 spectral intensity of the silicon substrate as a function of thickness are shown for three kinds of vicinal surfaces in the upper part of Fig. 10.6. According to this figure, the saturated normalized spectral intensities of intermediate states are in good agreement with those calculated for an abrupt compositional transition at the interface, which are shown by the dashed lines in this figure. [15] Because almost the same results are obtained for three kinds of vicinal surfaces, the atomic steps on the Si surface before the oxidation do not affect the abrupt compositional transition. The areal densities of Si¹⁺, Si²⁺, Si³⁺ and Si⁴⁺ as a function of thickness are shown in the bottom of Fig. 10.6. At vicinal (111) 0.02–0.03 degrees, the oxidation reaction occurs layer-by-layer at the interface, for a thickness less than 1.7 nm.

The effect of terrace width on the interface structures appears on the thickness dependence of the amount of Si¹⁺ in the thickness range from 0.7 to 1.0 nm because of the following two reasons: 1) in the thickness range from 0.7 nm to 1.0 nm, with progress of oxidation, the amount of Si¹⁺ increases, while that of Si³⁺ decreases. Therefore, in this thickness range the formation of Si¹⁺ interface proceeds with the progress of oxidation; 2) on the vicinal (111) surfaces there are double atomic steps of silicon, because silicon monohydrides can be only observed on these vicinal H-Si surfaces. [20] The oxidation of two Si layers from the top produces nearly 0.7 nm thick silicon oxide, which is equal to the period in thickness observed for the periodic changes in interface structures.

Then, until the thickness reaches 0.7 nm, the oxidation of the third silicon layer from the top cannot interfere with the oxidation on adjacent terrace on vicinal (111) 3 and 6 degrees. This results in the almost same results below the thickness of 0.7 nm for three kinds of vicinal surfaces. However, on vicinal (111) 6 degrees the size of atomically flat oxidized region which is produced by layer-by-layer oxidation reaction at the interface is limited by the terrace width of 3 nm, and the oxidation of the third silicon layer from the top starts to interfere with the oxidation on the adjacent terrace above the thickness of 0.7 nm, because the thickness dependence of the amount of $Si^{1+}$ for vicinal (111) 6 degrees starts to deviate from that for vicinal (111) 0.02–0.03 degrees above the thickness of 0.7 nm. On the other hand, at vicinal (111) 3 degrees, the size of atomically flat oxidized region at the thickness of 0.7 nm is smaller than the terrace width of 6 nm and the oxidation of the third silicon layer from the top starts to interfere with the oxidation on the adjacent terrace above the average

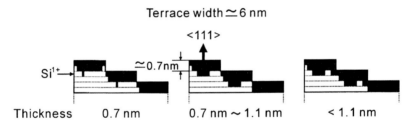

Fig. 10.7.  The oxidation process on a vicinal (111) 3 degree surface.

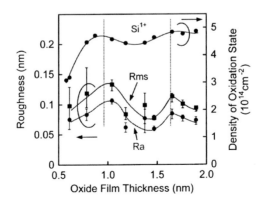

Fig. 10.8. Two kinds of surface microroughness of oxide formed on Si(111) and the density of $Si^{1+}$ shown as functions of the oxide film thickness.

thickness of 1.1 nm as shown in Fig. 10.7, because the thickness dependence of the amount of Si[1+] for vicinal (111) 3 degrees starts to deviate from that for vicinal (111) 0.02–0.03 degrees above the thickness of 1.1 nm. Therefore, the lateral size of an atomically flat oxidized region on Si(111) surface must be in the range of 3 to 6 nm at a thickness of 0.7 nm.

## 10.4  Oxidation-Induced Atomic-Scale Surface Roughness

Recently, it was found that the surface roughness changes in accordance with the changes in interface structure. [21] In order to investigate the correlation between surface and interface structures of ultrathin oxides formed on Si(111) surfaces, we measured non-contact mode atomic force microscope (NC-AFM) images and Si2p photoelectron spectra with progress of oxidation. In the experiment, the effect of atomic steps on the surface roughness was minimized by using vicinal (111) 0.017 degrees from <110> and 0.008 degrees from <112> with an average terrace area of about 1 $\mu$m × 2 $\mu$m.

Figure 10.8 shows averaged values of two kinds of surface roughness, Rms and Ra, of oxide film, measured at three positions on the oxide surface, as a function of thickness. It can be seen from this figure that the surface roughness changes periodically with the progress of oxidation and can be

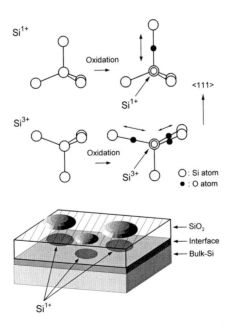

Fig. 10.9. The upper part shows the direction of volume expansion produced by the formation of Si[1+] and Si[3+] at the interface. The lower part shows a diagram of the relationship between the oxidation-induced protrusions and the interface structure consisting of Si[1+].

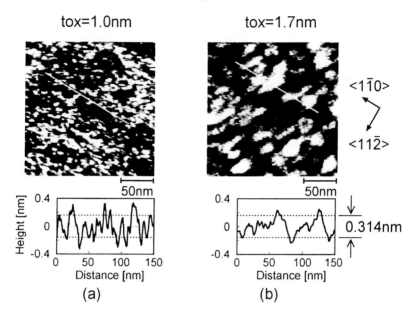

Fig. 10.10. AFM images and cross section of oxide films with thicknesses of (a) 1.0 nm and (b) 1.7 nm. Cross-sectional profiles were obtained along the lines shown in the AFM images.

correlated with the periodic changes in the amount of $Si^{1+}$ as explained in the following. As Fig. 10.9 illustrates, when $Si^{1+}$ is formed, the insertion of an oxygen atom between two Si atoms consisting of a Si-Si bond oriented along the <111> direction at the interface expands the oxide network only along the <111> direction, while when $Si^{3+}$ is formed the insertion of an oxygen atom between two Si atoms consisting of a Si-Si bond at the interface expands the oxide network mostly along the direction perpendicular to <111>. Therefore, the formation of $Si^{1+}$ at the interface results in an increase in the surface roughness accompanied by the formation of protrusions on the oxide surface.

Figure 10.10 shows the surface morphology of oxide films with thicknesses of 1.0 and 1.7 nm measured over an area of 200 nm × 200 nm. According to this figure, small protrusions with a lateral size of about 5 nm in diameter are present on the 1-nm-thick oxide film, while ones with a lateral size of about 20 nm in diameter are observed on the 1.7-nm-thick oxide film. These observations support the appearance of oxidation-induced protrusions discussed above. The lateral size of protrusions at a thickness of 1.0 nm is close to the lateral size of the atomically flat oxidized region determined from the effect of terrace width on the layer-by-layer oxidation described in chapter 3.2.

The same kind of study was also performed on a Si(100) surface. [22,23] Figure 10.11 shows the surface morphology of oxide films with thicknesses of

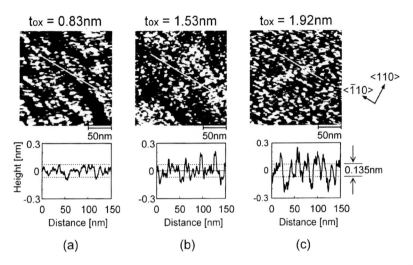

Fig. 10.11. AFM images and cross section of oxide films with thicknesses of (a) 0.83 nm, (b) 1.53 nm and (c)1.92 nm. Cross-sectional profiles were obtained along the lines in the AFM images.

0.83, 1.53 and 1.92 nm measured over an area of 200 nm × 200 nm. The upper part of Fig. 10.11 shows no significant change in surface morphology produced by the change in thickness from 0.8 to 1.9 nm, while the lower part of this figure shows a gradual increase in atomic-scale surface roughness. Figure 10.12 shows the height distributions on the three oxide surfaces shown in Fig. 10.11. These height distributions can be approximated by the Gaussian function. However, as can be seen in Fig. 10.11, the delta function which corresponds to atomically flat surfaces, also contributes slightly to these height distributions. Full with at half maximum (FWHM) of these Gaussian functions are shown as a function of the oxide film thickness in Fig. 10.13. In this figure, there is a clear difference in the thickness dependence of FWHM of a Gaussian function below and above the thickness of 1 nm, which is equal to the thickness of the structural transition layer. Above this critical thickness, the oxidation-induced stress is relaxed to form bulk $SiO_2$. Root mean square roughness, (Rms), are shown in Fig. 10.14 as a function of the thickness. According to this figure, surface roughness repeatedly increases and decreases with a period in thickness of 0.19 nm. This period in thickness is almost equal to the period in thickness observed for the oxidation of Si(100) surface in pure water at room temperature. [24] This periodic change in surface roughness is indicative of the layer-by-layer oxidation. Figure 10.15 gives a comparison between the changes in surface roughness with the progress of oxidation shown in Fig. 10.14 and the changes in interface roughness with progress of oxidation determined from the cross-sectional transmission electron microscope observation of the interface. [25]

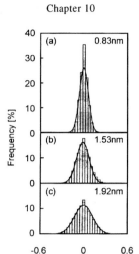

Fig. 10.12. Height distributions on three oxide surfaces with thicknesses of (a) 0.83 nm, (b) 1.53 nm and (c)1.92 nm shown in Fig. 10.11.

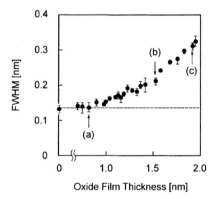

Fig. 10.13. FWHM of the Gaussian function, which approximates the height distribution, as a function of the oxide film thickness.

In order to correlate the surface roughness with the interface structure, the changes in interface structures were measured with the progress of oxidation. The upper part of Fig. 10.16 shows the Si2p spectral intensity of $Si^{4+}$ and that of intermediate states consisting of $Si^{1+}$, $Si^{2+}$ and $Si^{3+}$ normalized by the spectral intensity of Si substrate as a function of thickness. As the upper part of Fig. 10.16 shows, the saturated normalized spectral intensity of intermediate states

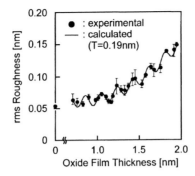

Fig. 10.14. Root mean square surface roughness of oxide formed on Si(100) and the density of Si[1+] shown as functions of oxide film thickness.

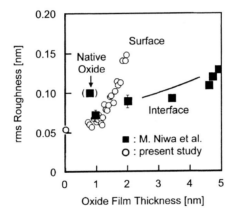

Fig. 10.15. Comparison between the changes in surface and interface roughness with oxide film thickness as a parameter.

is slightly larger than that calculated, which is shown by the dashed line in Fig. 10.16, for an abrupt interface [26] consisting only of $Si^{2+}$. As the lower part of Fig. 10.16 shows, the interface consists not only of $Si^{2+}$, but also of $Si^{1+}$ and $Si^{3+}$. Therefore, the interface structure must be a combination of the interface consisting of $Si^{2+}$, the interface [27] consisting of $Si^{1+}$ and $Si^{2+}$, the interface [28] containing $Si^{3+}$ and the interface [29] consisting of $Si^{1+}$ or $Si^{2+}$. Therefore, the layer-by-layer oxidation must occur locally. The existence of $Si^{1+}$ and $Si^{3+}$ at the $SiO_2/Si(100)$ interface must be necessary in order to minimize the oxidation-induced stress at the interface, which is larger than that at $SiO_2/Si(111)$ interface. [30]

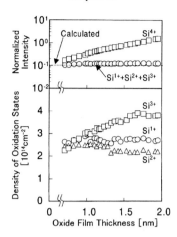

Fig. 10.16. Normalized spectral intensity of Si⁴⁺ and summation of intensities for all intermediate oxidation states are shown as a function of oxide film thickness. Dependence of areal densities of $Si^{1+}$, $Si^{2+}$, $Si^{3+}$ and $Si^{4+}$ on thickness are also shown.

## 10.5 Interface Electronic States and Their Correlation with Interface Structures

Figure 10.17 shows interface state densities at and near the midgap of Si with oxide film thickness as a parameter for oxide films formed on Si(111) surface. [31] For oxide film thicknesses of 1.35 and 1.98 nm, where the amounts of $Si^{3+}$ take their maximum values, the interface state densities at and near the midgap of Si are almost equal to $1 \times 10^{13}$ cm$^{-2}$eV$^{-1}$, while for oxide film thicknesses of 1.62 and 2.32 nm, where the amounts of $Si^{1+}$ take their maximum values, the interface state density near the midgap decreases drastically down to $1 \times 10^{12}$ cm$^{-2}$eV$^{-1}$. This drastic decrease in interface state density occurs only near the oxide film thickness of 1.62 and 2.32 nm. Therefore, the drastic decrease in interface state density near the midgap appears periodically with the progress of oxidation as a result of periodic changes in interface structures which are closely related to the formation of the $Si^{1+}$ interface. As described in section 10.4, the protrusions are produced on the oxide surface so as to relax stress produced by the formation of $Si^{1+}$ interface. Therefore, the protrusions on the oxide surface indicate the place where the interface state density is small. On the other hand, the volume expansion mostly along the interface plane induced by the formation of $Si^{3+}$ breaks a part of the Si-Si bonds just below the $Si^{3+}$ to produce dangling bonds with a density of about $10^{12}$ cm$^{-2}$ eV$^{-1}$.

Figure 10.18 shows interface state densities at and near the midgap of Si with oxide film thickness as a parameter for oxide films formed on a Si(100) surface. [32] For oxide film thicknesses of 1.65 and 1.83 nm, where the rms

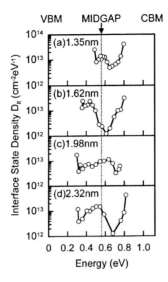

Fig. 10.17. Interfacestate densities at and near the midgap of Si for oxide films formed on the Si(111) surface with oxide film thickness as a parameter.

roughness takes its maximum value, the interface state densities at and near the midgap are almost equal to $1 \times 10^{13}$ cm$^{-2}$eV$^{-1}$, while for oxide film thicknesses of 1.54 and 1.73 nm, where the rms roughness takes its minimum value, the interface state density near the midgap decreases drastically down to $5 \times 10^{12}$ cm$^{-2}$eV$^{-1}$. Therefore, the drastic decrease in interface state densities near the midgap appears periodically with the progress of oxidation and is closely correlated with the smoothness of the oxide surface, which must reflect the smoothness of the interface. Therefore, the atomic step at the interface must be the structural origin of the interface states. This result contrasts with the results obtained for oxide films formed on a Si(111) surface.

## 10.6 *Valence Band Discontinuities at and near the SiO₂/Si Interface*

More than thirty years ago, Williams [33] made an energy band diagram of the SiO₂/Si system using internal photoemission. Later, a study [34] of the impurity-effect on the energy band discontinuity at the SiO₂/Si interface formed on a Si(100) surface found that the impurity-induced decrease in the conduction band (abbreviated to the C. B. hereafter) discontinuity and the impurity-induced increase in valence band (abbreviated to the V. B. hereafter) discontinuity are almost the same. Namely, they found that the V. B. discontinuity of 4.5 eV at the interface for the ultra-cleanly prepared oxide is smaller than that of 4.7 eV for the conventionally prepared oxide, while the C.

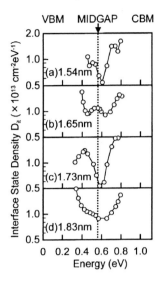

Fig. 10.18. Interfacestate densities at and near the midgap of Si for oxide films formed on the Si(100) surface with oxide film thickness as a parameter.

Table 10.1. Conduction and valence band discontinuity at $SiO_2$/Si(100) interface.

| Oxidation | Discontinuity [eV] | | Sum [eV] | $m^*/m_0$ | Reference |
|---|---|---|---|---|---|
| | C. B. | V. B | | | |
| Conventional | 3.2 | 4.7 | 7.9 | 0.42 | [50] |
| Superclean | 3.5 | 4.5 | 8.0 | 0.42 | [50] |
| a) | | 4.7 | | | [50] |
| b) | | 5.1 | | | [50] |
| | 3.25 | 4.99 | 7.74 | 0.34 | [51,52] |

a) Superclean oxidation followed by annealing in $H_2$/Ar, 450°C, 30min
b) Superclean oxidation followed by annealing in $N_2$ , 800°C, 60min

B. discontinuity of 3.5 eV at the interface for the ultra-cleanly prepared oxide is larger than that of 3.2 eV for the conventionally prepared oxide. The impurity effect on the conventionally prepared interface is equivalent to the effect of terminating dangling bonds with hydrogen atoms because almost the same V. B. and C. B. offset was obtained if the ultra-cleanly prepared oxide is annealed in the mixture of hydrogen and argon gases. Therefore, the increase

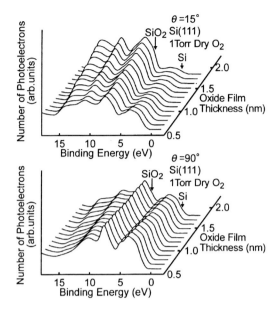

Fig. 10.19. Changes in valence band spectra of silicon oxide with the progress of oxidation for two photoelectron take-off angles.

in C. B. discontinuity and the decrease in V. B. discontinuity at the interface observed for the ultra-cleanly prepared oxide must have a correlation to the dangling bond-induced dipole layer. [35] The results are summarized in Table 10.1 It should be noted that an extremely large increase in V. B. discontinuity was observed for the annealing of the ultra-cleanly prepared oxide in a nitrogen atmosphere at 800°C. The impurity-induced stabilization of $SiO_2$ on Si(100) [36] must also be caused by the termination of dangling bonds by hydrogen atoms. It will be shown in the following that V. B. discontinuities exist not only at the interface, but also in the oxide near the interface. [37,38]

The changes in V. B. spectra with the progress of oxidation measured for oxides formed on Si(111) surface are shown in Fig. 10.19 and analyzed by the following two kinds of analytical procedure. The first analytical procedure is to take the difference in V. B. spectrum measured for two thicknesses close enough to each other so as to eliminate the V. B. spectrum of Si substrate. From this analysis, the V. B. spectrum of oxide surface can be obtained: if we assume that the V.B. discontinuity does not depend on the interface structure, but depends on the distance from the interface, the change in V. B. spectrum produced as a result of a small increase in the oxide film thickness corresponds to the V. B. spectrum of the oxide near the surface. The second analytical procedure is to take the difference in V. B. spectra measured for two

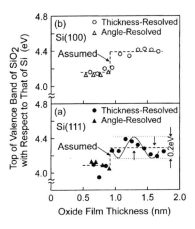

Fig. 10.20. Changes in the top of the valence band edge are shown with the progress of oxidation determined for two kinds of analytical procedures described in the text.

photoelectron take-off angles, that is, 15 and 90 degrees, so as to eliminate the V. B. spectrum of Si substrate. Because the electron escape depth in silicon dioxide for these two photoelectron take-off angles is 34 and 0.9 nm, respectively, the relative contribution of silicon oxide and silicon to the V. B. spectrum is different for two photoelectron take-off angles. Therefore, from the difference in two spectra for two take-off angles, the V. B. spectrum of silicon oxide occurring mostly within 0.9 nm from the oxide surface can be obtained.

Figure 10.20(a) shows the difference in energy between the top of the V. B. of the surface of oxides formed on Si(111) surface and that of silicon as a function of thickness determined from the first procedure on the assumption that the density of states near the V. B. edge follows parabolic dependence on the binding energy. Here, the thickness is an average of two thicknesses, where the difference between two spectra is taken. According to this figure, the top of the valence band of the oxide surface increases by about 0.2 eV near the thickness of 0.9 nm. In order to explain the spectral difference, it is assumed that the top of the V. B. of the oxide surface changes abruptly at the thickness of 0.9 nm as indicated by the dashed line in Fig. 10.20(a). This assumption implies that the oxide layer consists of a transition layer and a bulk layer (referred to as the two layer model hereafter). The energy of the top of the V. B. of the silicon oxide within 0.9 nm from the interface can be also determined from the second procedure and is also shown in Fig. 10.20(a). The critical thickness of 0.9 nm above which the energy level of top of V. B. changes abruptly is very close to the thickness of the structural transition layer determined from the analysis of infra-red absorbance and X-ray reflectance.

[3,4] Therefore, the present results suggest that the oxidation-induced stress in the interfacial transition layer produces a change in the energy level of the top of the V. B. of the silicon oxide near the interface. The difference in V. B. spectra obtained from the second procedure can be explained if the two layer model described above and the non-uniform oxide film expressed by a Gaussian function with FWHM of 0.5 nm are considered. If we examine Fig. 10.20(a) closely, it will be seen that the difference between the experimental data and the dashed line exhibits the periodic changes in the energy level of the top of the V. B. with progress of oxidation, which exactly corresponds to the periodic changes in the interface structures. [15] The difference in energy between the top of the V. B. of the oxide surface and that of silicon determined for Si(100) is also shown in Fig. 10.20(b). [38] The critical distance from the interface, above which the energy level of V. B. edge of oxide agrees with that of bulk SiO$_2$, are almost equal to each other for Si(111) and Si(100). The saturated values of V. B. discontinuities in Fig. 10.20 almost agree with those obtained by Alay *et al.* [39,40]

## 10.7 Summary

The surface, interface and valence band structures of ultrathin silicon oxides at the initial stage of oxidation are reviewed. On a Si(111) surface, the periodic changes in interface structures appear to be the result of the bonding nature of the Si crystal at the interface. Abrupt compositional transitions can be realized on Si(111) and (100) surfaces and are weakly affected by the atomic steps on the initial surface. The lateral size of the atomically flat oxidized region on the Si(111) surface is 3–6 nm at the thickness of 0.7 nm and roughly agrees with that observed using the non-contact mode atomic force microscope. The surface roughness of the oxide surface on Si(111) surface was found to change periodically with the progress of oxidation. This change can be correlated to the changes in interface structures. Namely, the formation of Si$^{1+}$ results in the formation of protrusions on the oxide surface. Periodic changes in surface micro-roughness were also found on the Si(100) surface; these must arise from the monolayer by monolayer growth of thermal oxide. The interface state densities at and near the midgap of Si exhibit periodic changes with the progress of oxidation in accordance with the periodic changes in the interface structures. The top of the valence band of silicon oxide within 0.9 nm from the interface was found to be different from that of bulk SiO$_2$. Furthermore, it was found that on the Si(111) surface, the valence band discontinuity at the interface is affected by the interface structure.

## Acknowledgments

One of the authors (T.H.) expresses his hearty thanks to Dr. Masatake Katayama of SEH Isobe R & D Center for supplying the silicon wafers used in the present study.

[References]

1)  H. Sasaki Momose, M. Ono, T. Yoshitomi, T. Ohguro, S. Nakamura, M. Saito and H. Iwai
    (1994), IEDM 94-593.

2)  M. Fukuda, W. Mizubayashi, A. Kohno, S. Miyazaki and M. Hirose, *Jpn. J. Appl. Phys.* **37**,
    L1534 (1998).

3)  K. Ishikawa, H. Ogawa, S. Oshida, K. Suzuki and S. Fujimura, *Ext. Abstr. of Int. Conf. on
    Solid State Devices and Materials*, Osaka, 1995, p. 500.

4)  Y. Sugita, N. Awaji and S. Watanabe, *Ext. Abstr. of Intern. Conf. on Solid State Devices
    and Materials*, Yokohama, 1996, p. 380.

5)  T. Hattori, *Critical Rev. Solid State Mat. Sci.* **20**, 339 (1995).

6)  T. Takahagi, I. Nagai, A. Ishitani and H. Kuroda, *J. Appl. Phys.* **64**, 3516 (1988).

7)  M. Sakuraba, J. Murota and S. Ono, *J. Appl. Phys.* **75**, 3701 (1994).

8)  G. S. Higashi, R. S. Becker, Y. J. Chabal and A. J. Becker, *Appl. Phys. Lett.* **58**, 1656 (1991).

9)  H. Bender, S. Verhaverbeke, M. Caymax, O. Vatel and M. M. Hynes, *J. Appl. Phys.* **75**,
    1207 (1994).

10) T. Ohmi, M. Morita, A. Teramoto, K. Makihara and K. S. Tseng, *Appl. Phys. Lett.* **60**, 2126
    (1992).

11) U. Gelius, B. Wannberg, P. Baltzer, H. Fellner-Feldegg, G. Carlsson, C.-G. Johansson, J.
    Larsson, P. Munger and G. Vergerfos, *J. Electron Spectrosc. & Relat. Phenom.* **52**, 747
    (1990).

12) P. Guthner, *J. Vac. Sci. & Technol.* **B14**, 2428 (1996).

13) W. M. Lau and X.-W. Wu, *Surf. Sci.* **245**, 345 (1991).

14) H. Nohira, Y. Tamura, H. Ogawa and T. Hattori, *IEICE Trans. Electron.* **E75-C**, 757
    (1992).

15) K. Ohishi and T. Hattori, *Jpn. J. Appl. Phys.* **33**, L675 (1994).

16) G. Hollinger and F. J. Himpsel, *Appl. Phys. Lett.* **44**, 93 (1984).

17) T. Aiba, K. Yamauchi, Y. Shimizu, N. Tate, M. Katayama and T. Hattori, *Jpn. J. Appl. Phys.*
    **34**, 707 (1995).

18) K. Ohishi and T. Hattori (unpublished).

19) A. Omura, H. Sekikawa and T. Hattori, *Appl. Surf. Sci.* **117/118**, 127 (1997).

20) I.-W. Lyo, Ph. Avouris, B. Schubert and R. J. Hoffmann, *Phys. Chem.* **94**, 4400 (1990).

21) M. Ohashi and T. Hattori, *Jpn. J. Appl. Phys.* **36**, L397 (1997).

22) T. Hattori, M. Fujimura, T. Yagi and M. Ohashi, *Appl. Surf. Sci.* **123/124**, 87 (1998).

23) M. Fujimura, K. Inoue, H. Nohira and T. Hattori, *Appl. Surf. Sci.* **162/163**, 62(2000).

24) Y. Yasaka, S. Uenaga, H. Yasutake, M. Takakura, S. Miyazaki and M. Hirose, *Mater. Res.
    Soc. Symp. Proc.* **259**, 385 (1992).

25) M. Niwa, K. Okada and R. Sinclair, *Appl. Surf. Sci.* **100/101**, 425 (1996).

26) S. T. Pantelides and M. Long, *The Physics of SiO$_2$ and Its Interface*, S. T. Pantelides, ed.,
    Pergamon, 1978, New York, p. 339.

27) F. Herman, I. P. Batra and R. V. Kasowski, *The Physics of SiO$_2$ and Its Interface*, S. T.
    Pantelides, ed., Pergamon, 1978, New York, p. 333.

28) M. M. Banaszak Holl, S. Lee and F. R. McFeely, *Appl. Phys. Lett.* **65**, 1097 (1994).

29) A. Pasquarello, M. S. Hybertsen and R. Car, *Phys. Rev. Lett.* **74**, 1024 (1995).

30) I. Ohdomari and H. Akatsu, *Solid State Electronics* **33**, Suppl., 265 (1990).

31) N. Watanabe, A. Omura, H. Nohira and T. Hattori, *Ext. Abstr. of Intern. Conf. on Solid State
    Devices and Materials*, Hiroshima, 1998, p. 130.

32) Y. Teramoto, N. Watanabe, M. Fujimura, H. Nohira and T. Hattori, *Appl. Surf. Sci.* **159/
    160**, 67 (2000).

33) R. Williams, *Phys. Rev.* **A140**, 569 (1965).

34) T. Ohmi, M. Morita and T. Hattori, *The Physics and Chemistry of SiO$_2$ and the Si-SiO$_2$*

*Interface*, Plenum Press, 1988, New York, p. 413.

35)  T. Yoshida, D. Imafuku, J. L. Alay, S. Miyazaki and M. Hirose, *Jpn. J. Appl. Phys.* **34**, L903 (1995).

36)  C. Heimlich, M. Kubota, Y. Murata, T. Hattori, M. Morita and T. Ohmi, *Vacuum* **41**, 793 (1990).

37)  H. Nohira and T. Hattori, *Appl. Surf. Sci.* **117/118**, 119 (1997).

38)  H. Nohira, A. Omura, M. Katayama and T. Hattori, *Appl. Surf. Sci.* **123/124**, 546 (1998).

39)  J. L. Alay, M. Fukuda, C. H. Bjorkman, K. Nakagawa, S. Sasaki, S. Yokoyama and M. Hirose, *Jpn. J. Appl. Phys.* **34**, L653 (1995).

40)  J. L. Alay, M. Fukuda, K. Nakagawa, S. Yokoyama and M. Hirose, *Ext. Abstr. of Intern. Conf. on Solid State Devices and Materials*, Osaka, 1995, p. 28.

# Chapter 11

# Characterization of Molecular Films by a Scanning Probe Microscope

*11.1    Local Area Visualization of Organic Ultra-Thin Films by the Scanning Probe Microscope*

*11.1.1   Introduction*

The molecular alignment on solid substrates is important both to fundamental issues in physics and chemistry and to practical applications such as optoelectronic molecular devices. It is well-known that the orderings of adsorbed molecules in the anchoring region near the substrate are strongly affected by the substrate surface structure and the balance of molecule-molecule and molecule-substrate interactions. This is especially evident in the case of liquid crystal molecules, for example, where such an alignment technique with boundary effects at the substrate surface has been widely used to produce director configurations in the monodomain cells such as flat-panel displays and optical shutters. However, the actual mechanism of their orientation at the interface has long been unclear, because there have been technical difficulties in the analysis of the anchoring structures at a molecular level. Since its invention, the scanning probe microscope (SPM) has been successfully applied to direct visualization of the interfacial structure with molecular resolution, and used to study positional and orientational order in organic molecules on solid substrates. [1] In fact, high-resolution scanning tunneling microscope (STM) images now allow us to make a real-space analysis of the molecular alignment at the interface and to discuss details of the growth mechanisms of molecular films on an individual molecular scale. This section will briefly review local area visualizations that have been performed with the STM and allied microscopies on organic ultra-thin films, rendering features from the nanometer to the submicron scale.

## 11.1.2 STM imaging of molecular films

Since the STM method itself was developed in a clean surface science field, imaging of organic molecules was firstly performed under well-defined ultra high vacuum (UHV) conditions for the low-molecular weight adsorbates on single crystalline metal substrates. Especially in the earlier studies in the late 80's, co-adsorption of benzene with CO, for example, from the vapor phase on clean Rh(111) leads to a stable surface structure under UHV, resulting in atomic resolution STM images at room temperature. [2] Such UHV STM imaging has also been extensively utilized to study naphthalene chemisorbed on a Pt(111) [3] and isolated Cu-phthalocyanine molecules on various metal substrates. [4] During the early 90s, unambiguous and reproducible STM images of organic molecules were only available in the UHV STM system until the first STM imaging of liquid crystals in air was reported [5], because the apparent success of the STM in obtaining molecular images was accompanied by misleading "artifacts" or "failures" which gave no recognizable results nor new insights. [6] After overcoming the "artifacts" era, STM imaging (especially in air) has been carefully carried out and extended to a wide variety of molecular films including ordered monolayers [7], conducting polymers [8], fullerenes [9] and biological molecules (Fig. 11.1). [10] The systematic studies of the STM imaging in liquid crystals [11] and self-assembled monolayers (SAMs) [12] as typical examples of the ordered monolayers will be described separately in the following section 11.2.

The application of the STM to organic molecular films has been a challenging task, basically requiring suitable conductive and atomically-flat single crystalline substrates and immobilization of the deposited molecules for stable STM imaging, while the most serious limitation still being encountered is the problem of image interpretation. [13] One of the most typical sample structures which remain stationary under the scanning probe is an organic monolayer with a two-dimensionally ordered orientation analogous to epitaxial growth on an ordered substrate. The two-dimensional molecular ordering at the substrate-solution interface is not unique for liquid crystal and self-assembly molecules, but can be also observed for other molecules, e.g. alkane derivatives [14] and Langmuir-Blodgett films. [15] These STM images of mainly straight-chain hydrocarbons constitute one of the most straightforward examples of imaging of electric insulators with a tunneling current. On the other hand, STM has been also applied to study the atomic and electronic structures of electrically conducting and superconducting organic molecular crystals such as TTF-TCNQ (tetrathiafulvalene-tetracyanoquinodimethane) [16] and the BEDT-TTF (bis(ethylenedithio)-tetrathiafulvalene) family. [17] This has been motivated by the desire for a better understanding of the relationship between the molecular structure and conductivity, and the tunneling phenomenon itself. [13] As mentioned above, while tunneling and imaging mechanisms in organic molecular films should themselves be clarified through

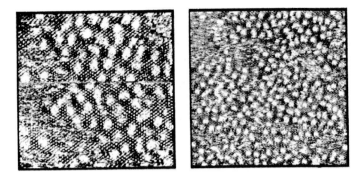

Fig. 11.1. STM images of biological macromolecule cytochrome c on an iodine-modified surface of Au(111). The small hexagonal dots correspond to iodine atoms and the large dots represent cytochrome c. (27 × 27 nm and 46.5 × 46.5 nm)

studies using conducting and nonconducting materials, the ability to "see" single molecules is providing "never-ever-seen" information for molecular film studies even though it is still partially based on empirical findings at present. The impact of STM on organic molecular films seems to have slowed down at the moment, but in fact other organic molecular systems are still challenging further innovative research.

### 11.1.3 AFM imaging of molecular films

In contrast to the examples of the STM visualization mentioned above, most organic molecular films are electrically insulating and should therefore be investigated by the atomic force microscope (AFM) and related techniques using force interactions between the scanned probe and the sample surface such as van der Waals, frictional, electrostatic, capillary and magnetic forces. The early AFM studies of organic molecules were mainly on macromolecules [18], because the resolution itself was not sufficient to visualize the individual molecular structures to the same extent as STM images. However, the capability of the AFM has been improved drastically and has reached the resolution needed to distinguish between different molecular orientations with aspect ratios that correspond to the dimensions of individual head groups in membranes. [19] More recent AFM studies of molecular films show that resolution on the scale of individual molecules and atoms can be realized especially for LB [20] and SAM samples. This progress is much indebted to the development of specially sharpened cantilevers, precise control of the feedback, separation of multiple signals, and suitable sample preparation with a well-controlled scanning environment.

The AFM is now utilized in a much wider area than STM, from basic biology through to semiconductor production lines, because of the lower restrictions on the sample structures. In addition, further improvement of force measurements have promoted wider applicability such as to the friction force microscope (FFM, or lateral force microscope (LFM)), the magnetic/ electrostatic force microscope (MFM/EFM), and the phase contrast microscope (PCM). The ability of the AFM to visualize or characterize molecular film features has been well demonstrated, and one of such challenges was the differentiation of subtle features by chemically modified domains and compositional changes within a film. For example, the localized properties of adhesion, elasticity, and friction have been probed with the AFM within single domains in LB films composed of mixtures of molecules with hydrocarbon and fluorocarbon moieties. [21]

Although the restrictions on the structure are lower in general AFM samples, similar serious problems to those in STM have often been encountered in trying to obtain unambiguous AFM images of soft biological macromolecules. These include reliable immobilization methods for those molecules on atomically flat substrates, controlling the force acting on the sample surface, and the identification of images against "artifacts". Especially when scanning, the force exerted by the probe itself must be questioned. Even if the macromolecules form an ideal monolayer in two dimensions, the surface of the soft biological specimen is easily deformed by the strong force applied between probe and specimen, even in the order of $10^{-9}$ N, making the production of high-quality images difficult. [22] In fact, clear AFM images of biological macromolecules have so far been restricted mainly to naturally formed two-dimensional protein crystals, which are sufficiently stable. [23] In order to overcome such difficulties, AFM imaging of artificially formed two-dimensional arrays of water-soluble protein molecules bound to a charged polypeptide monolayer on a silicone surface has been performed. [24] The imaging itself has been discussed from the viewpoint of controlling the electrostatic force between the probe and the sample surface, optimized by the cancellation of electrostatic charges at the interface of the negatively charged protein and positively charged polypeptide monolayer, which provides one of the most effective factors in attaining low-force AFM imagings. By introducing such a "self-screening effect" of the surface charges of the biological macromolecule itself, the forces between probe and sample layer could be kept sufficiently smaller than $10^{-10}$ N, resulting in reproducible nondestructive imaging of the water-soluble molecules in solution (Fig. 11.2).

### 11.1.4 SNOM imaging of molecular films

Real time visualization of biological molecular behavior has been performed using optical phase contrast or the differential interference contrast microscope. Resolution of these systems is limited by the wavelength of light. More detailed submicrometer studies of the ultrastructure have been carried out using scanning and transmission electron microscope (SEM and TEM)

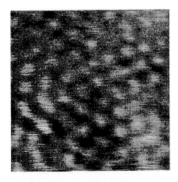

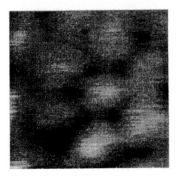

Fig. 11.2. (a, upper) Unfiltered AFM image of a regular array of water-soluble protein ferritin molecules bound to a PBLH layer on a silicon wafer. The imaging was carried out in pure water using a fluid cell (125 × 125 nm). (b, lower) Higher magnification AFM image showing the hexagonal packing of ferritin molecules (45 × 45 nm).

techniques. However, live molecules cannot be examined in real time using these methods because SEM and TEM systems require molecular immobilization and visualization under high vacuum. On the other hand, AFM has been used to examine the morphology of live and fixed macromolecules. These systems have high resolutions but may deform the soft matter, because of the direct contact between the probe and samples. During the last decade, the scanning near-field optical microscope (SNOM) has been developed to increase the resolution of optical imaging beyond that of the traditional optical microscope [25], and more recently, the first *in situ* SNOM observation of live-cell dynamics to examine rhythmically beating cardiac myocytes in culture has been performed. [26] By scanning over the surface of cells, three-dimensional SNOM images can be obtained in solution. While scanning, the probe can be stopped at any point to optically record the localized contractile activity of submicrometer areas of individual cells. The contraction profiles have been seen to change dramatically within adjacent areas, suggesting that the SNOM system is capable of detecting submicrometer features that directly influence the recorded contraction profiles. There is evidence from acoustic microscopy that submembrane shortened actin filaments can act as detectable scatterers in

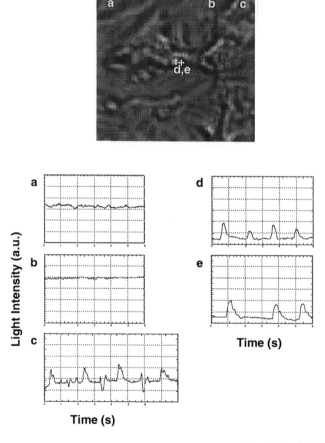

Fig. 11.3. SNOM images of live cardiac myocytes ($60 \times 60$ $\mu$m, $90 \times 90$ pixels). White crosses indicate recording sites where the scan was stopped to monitor the cell activity in real time. Sites d and e are two adjacent pixels, where crosses are superimposed. The lower five figures show the recording signals taken at the points a to e, plotting the dynamic activity of the cell in the crossed sites. Recording time resolution is 0.03 seconds, and the scanning was restarted from the same point. The whole image scan took about 5 minutes. The time scales are independent and the two signals are not synchronized.

contractile cells. There is also evidence that the SNOM system can resolve submembrane cytoskeletal elements in fixed cells. The spatially dependent amplitudes of contractions then become significant because of the SNOM's ability to resolve the behavior of individual submembrane actin bundles. Combining SNOM imaging and real-time recording in localized areas will provide a new noninvasive method for studying the dynamics of live biological molecular films under physiological conditions (Fig. 11.3).

## 11.2  Application to Anchoring Phase Studies

### 11.2.1  Introduction

A series of organic molecules that has been most extensively imaged to date is liquid crystals [27] and self-assembled monolayers (SAMs) [28], since they spontaneously form highly ordered anchoring structures on solid substrates. In this section, such "anchoring phase" studies, an example of the successful application to organic materials science, will be described on the basis of the highly reproducible STM images.

The most typical liquid crystal system used in STM is a homologous series of n-alkylcyanobiphenyls (mCBs: m = number of carbon atoms in the alkyl chains). [29] Those anchoring structures can be divided into two clear categories (a "single-row" type and a "double-row" type) by STM. It has been proposed that the single-row is the anchoring structure of a nematic phase, while the double-row is the one of a smectic phase. Furthermore, a new phase transition that is only available at the boundary has been confirmed for the first time by STM studies of various bulk compositions of 8CB-12CB binary mixtures. [30] The correlation between those anchoring structures and the bulk phase diagrams has been discussed from the viewpoint of anchoring phase formation at a molecular level. [31]

In the case of STM imaging for SAMs, alkanethiols and dialkyl disulfides on Au(111) have been extensively studied because of their tendency to form highly ordered and densely packed monolayers having high stability. [32] In spite of the number of such studies, there remain some intrinsic questions regarding adsorption processes and states of dialkyl disulfide SAMs on the gold surface. For example, does the S-S bond cleavage of disulfide actually occur in the course of the formation of monolayers, and do monolayers thereafter consist of gold-bound thiolates (S-Au, chemisorption) or the dimer-like form (S-S, physisorption) of sulfur head groups, or if the dissociative adsorption of disulfides on gold occurs, what is the adsorption behavior of thiol moieties after the S-S bond cleavage on the gold surface? Although STM is a very powerful technique for obtaining information on the surface structures of adsorbates on substrates, STM has to date given no clear evidence from which to obtain answers to these questions. [33]

## 11.2.2 "Single-row" and "double-row" anchoring phases of liquid crystals

Figure 11.4 shows typical STM images of *m*CB molecules; (a, left) 8CB, (b, middle) 8CB-12CB binary mixtures, and (c, right) 12CB adsorbed on MoS$_2$. Individually distinguishable rodlike patterns and regular alignment can be seen. Each rodlike pattern shows the individual liquid crystal molecules. In numerous studies of this type on MoS$_2$, the anchoring structures are divided into two clear categories; a "single-row" type (7CB, 8CB (Fig. 11.4(a)), 9CB, and 11CB) in which cyanobiphenyl head groups and alkyl tails alternate in each row and a "double-row" type (10CB and 12CB (Fig. 11.4(c))) in which cyano groups are facing one another. In the case of the binary mixtures, on the other hand, it has been realized that there exists an inhomogeneous (mixed) double-row type consisting of 8CB and 12CB (Fig. 11.4(b)).

While there exist various interactions for the formation of the anchoring phases, one possible model for the origin of the two phases from the viewpoint of their phase sequences in the bulk has been proposed. Firstly, while 10CB and 12CB only have a smectic (S) liquid crystal phase between isotropic (I) and solid crystal (C) phases, STM images of both molecules exhibit the same double-row structure nearly equivalent to the bulk S phase ordering. It is natural to assume, therefore, that the double-row is formed at the interface when the phase transition takes place from I to the ordered S phase in the bulk. This double-row then can be attributed to the anchoring phase of the S phase. Secondly, while 7CB has only a nematic (N) liquid crystal phase, the STM

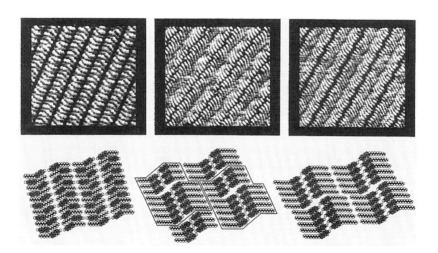

Fig. 11.4. STM images and models showing the anchoring structures of *m*CBs on MoS$_2$. (a, left) Homogeneous 8CB single-row (15 × 15 nm). (b, middle) Inhomogeneous (mixed) double-row (20 × 20 nm). (c, right) Homogeneous 12CB double-row (20 × 20 nm).

image exhibits a single-row structure on $MoS_2$. Comparing the results for 7CB (phase sequence I-N-C) with those for 10CB and 12CB (I-S-C), the single-row can be attributed to the anchoring phase of the N phase. In the same manner, if the liquid crystals have a phase sequence I-N-S-C, the molecules should be ordered in the N phase which actually appears first when the temperature is decreased during the sample preparation. In fact, the STM images of pure 8CB samples, for example, which possess an N phase in a higher temperature region than an S phase in the bulk (I-N-S-C), exhibit a single-row phase on $MoS_2$. From this point of view, the single-row and double-row are attributed to the formation of anchoring phases from the N and the S phases, respectively (Fig. 11.5). [29] The STM imaging of various bulk compositions from 0 to 100 mol% 8CB-12CB binary mixtures showed the same anchoring phase transitions, which supports the interpretation of the anchoring structures given above. [30] The field of characterization of molecular films, which had been done macroscopically and quite empirically before the introduction of the STM, has been opened up by these findings.

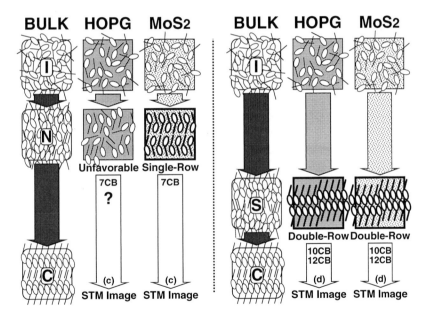

Fig. 11.5. Phase sequences in bulk and anchoring region on $MoS_2$ and graphite (HOPG). (a, left) 7CB with I-N-C sequence. Energetically favorable anchoring phase does not form on graphite because of no S phase, while single-row forms during I-N bulk transition on $MoS_2$. (b, right) 10 and 12CBs with I-S-C sequence. The same double-row forms both on $MoS_2$ and graphite during I-S bulk transition.

### 11.2.3 Substrate dependence: MoS₂ and graphite (HOPG)

STM imaging of organic molecular films under ambient conditions had been carried out mainly on graphite (highly oriented pyrolytic graphite: HOPG) before MoS₂ was introduced to the study of liquid crystals. [34] In fact, it has been well accepted that 8CB, 10CB and 12CB exhibit smectic-like double-row ordering on graphite, while those molecules are so-called smectic liquid crystals. From the viewpoint of interpretation of the anchoring phases, however, this result on graphite suggests that 8CB did not form an energetically favorable anchoring phase on graphite in the N phase (single-row), but rather in the S phase (double-row). This difference can be explained and attributed to the different surface structures of the substrates as follows. [31]

In the case of graphite, the cleaved surface consists of a condensed ring of six carbon atoms. In addition, it is well-known that the hydrogen orbitals of *n*-alkanes can fit into the holes of the carbon rings on graphite with nearly commensurate condition, resulting in only three degrees of freedom for the stable anchoring of alkyl chains. [35] The alkyl chains are then most stable when the angle between chains is 0° or 60°. On the other hand, the cleaved surface of MoS₂ consists of a hexagonal lattice of sulfur atoms and there are at

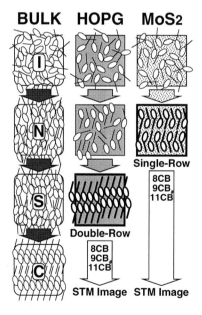

Fig. 11.6. Substrate dependence of anchoring phase sequences. Anchoring phase transition takes place during I-N bulk transition on MoS₂, while it takes place during N-S bulk transition on graphite. # It turns out that N-S transition of 11CB was caused by impurity (from data sheet distributed by BDH Ltd. (Poole, UK)).

least six degrees of freedom for the anchoring of alkyl chains with multiples of 30°. The angle between the alkyl chains in the single-row is 0° for every second molecule and approximately 30° for neighboring molecules. For this reason, since the 30°-required single-row is unstable on graphite, 8CB is not adsorbed stably in the N phase, which appears first with decreasing temperature, but rather is adsorbed in the subsequent S phase. In fact, STM images of 7CB on graphite exhibit neither single-row nor double-row structures. This result can be explained as 7CB forming only an N phase on graphite, resulting in instability of the single-row on graphite (Fig. 11.6).

### 11.2.4 Self-assembled monolayers: chemisorption and physisorption

A number of earlier STM studies were performed with the aim of solving the adsorption processes of dialkyl disulfide using asymmetric or unsymmetric disulfides. However, these studies were mainly performed on the fully covered monolayers derived from compounds on Au(111). This is a limitation for finding a solution to the problem because the exchange process, from the shorter alkyl chain thiolates to the longer alkyl chain ones during increasing surface coverage [36], different lateral diffusion rate of adsorbed molecules on the surface, and interaction differences between adsorbing molecules and the solvent induced by two different alkyl parts, can affect the SAM formation and STM images. [37] In order to overcome this experimental restraint, asymmetric

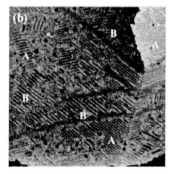

Fig. 11.7. STM images of 11-hydroxyundecyl octadecyl disulfide (HUOD) SAMs on Au(111) obtained after (a) 1 min and (b) 3 min deposition in 0.25 $\mu$M ethanol solutions of HUOD. (a) The STM image showed an ordered nucleation of these molecules on the gold surface during the initial SAM growth stage. The height of the nuclei reached about 0.5 nm, while the monatomic step height of Au(111) was about 0.25 nm. The scan size was 205 nm × 205 nm, and imaging conditions were 500 mV (sample positive) and 0.3 nA in the constant current mode. (b) Two phase-separated domains (A and B), formed as the result of the S-S bond cleavage of disulfide, were clearly observed in this image. The scan size was 130 nm × 130 nm, and imaging conditions were 300 mV (sample positive) and 0.2 nA in the constant current mode.

11-hydroxyundecyl octadecyl disulfide ($CH_3(CH_2)_{17}S$-$S(CH_2)_{11}OH$, HUOD), which for example, has different alkyl chain lengths and terminal groups, was introduced to the STM imaging. It is well-known that striped phases where molecules are oriented parallel to the surface are usually observed in the low surface coverage region during the initial SAM growth stage, or on a surface which is prepared using an extremely dilute solution of organosulfurs. [38] The periodicity of the striped phases depend strongly on the alkyl chain length of organosulfur compounds used for the formation of SAMs on metal surfaces. STM imaging of the striped phases formed during the process of the SAM growth stage from a solution containing asymmetric molecules has an important role in revealing adsorption states such as chemisorption and physisorption. In fact, STM imaging has clearly exhibited two types of phase-separated domains having different corrugation periodicities consistent with the lengths of $CH_3(CH_2)_{17}S$ and $HO(CH_2)_{11}S$ molecules, respectively. This is the first time that there has been a direct observation of the dissociative adsorption of disulfides on the nanometer scale. The self-assembly process of such asymmetric molecules physisorbed on graphite (HOPG), on the other hand, is mainly governed by a hydrogen bond with a hydroxyl group facing another hydroxyl group of adjacent molecules without any S-S bond cleavage of the disulfide group unlike the adsorption process on gold. By introducing different substrates, STM has provided a new, simple and insightful way of comparing monolayers chemisorbed on Au(111) with those physisorbed on graphite.

The STM image in Fig. 11.7 clearly exhibits the disordered intermediate phase and two types of striped-phase domains with phase separation of the SAM sample obtained after 3-min deposition. The corrugation periodicity values of the striped phases in regions A and B are 1.80 and 2.53 nm, which show good agreement with the length of $HO(CH_2)_{11}S$ and $CH_3(CH_2)_{17}S$ molecules, respectively. This finding strongly suggests that the S-S bond cleavage of asymmetric disulfide had already occurred when the HUOD molecules were adsorbed on the surface in the reaction with gold at room temperature, and that the direct adsorption of the original disulfide cannot occur on the surface during the SAM formation. This conclusion can also be drawn from the fact that if there is no S-S bond cleavage during the adsorption process of disulfide, the striped periodicity should show the value of an entire molecular length of HUOD molecule. In this respect, STM imaging is the first direct observation of the dissociative adsorption of disulfides on Au(111) on the nanometer scale. The disordered phase between striped phases, on the other hand, can be attributed to the phase transition from the striped phase to the standing-up phase as the surface coverage increases.

To confirm the result of the S-S bond breaking on Au(111), the molecular arrangements of HUOD molecules physisorbed on graphite were examined using the STM. The main difference in the self-assembly process of HUOD molecules on gold and graphite is the interactions (chemisorption or physisorption) between the molecule and the substrate, as discussed above. Organic molecules on the graphite surface usually lie flat on the surface in a similar way to the striped phases observed on gold. The STM image of a large scan area in Fig. 11.8(a) shows well-ordered molecular arrangements of HUOD molecules that are oriented parallel to the graphite surface. The high-resolution STM image in Fig. 11.8(b) displays the individual disulfide groups, indicated by spots in bright lines, and about $23°$ tilted orientation of alkyl chains. The bright lines corresponding to the disulfide group in this STM image may be due to the large molecular polarizability and an increase in the local density of states (LDOS) near the Fermi level of the surface. [39] These bright lines are paired by a hydrogen bond through a hydroxyl group of the alkyl chain, facing another hydroxyl group of adjacent molecules. Here, two longer alkyl moieties are located in the large dark area (region A) between two bright lines, and two shorter hydroxylated alkyl moieties are located in the small dark area (region B). The three bright lines shown in region C, on the other hand, may be due to the incorporation of di(11-hydroxyundecyl)disulfide $((HO(CH_2)_{11}S)_2)$ molecules as an impurity into the monolayers during the monolayer formation on graphite. These results suggest that the hydrogen bond between molecules is an important factor in determining the molecular arrangements in this system. The STM image provides supports for the finding that the self-assembly process on graphite proceeds without any disulfide bond rupture as expected.

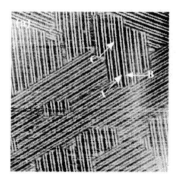

Fig. 11.8. STM images exhibiting the striped phase of 11-hydroxyundecyl octadecyl disulfide (HUOD) physisorbed on graphite. Bright rows correspond to the disulfide groups. Molecular arrangements were strongly dependent on the hydrogen bond derived from the hydroxyl group positioned at the end of the shorter alkyl chain among the two different alkyl chain lengths attached to the disulfide group. (a) The scan size was 160 nm × 160 nm, and imaging conditions were 1500 mV (sample positive) and 0.12 nA in the constant current mode. (b) The scan size was 16 nm × 16 nm, and imaging conditions were 1300 mV (sample positive) and 0.12 nA in the constant current mode.

On the basis of such STM studies, it has been confirmed for the first time from a nanoscopic viewpoint that the self-assembly of disulfide on gold proceeds along with the bond breaking of disulfide by a chemical reaction between the disulfide group and gold without the direct adsorption of the disulfide group on the gold surface. In addition, the SAM formation on the gold surface was not strongly influenced by a hydrogen bond, at least not in the initial growth stage, because there were no striped phases induced by a hydrogen bond between molecules, while phase separation had already begun in the striped phase by surface diffusion. Thus, observation of the striped phases before the formation of the standing-up SAM with a well-designed model compound will provide new guidelines for probing and interpreting new and fundamental aspects such as surface reactions, interactions, mobilities and orientation mechanisms of molecules, and for monitoring target molecules on a surface. Comparing STM images of the monolayer ordering on chemically reactive substrates with those on inert surfaces is a simple and insightful method which can be widely applied to various organic molecular film systems. [40]

### 11.3    Local Area Characterization of Organic Ultra-Thin Films by a Scanning Probe Microscope

#### 11.3.1  Introduction

The scanning probe microscopes (SPM) such as the scanning tunneling microscope (STM) and the atomic force microscope (AFM), and their related

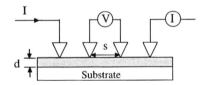

(a) A Classical Four-Point Probe Measurement

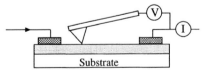

(b) Scanning Potentiometry Using
Conductive AFM Probe

Fig. 11.9. Scanning potentiometry. (a) A classical four-point probe measurement, (b) Scanning potentiometry using conductive AFM probe.

methods are being used in the development of powerful surface analysis techniques enabling nanometer resolution. As a consequence, various characterization methods combined with SPM have been introduced. [41] One important method is scanning potentiometry and scanning tunneling spectroscopy (STS), which permits electrical characterization on an atomic scale of properties such as electrical transport, local density of states (DOS), and vibrational modes in small-scale structures. It has also been applied to many kinds of materials, including organic molecules. Use of it in molecular switching devices and high density data storage systems is also arousing great interest. However, there is no precise understanding of it yet, because the surface states of the SPM probe and the samples are strongly reflected in the experimental data, and either erratic or unstable results are often obtained due to thermal drifts, piezoelectric hysterisis, and unknown tip-sample interactions. In this section, some electrical characterization methods using SPM are described.

### 11.3.2 Scanning potentiometry and scanning capacitance measurements

The voltage probe method using SPM has been extensively studied by researchers doing fundamental research in nanoelectronics. Surface potentiometry can be realized by using a conductive AFM probe and the basic analysis of this technique is the same as in a classical four-point probe measurement. Figure 11.9 shows a typical four-probe arrangement and scanning potentiometry using a conducting AFM probe. Current I passes between the outer two probes and the voltage V is the potential drop between the inner two probes. When a collinear four-point probe with equal separation s between the points is used, the following expression is obtained.

$$\rho = 2\pi s(V/I). \tag{11.1}$$

The equation represents the average resistivity $\rho$ of the film between the inner probes. Note that the four-point probes need not be either equidistant or collinear; in this case, however, the expressions are more complicated. [42] Consequently, if we employ an AFM probe instead of two inner probes (Fig. 11.9), the potential variations or the local area resistivity of the sample can be measured. Another merit of this method is that the constant and low stiffness of the AFM probe reduces probe-sample interaction effects such as sample damage, vibration, piezoelectric effects and so on.

A typical experimental result obtained on a TTF-TCNQ evaporated film is shown in Fig. 11.10. Simultaneously obtained topographic and potential images using an Au coated AFM probe are indicated in Figs. 11.10(a) and (b), respectively. The observed potential image corresponds to the AFM topographic image and these correlations are also examined under the different conditions of potential polarity and raster direction of the sample. In this case, TTF-

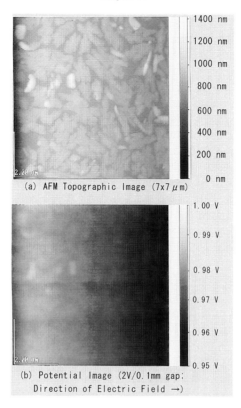

(a) AFM Topographic Image (7x7 μm)

(b) Potential Image (2V/0.1mm gap;
    Direction of Electric Field →)

Fig. 11.10. Topographic and potential images of a TTF-TCNQ evaporated film.

TCNQ evaporated films form charge-transfer complex domains with a high conductivity immersed in a low conductive amorphous-like film. The results indicate that the conductivity difference in the domain structure of the film reflects the potential image.

Scanning potentiometry using STM is also possible. Muralt and Pohl proposed the first tunneling potentiometry apparatus and experimental results of this method on a granular gold film and a semiconductor heterostructure have been reported. [43,44]

Capacitance microscopy makes it possible to measure the electrical properties of both conducting and insulating samples as well as the surface microtopography. [45] The high-resolution capacitance and potential measurement of organic or inorganic microstructures such as photoresists and pn junction have been achieved by using a SPM capacitance probe. [46,47] Scanning capacitance measurements are also important in investigating of the local work function or dielectric nature of the materials. The Maxwell stress

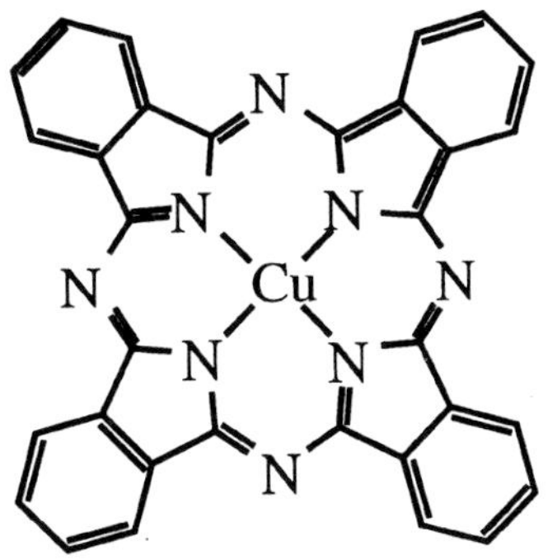

# Copper phthalocyanine (CuPc)

Fig. 11.11. Chemical structure of CuPc molecule.

microscope and Kelvin vibrating capacitor method have been described. [48–50]

These electrical measurements using SPM allow simultaneous investigation of the topography of the surface and their electrical properties with microscopic resolution, giving an insight into conduction through domain structures, heterostructures, and interfaces.

## 11.3.3 STS of copper phthalocyanine thin films

In a conventional STS system, the tunneling conductance reflects the local DOS of the probe and sample. If we measure the tunneling spectrum using a metal probe (that is the DOS of the probe is assumed not to influence the shape of the overall DOS), the observed STS mainly contains information on the sample. The tunneling conductance was predicted by Tersoff and Hamman, [51] in the following equation:

$$dI/dV = \rho_s(r_0, eV) \qquad (11.2)$$

where $\rho_s$ is the local density of states of the sample, $r_0$ is the position of the center of curvature of the probe, and $V$ is the applied voltage. On the other hand, the STS using a semiconductor probe is strongly influenced by the DOS of both the probe and sample. In this case, the tunneling conductance will be the following:

$$dI/dV = \rho_t\rho_s(r_0, eV) \qquad (11.3)$$

where $\rho_t$ is the DOS of the semiconductor probe. The STS measurement has

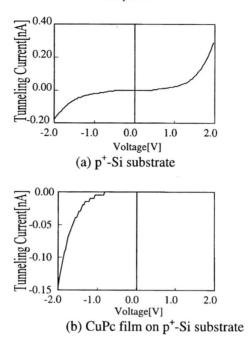

Fig. 11.12. I-V curves of a p⁺-Si substrate and CuPc films on p⁺-Si substrate.

also been applied to determine the band structure of conducting polymer films. [52,53]

The STS of copper phthalocyanine (CuPc) thin films were measured as follows. CuPc films in which the thickness of the CuPc was approximately 70 nm were vacuum evaporated on p⁺-Si substrates. Figure 11.11 shows the chemical structure of CuPc. CuPc shows p-type semiconducting properties and has a band gap of 1.6 eV. [54] The STS measurements were performed using a rearranged STM system (JSTM-4100S: JEOL). The STS probes used here were Pt-Ir, cleaved p⁺-Si and CuPc coated Pt-Ir probes. Si probes were fabricated by freshly cleaving a Si wafer and using the edge as the probe. Before introducing the Si probe to the STS system, the sharpness of the probe was examined by a scanning electron microscope. The CuPc coated Pt-Ir probe was prepared by evaporating CuPc on the top of Pt-Ir tip. All the measurements were carried out under vacuum at room temperature.

Figure 11.12 shows I-V curves of a p⁺-Si substrate and CuPc films deposited on p⁺-Si substrate obtained with a Pt-Ir probe. Although a symmetric I-V curve was observed in the p⁺-Si sample, the CuPc sample showed asymmetric I-V characteristics. Similar results for I-V curves have been reported in CuPc films and have been explained as being due to changes in the density of states

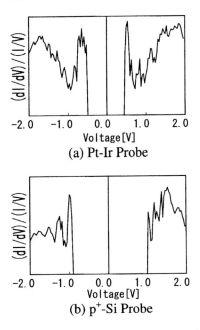

Fig. 11.13. Differential conductance curves, (dI/dV)/(I/V), of p$^+$-Si substrate. (a) Pt-Ir probe and (b) p$^+$-Si probe.

arising from the molecular energy levels or the formation of a rectifying Schottky barrier at the interface. [55] To examine the cause of the asymmetry, we performed STS measurements by changing the probe and sample materials. Figures 11.13 and 11.14 show the differential conductance curves, (dI/dV)/(I/V), of the p$^+$-Si sample and the CuPc films on p$^+$-Si substrate. The results obtained by (a) a Pt-Ir probe and (b) a p$^+$-Si probe are shown in each figure. The STS of p$^+$-Si sample measured by the Pt-Ir probe shows a clear band gap of 1.1 eV for Si. However, the STS measured by the Si probe are influenced by the DOS of Si. On the other hand, the STS of a CuPc film on a p$^+$-Si substrate measured by the Pt-Ir probe shows a strong asymmetric characteristic (Fig. 11.14(a)) and that measured by the p$^+$-Si probe shows a gap of approximately 2 eV. This value is much larger than the band gap of CuPc (1.6 eV). These results indicate that the asymmetry of STS and the difference in band gap is due to the influence of the probe DOS and the formation of a rectifying barrier at the probe and sample interface.

To examine this more precisely, we also measured STS using a CuPc coated Pt-Ir probe. The STS characteristics of a metal (evaporated Al on p-Si) and a p$^+$-Si sample are shown in Fig. 11.15. Comparing Fig. 11.14 with Fig. 11.15, shows that similar results were obtained, that is, the STS of the p$^+$-Si

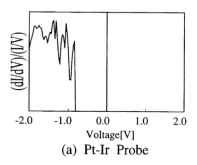

(a) Pt-Ir Probe

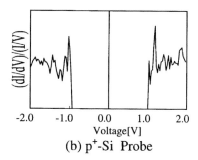

(b) p⁺-Si Probe

Fig. 11.14. Differential conductance curves, (dI/dV)/(I/V), of CuPc films on p⁺-Si substrate. (a) Pt-Ir probe and (b) p⁺-Si probe.

sample shows a band gap of about 2 eV and that of the Al coated Si sample exhibits a strong asymmetric characteristic. These results also demonstrate that an extremely small-size rectifying barrier formed near the probe affects the STS characteristics and the samples with the asymmetric STS behave like a MIS (metal insulator (tunneling barrier) semiconductor) diode.

### 11.3.4 Inelastic electron tunneling spectroscopy of organic thin films

The above discussion of STS describes is the elastic tunneling process in which the electrons tunnel through the barrier without giving up their energy. In addition to the elastic tunnel, there are inelastic tunneling processes which occur when the tunneling electrons transfer energy to various excitations in the vicinity of the barrier. If we measure the current as a function of voltage, these two components will be seen to be an accompanying step in dI/dV and thus a peak in $d^2I/dV^2$. The spectroscopy conducted by plotting $d^2I/dV^2$ or $d^2V/dI^2$ as a function of V is called inelastic electron tunneling spectroscopy (IETS) and the peak voltages correspond to the excitation energies of impurities or vibration modes of molecules of the barrier region. The IETS measurement can provide valuable information about the structure, orientation, electronic states

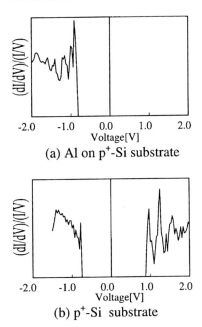

(a) Al on p$^+$-Si substrate

(b) p$^+$-Si substrate

Fig. 11.15. STS characteristics using CuPc coated Pt-Ir probe. (a) Al on p$^+$-Si and (b) p$^+$-Si sample.

and interactions of molecules in organic thin films [56–59].

In particular, the IETS using an STM tip is attractive not only as a local characterization but also as a high-density storage system, because this system provides both spatial resolution as with STM and spectral information as with IETS. Smith *et al.* have shown that an STM operating in liquid helium is sensitive to phonons and have observed a well-defined spectrum of peaks whose energies corresponded to the vibrational modes of sorbic acid molecules. [60]

We investigated the STS of a polydiacetylene (PDA) monolayer film at low temperature, 120 K. In the experiment, a Langmuir-Blodgett (LB) film of diacetylene (DA) monomer was deposited onto a highly oriented pyrolytic graphite (HOPG) substrate. The chemical structures of DA and PDA are shown in Fig. 11.16. The changes of IETS by applying voltage pulses were measured. Figure 11.17 shows the dI/dV and d$^2$I/dV$^2$ as a function of V. The second-derivative curve shows several IETS peaks corresponding to the vibrational mode frequencies of the PDA molecule.

In summary, electric measurements using the SPM probe are a powerful means for investigating the characterization of organic thin films, because the data obtained by these techniques provide not only local-area electrical properties such as electrical conductivity, dielectric constant and work function,

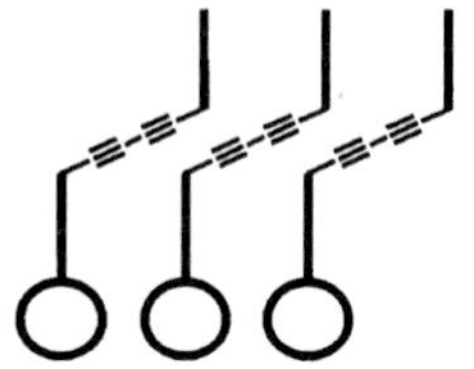

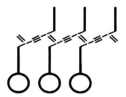

$$C_{12}H_{25} - C \equiv C - C \equiv C - (CH_2)_8 - COOH$$

(a)  Diacetylene Monomer (DA)

(b)  Polydiacetylene  (PDA)

Fig. 11.16. Chemical structure of (a) diacetylene (DA) monomer and (b) polydiacetylene (PDA).

but also vibrational modes at the molecular scale. The information, however, sometimes does not directly reflect the real properties due to unwanted interactions between tip and sample, or sample and substrate. Since the electrical measurements using SPM and related techniques are now being carried out, even though it is still uncertain how the mechanism works and how to analyse it, it is essential to establish the true characteristics for an ideal interaction between electrons and molecules. It is also important to verify the data by starting with well-defined and characterized test structures or comparing the data with results obtained from different experimental techniques.

*11.4  Application to Molecular Devices*

*11.4.1  Introduction*

There are a lot of reports in the literature on reproducible switching in organic films and bulk crystals. [61–65] It would appear that the switching mechanism is electric, not thermal, in nature. Though the details are complex

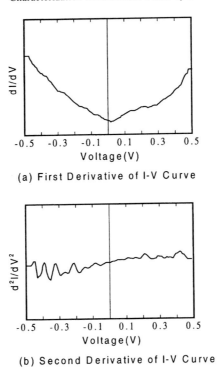

(a) First Derivative of I-V Curve

(b) Second Derivative of I-V Curve

Fig. 11.17. STS and IETS of UV-irradiated DA film measured by STM tip.

and are not yet completely settled, switching phenomena at molecular level would provide a insight into molecular electronic devices.

On the other hand, if these changes in molecular level can be induced by applying voltage pulses through the STM probe, high-density recording can be expected with the utilization of IETS probing beneath the STM tip. Several attempts have been made to create high-density memory or switching devices with STM and/or AFM [66–72]. To our knowledge, however, most of them deal with the detection of the surface topographic or conductance change in a domain that the probe can approach. From this point of view, memory media using organic thin films and electron spectroscopy based on IETS have the potential to be a high-density storage system. In other words, if complex changes of the IETS pattern correspond to the various changes in organic thin films induced by the voltage pulses, multilevel or multiplex recordings can be realized. [73]

We describe here the basic experimental results and the application to molecular devices using STM system such as molecular switching devices and high-density recording systems.

### 11.4.2 Electric devices with molecular films

Research on the local characterization techniques is now focusing on nanodevices such as single electron transistors, ultrahigh-density memory devices, and molecular electronic devices. Recently, novel electrical switching and diode properties have been reported in organic thin film. [65–67] Most of these characteristics were observed in a conventional metal/organic film/metal planar structure. The planar junctions sometimes suffer from the disadvantage that the interpretation of results can be confused by unwanted effects such as nonideal solid tunneling barrier, spatially averaged information over the junction, etc. From this point of view, electrical measurements using an SPM tip have important implications for not only the basic investigation at the molecular level but also the ultimate size-limit of electronic devices.

Electrical switching phenomena of lead phthalocyanine (PbPc) films in the Au/PbPc/Au sandwich type cells have been reported by several authors. [62,63] The PbPc molecule in the solid state has a nonplanar "shuttle-cock" form and the triclinic crystal has two stacking structures, drawn diagrammatically in Figs. 11.18(a) and (b). Although the electrical switching has been explained in terms of the structural change in the stacking domains, a detailed explanation, in particular the theory at a molecular level, has not been yet provided.

The current-voltage (I-V) characteristics of thin PbPc film were investigated using an STM probe. The samples were fabricated by evaporating PbPc films (–300 nm thick) on Au electrodes. The I-V measurements were performed as follows. The probe was positioned above some point in the sample surface, and the distance between the probe and the sample was held

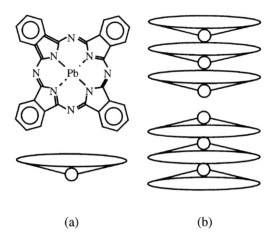

(a)                                (b)

Fig. 11.18. Chemical structure of (a) PbPc molecule and (b) two types of stacking structure.

constant, with the feedback loop activated. Then the feedback loop was disabled during the I-V measurement, while the Z-piezo voltage was held constant. Figure 11.19 shows the I-V characteristics for a PbPc evaporated film sample. The I-V curves in the voltage range of less than 3 V show reproducible diode characteristics (high-resistance off-state in Figs. 11.19(a) and (b)). At around a negative voltage of –4 V, however, the current shows a rapid increase and the diode characteristics also disappears (low-resistance on-state; Fig. 11.19(c)). Although this low-resistance state is maintained in the voltage scan from –2 to +2 V, switching from the on-state to the off-state takes place when the applied voltage exceeds +3.5 V. However, these switching phenomena have not been observed in other phthalocyanine (H₂Pc and CuPc) samples. These results indicate that the electrical switching occurs at the local area and originates from the intrinsic properties of PbPc films.

### 11.4.3 High-density recording systems using IETS

As a preliminary experiment, the IETS of Al electrode/LB monolayer film/SiO₂/Si samples were investigated. The sample structure is shown in Fig. 11.20. LB monolayer films of DA were deposited on the highly doped Si wafer with a very thin SiO₂ film (–2 nm thick). The IETS variations were studied by applying voltage pulses. During the IETS measurement, the samples were cooled to 20 K in a low-temperature refrigerator.

Electron beam or UV light exposure induces polymerization of the DA film. A reversible color phase transition of the PDA film is also caused by temperature change, [74] that is, chromism phenomena. Consequently, complicated phenomena of DA films under the STM tip are expected. Figure

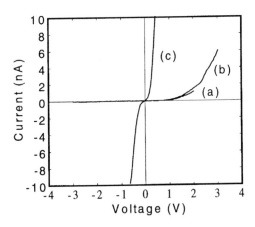

Fig. 11.19. Electrical switching phenomena of PbPc film observed under the STM probe. (a) and (b) are high-resistance off-states, (c) is a low-resistance on-state.

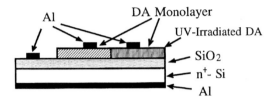

Fig. 11.20. Sample structure for IETS measurement.

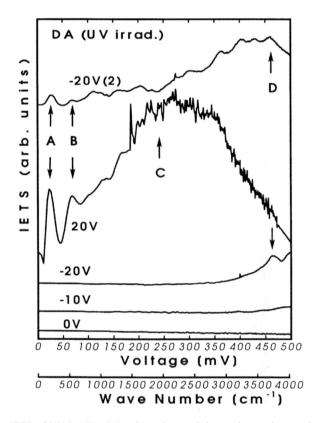

Fig. 11.21. IETS of UV-irradiated DA film after applying various voltage pulses to the Al electrode.

11.21 shows the IETS of UV-irradiated DA film before and after applying one set of voltage pulses. Although no obvious change was observed at low bias voltages after applying a negative voltage pulse, the IETS peaks labeled C and D in Fig. 11.21 were modified reversibly by applying voltage pulses of 20 V

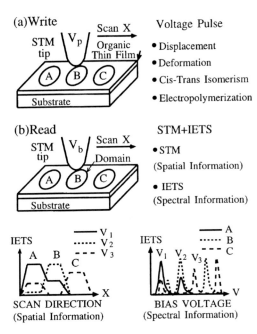

(a)Write    Scan X    Voltage Pulse

STM $V_p$    Organic
tip          Thin Film

• Displacement
• Deformation
• Cis-Trans Isomerism
• Electropolymerization

Substrate

(b)Read    Scan X    STM+IETS

STM $V_b$    Domain
tip

• STM
(Spatial Information)

• IETS
(Spectral Information)

Substrate

IETS —— $V_1$
······ $V_2$
- - - $V_3$
A  B
C

SCAN DIRECTION
(Spatial Information)
X

IETS —— A
······ B
- - - C
$V_1$  $V_2$  $V_3$

BIAS VOLTAGE
(Spectral Information)
V

Fig. 11.22.  High-density recording system using STM tip. (a) Writing by voltage pulse, (b) Reading by IETS.

and –20 V. These results may be related to the chromism properties of PDA films and are attractive possibilities for a rewritable memory system. However, some peaks (A, B in Fig. 11.21) remain in the IETS after applying the second –20 V pulse. In particular, the B peak at 60 mV only changes its peak height, but the A peak at 23 mV shifts toward the higher voltage of 30 mV. These results demonstrate that the IETS pattern provides information about different voltage pulses.

From these basic results of IETS measurements, a new high-density recording system using the IETS of organic ultrathin films is expected to be possible. The IETS using the STM tip can detect changes in nanometer to submicron-size domains of organic ultrathin films.

Figure 11.22 gives a schematic diagram of this system. [73] A similar concept is photochemical hole burning (PHB) [75], which utilizes the spectral information at one probing point in an optical method. The storage density of PHB is estimated as being about $10^{12}$ bit/cm$^2$. In the PHB system, the recording spot size of about 1 mm is restricted by the scanning wavelength of a laser beam. On the other hand, the IETS system using the STM tip has high spatial resolution, even though multiplexing is low in the voltage signal. For device application, however, the greatest disadvantage in both the IETS and PHB is the low-temperature operation required to minimize the linewidth broadening due to the thermal effect. The linewidth of IETS peaks in a standard metal/insulator/metal junction is determined by the modulation voltage and the thermal smearing of the Fermi surface of metal electrodes. From this point of view, the most effective method would be to minimize the thermal smearing of the Fermi level of the metal or semiconductor STM tip, which means that cooling of the STM tip, not of the sample itself, is essential. To confirm the potential for high-density recording, it is necessary to obtain experimental results using a low-temperature and stable STM system.

Our focus here has been on new technologies of characterization, and device applications of organic thin films for electronics, especially, electric characterizations using the scanning probe microscope and the application to high-density recording system have been discussed. To sum up, organic thin films have many possibilities for optical and electrical applications, and they will become one of the key materials for molecular electronic devices and even information processing, in the near future.

## [References]

1)   J. E. Frommer, *Angew. Chem. Int. Ed. Engl.*, **31**, 1298 (1992)
2)   H. Ohtani, R. J. Wilson, S. Chiang and C. M. Mate, *Phys. Rev. Lett.*, **60**, 2398 (1988)
3)   V. M. Hallmark, S. Chiang, J. K. Brown and Ch. Woell, *Phys. Rev. Lett.*, **66**, 48 (1991)
4)   P. H. Lippel, R. J. Wilson, M. D. Miller, Ch. Woell and S. Chiang, *Phys. Rev. Lett.*, **62**, 171 (1989)
5)   J. S. Foster and J. E. Frommer, *Nature*, **333**, 542 (1988)
6)   C. Clemmer and T. Beebe, *Science*, **251**, 640 (1991)
7)   Y. Kim and A. Bard, *Langmuir*, **8**, 1096 (1992)
8)   R. Yang, K. Naoi, D. Evans, W. Smyrl and W. Hendrickson, *Langmuir*, **7**, 556 (1991)
9)   R. Wilson, G. Meijer, D. Bethune, R. Johnson, D. Chambliss, M. de Vries, H. Hunziker and H. Wendt, *Nature*, **348**, 621 (1990)
10)  A. Engel, *Ann. Rev. Biophys. Chem.*, **20**, 79 (1991)
11)  D. P. E. Smith, J. Hörber, Ch. Gerber and G. Binnig, *Science*, **245**, 43 (1989)
12)  A. Ulman, *Chem. Rev.*, **96**, 1533 (1996)
13)  N. A. Kato, M. Hara, H. Sasabe and W. Knoll, *Nanotechnology*, **7**, 122, (1996)
14)  G. McGonigal, R. Bernhardt and D. Thomson, *Appl. Phys. Lett.*, **57**, 28 (1990)
15)  H. Fuchs, S. Akari and K. Dransfeld, *Z. Phys. B Condens. Matter*, **80**, 389 (1990)
16)  T. Sleator and R. Tycko, *Phys. Rev. Lett.*, **60**, 1418 (1988)
17)  M. Yoshimura, H. Shigekawa, H. Yamochi, G. Saito, Y. Saito and A. Kawazu, *Phys. Rev. B*, **44**, 1970 (1991)
18)  D. Drake, C. Prater, A. Weisenhorn, S. Gould, T. Albrecht, C. Quate, D. Cannell, H. Hansma and P. Hansma, *Science*, **243**, 1586 (1989)
19)  M. Egger, F. Ohnesorge, A. Weisenhorn, S. Heyn, D. Drake, C. Prater, S. Gould, P. Hansma and H. Gaub, *J. Struct. Biol.*, **103**, 89 (1990)
20)  E. Meyer, L. Howald, R. Overney, H. Heinzelmann, J. Frommer, H. Guentherodt, T. Wagner, H. Schier and S. Poth, *Nature*, **349**, 398 (1991)
21)  R. Overney, E. Meyer, J. Frommer, D. Brodbeck, L. Howald, H. Guentherodt, M. Fujihira, H. Takano and Y. Gotoh, *Nature*, **359**, 133 (1992)
22)  A. L. Weisenhorn, P. K. Hansma, T. R. Albrecht and C. F. Quate, *Appl. Phys. Lett*, **54**, 2651 (1989)

23) G. Devaud, P. S. Furcinitti, J. C. Fleming, M. K. Lyon and K. Douglas, *Biophys. J.*, **63**, 630 (1992)

24) S. Ohnishi, M. Hara, T. Furuno, T. Okada and H. Sasabe, *Biophys. J.*, **65**, 573 (1993)

25) E. Betzig, J. K. Trautman, T. D. Harris, J. S. Weiner and R. L. Kostelak, *Science*, **251**, 1468 (1991)

26) R. Micheletto, M. Denyer, M. Scholl, K. Nakajima, A. Offenhaeusser, M. Hara and W. Knoll, *Appl. Opt.*, **38**, 6648 (1999)

27) D. P. E. Smith, H. Hörber, Ch. Gerber and G. Binnig, *Science*, **245**, 43 (1989)

28) G. E. Porier and M. J. Tarlov, *Langmuir*, **10**, 2853 (1994)

29) Y. Iwakabe, M. Hara, K. Kondo, K. Tochigi, A. Mukoh, A. Yamada, A. F. Garito and H. Sasabe, *Jpn. J. Appl. Phys.*, **30**, 2542 (1991)

30) Y. Iwakabe, K. Kondo, S. Oh-hara, A. Mukoh, M. Hara and H. Sasabe, *Langmuir*, **10**, 3201 (1994)

31) Y. Iwakabe and M. Hara, *Mol. Cryst. Liq. Cryst.*, **267**, 129 (1995)

32) G. E. Porier and E. D. Pylant, *Science*, **272**, 1145 (1996)

33) N. Nishida, M. Hara, H. Sasabe and W. Knoll, *Jpn. J. Appl. Phys.*, **35**, 5866 (1996)

34) M. Hara, Y. Iwakabe, K. Tochigi, H. Sasabe, A. F. Garito and A. Yamada, *Nature*, **344**, 228 (1990)

35) G. McGonigal, R. Bernhardt and D. Thomson, *Appl. Phys. Lett.*, **57**, 28 (1990)

36) K. Kajikawa, M. Hara, H. Sasabe and W. Knoll, *Jpn. J. Appl. Phys.*, **36**, L1116 (1997)

37) J. B. Schlenoff, M. Li and H. Ly, *J. Am. Chem. Soc.*, **117**, 12528 (1995)

38) G. E. Porier, *Langmuir*, **15**, 1167 (1999)

39) B. Venkataraman, G. W. Flynn, J. L. Wilbur, J. P. Folkers and G. M. Whitesides, *J. Phys. Chem.*, **99**, 8684 (1995)

40) J. Noh and M. Hara, *Langmuir*, **16**, 2045 (2000)

41) For example; *Scanning Tunneling Microscopy I*, ed. by H.-J. Guntherrodt, R. Wiesendanger (Springer-Verlag, Berlin, Heiderberg, New York, 1992); *Scanning Tunneling Microscopy III*, 2nd ed., ed. by R. Wiesendanger, H.-J. Guntherrodt (Springer-Verlag, Berlin, Heiderberg, New York, 1996)

42) *Semiconductor Measurements and Instrumentation*, W. R. Runyan (International Student Edition, Texas Instruments Electronic Series, McGraw-Hill, Kogakusya, Tokyo, 1975)

43) P. Muralt and D. W. Pohl, *Appl. Phys. Lett.*, **48**, 514 (1986)

44) P. Muralt, H. Meier, D. W. Pohl and H. W. M. Salemink, *Appl. Phys. Lett.*, **50**, 1352 (1987)

45) J. R. Matey and J. Blanc, *J. Appl. Phys.*, **57**, 1437 (1985)

46) Y. Martin, D. W. Abraham and H. K. Wickramasinghe, *Appl. Phys. Lett.*, **52**, 1103 (1988)

47) C. C. Williams, W. P. Hough and S. A. Rishton, *Appl. Phys. Lett.*, **55**, 203 (1989)

48) H. Yokoyama, M. J. Jeffery and T. Inoue, *Jpn. J. Appl. Phys.*, **32**, L1845 (1993)

49) S. Cunningham, I. A. Larkin and J. H. Davis, *Appl. Phys. Lett.*, **73**, 123 (1998)

50) Y. Majima, S. Miyamoto, Y. Oyama and M. Iwamoto, *Jpn. J. Appl. Phys.*, **37**, 4557 (1998)

51) J. Tersoff and D. R. Hamman, *Phys. Rev.*, *B*, **31**, 805 (1984)

52) R. Yang, W. H. Smyrl, D. F. Evans and W. A. Hendrickson, *J. Phys. Chem.*, **96**, 1428 (1992)

53) S. Wakabayashi, H. Kato, M. Tomitori and O. Nishikawa, *J. Appl. Phys.*, **76**, 5595 (1994)

54) A. J. Ikushima, T. Kanno, S. Yoshida and A. Maeda, *Thin Solid Films*, **273**, 35 (1996)

55) M. Pomerantz, A. Aviram, R. A. McCorkle, L. Li and A. G. Schrott, *Science*, **255**, 1115 (1992)

56) R. C. Jaklevic and J. Lambe, *Phys. Rev. Lett.*, **17**, 1139 (1966)

57) *Principles of Electron Tunneling Spectroscopy*, ed. by E. L. Wolf (Oxford University Press, Oxford, 1985)

58) K. Kudo, C. Okazaki, S. Kuniyoshi and K. Tanaka, *Jpn. J. Appl. Phys.*, **30**, 1452 (1991)

59) S. Kuniyoshi, K. Kudo and K. Tanaka, *Thin Solid Films*, **210/211**, 531 (1992)

60) D. P. E. Smith, M. D. Kirk and C. F. Quate, *J. Chem. Phys.*, **86**, 6034 (1987)

61) R. S. Potember, T. O. Poehler and D. O. Cowan, *Appl. Phys. Lett.*, **34**, 405 (1979)

62)  Th. Frauenheim, C. Hamann and M. Muller, *Phys. Status Solidi*, **86**, 735 (1984)

63)  Y. Machida, Y. Saitoh, A. Taomoto, K. Nichogi, K. Waragai and S. Asakawa, *Jpn. J. Appl. Phys.*, **28**, 297 (1989)

64)  A. R. Elsharkawi and K. C. Kao, *J. Phys. Chem. Sol.*, **38**, 95 (1977)

65)  K. Sakai, H. Matsuda, H. Kawada, K. Eguchi and T. Nakagiri, *Appl. Phys. Lett.*, **61**, 3032 (1992)

66)  C. M. Fischer, M. Burghard, S. Roth and K. v. Klitzing, *Appl. Phys. Lett.*, **66**, 3331 (1995)

67)  M. Pomerantz, A. Aviram, R. A. McCorkle, L. Li and A. G. Schrott, *Science*, **255**, 1115 (1992)

68)  K. Takimoto, H. Kawade, E. Kishi, K. Yano, K. Sakai, K. Hatanaka, K. Eguchi and T. Nakagiri, *Appl. Phys. Lett.*, **53**, 1274 (1988)

69)  H. Nejoh, *Nature*, **353**, 640 (1991)

70)  H. J. Mamin and D. Rugar, *Appl. Phys. Lett.*, **61**, 1003 (1992)

71)  R. C. Barrett and F. Quate, *J. Appl. Phys.*, **70**, 2725 (1991)

72)  N. Gemma, H. Hieda, K. Tanaka and S. Egusa, *Jpn. J. Appl. Phys.*, **34**, L859 (1995)

73)  K. Kudo, S. Kuniyoshi and K. Tanaka, *Jpn. J. Appl. Phys.*, **34**, 3782 (1995)

74)  D. Bloor and R. R. Chance, *Polydiacetylenes* (Martinus Nijhoff, Boston, 1985)

75)  G. Castro, D. Haarer, R. M. Macfarlane and H. P. Trommsadorff, U.S. Patent, No. 4, 1010, 976 (1978)

# INDEX